T0156088

Lecture Notes on Mathematical Modelling in the Life Sciences

The rapid pace and development of new methods and techniques in mathematics and in biology and medicine creates a natural demand for up-to-date, readable, possibly short lecture notes covering the breadth and depth of mathematical modelling, mathematical analysis and numerical computations in the life sciences, at a high scientific level.

The volumes in this series are written in a style accessible to graduate students. Besides monographs, we envision the series to also provide an outlet for material less formally presented and more anticipatory of future needs due to novel and exciting biomedical applications and mathematical methodologies.

The topics in LMML range from the molecular level through the organismal to the population level, e.g. gene sequencing, protein dynamics, cell biology, developmental biology, genetic and neural networks, organogenesis, tissue mechanics, bioengineering and hemodynamics, infectious diseases, mathematical epidemiology and population dynamics.

Mathematical methods include dynamical systems, partial differential equations, optimal control, statistical mechanics and stochastics, numerical analysis, scientific computing and machine learning, combinatorics, algebra, topology and geometry, etc., which are indispensable for a deeper understanding of biological and medical problems.

Wherever feasible, numerical codes must be made accessible.

King-Yeung Lam • Yuan Lou

Introduction to Reaction-Diffusion Equations

Theory and Applications to Spatial Ecology and Evolutionary Biology

King-Yeung Lam
Department of Mathematics
The Ohio State University
Columbus, OH, USA

Yuan Lou
School of Mathematical Sciences
Shanghai Jiao Tong University
Shanghai, China

This work was supported by the National Science Foundation [DMS-1853561].

ISSN 2193-4789 ISSN 2193-4797 (electronic)
Lecture Notes on Mathematical Modelling in the Life Sciences
ISBN 978-3-031-20421-0 ISBN 978-3-031-20422-7 (eBook)
https://doi.org/10.1007/978-3-031-20422-7

Mathematics Subject Classification (2020): 35B40, 35K57, 47H07, 92D15, 37L30, 37C65

This Springer imprint is published by the registered company Springer Nature Switzerland AG
The registered company address is: Gewerbestrasse 11, 6330 Cham, Switzerland

For our families

Wendy and Hazel, Jianling and Ankai

Preface

This set of lecture notes is based on the mini-courses given by the authors at the Center for Partial Differential Equations, East China Normal University in 2013, Séminaire de Mathématiques Supérieures at the University of Alberta in 2016, the Center for Applied Mathematics at Guangzhou University in 2021 and 2022, and most recently, at Institut Henri Poincaré, Paris in 2022. Several modern mathematical theories have found broad and profound applications at the intersection of reaction-diffusion equations and mathematical biology in recent years. Our goal is to present, in a self-contained manner, some of these theories and tools to interested readers, especially beginning graduate students. We also want to connect these theories with biological concepts and to illustrate their usefulness in tackling various mathematical problems motivated by ecology and evolution. The selection of the material is subjective and has been influenced by the authors' own research interests.

Population distributions change dynamically in space either by movement or dispersal. The question of how different species survive and interact in space, and how they choose their dispersal strategies, generates many fascinating problems in biology. Since the seminal work of Fisher, Kolmogorov–Petrovsky–Piskunov, Okubo, Murray, Turing, Skellam, Aronson and Weinberger, and many others, the theory of reaction-diffusion equations has become a major mathematical tool in the field of mathematical biology. The interplay between reaction-diffusion equations and biology works in both directions: On the one hand, mathematical models are used to quantitatively describe and study biological concepts such as population persistence and critical patch size, competitive and cooperative interactions, and asymptotic speed of invasion, among others. On the other hand, the biological point of view suggests new and often challenging mathematical problems, and promote new areas for mathematical analysis and new ways of thinking. For instance, research developments in the theory of monotone dynamical systems, pattern formation, and traveling wave solutions have been propelled by their intersection with biology, to name a few. The purpose of this work is to introduce the readers to several aspects of the mathematical theory of reaction-diffusion equations as guided by their applications in ecology and evolution.

By now there is a tremendous research literature devoted to reaction-diffusion equations in biology. Although mathematical tools such as the comparison principle, the concept of principal eigenvalue, and the theory of monotone dynamical systems are relatively "user-friendly", it is a nontrivial task to understand their proofs. One of the motivations in writing this set of lecture notes grew out of our practical need to teach the theory to our graduate students. The materials we wish to cover are often scattered in the literature, and some areas are even still under development. As such, there are very limited sources where the theory is developed in a self-contained manner that are accessible to beginners. The monographs by Ye et al.[1], Cantrell and Cosner[2] and Perthame[3] on the theory of reaction-diffusion equations, are some excellent sources. So are the monographs of Zhao[4], and Smith and Thieme[5] on dynamical systems approaches to problems in biology. However, their focus is different from ours, which is outlined below.

This set of lecture notes is divided into four parts. Part I concerns the theory of linear elliptic and parabolic equations; Parts II and III cover applications to ecological and evolutionary dynamics, respectively; while Part IV contains the proofs of the Krein–Rutman Theorem and elements of the theory of monotone dynamical systems. We strive to provide a self-contained account of the theory in Parts I and IV, and to illustrate how to use them to study reaction-diffusion models from mathematical biology in Parts II and III. Below we describe the contents in more details.

Part I, consisting of Chapters 1 to 4, is devoted to the linear theory of elliptic and parabolic PDEs, with emphasis on the maximum principle. Chapter 1 begins with scalar parabolic equations with oblique boundary conditions, which include the biologically interesting cases of Neumann, and no-flux boundary conditions. We introduce a concept of super- and subsolutions in a generalized sense which enables the gluing of classical super- or subsolutions. This is a simplified version of the usual weak super- and subsolutions for divergence form operators, and the solution concept in the viscosity sense. Next, we derive the existence of the principal eigenvalue by applying the Krein–Rutman Theorem. We will also determine the limit of the principal eigenvalue as the diffusion rate tends to zero or infinity. A characterization of the maximum principle based on the existence of a strict positive supersolution is also proved. Chapter 2 is devoted to the principal eigenvalue of periodic-parabolic problems, where we study the effects of both spatial and temporal heterogeneity by giving various asymptotic estimates of the principal eigenvalues, and show the monotone dependence of the principal eigenvalue in frequency. The corresponding

[1] Q. X. Ye, Z. Y. Li, Y. P. Wu and M. X. Wang, *Introduction to reaction-diffusion equations*, Foundations of Modern Mathematics Series, Science Press, Beijing, 2011.

[2] R. S. Cantrell and C. Cosner, *Spatial ecology via reaction-diffusion equations*, Wiley Series in Mathematical and Computational Biology, John Wiley & Sons, Ltd., Chichester, 2003.

[3] B. Perthame, *Parabolic equations in biology*, Lecture Notes on Mathematical Modelling in the Life Sciences. Springer, Cham, 2015.

[4] X.-Q. Zhao, *Dynamical systems in population biology*, CMS Books in Mathematics/Ouvrages de Mathématiques de la SMC, Springer, Cham, second ed., 2017.

[5] H. L. Smith and H. R. Thieme, *Dynamical systems and population persistence*, vol. 118 of Graduate Studies in Mathematics, American Mathematical Society, Providence, RI, 2011.

theory for the cooperative systems of equations is contained in Chapter 3. Chapter 4 concerns the notion of the principal Floquet bundle, which is a generalization of the principal eigenvalue to space-time varying environments that are not necessarily periodic in time. We present its basic theory and a recent result concerning the smooth dependence of the principal Floquet bundle on the coefficients of parabolic operators. In subsequent chapters, the consequences of these analytical results on the persistence and stability questions in single and multiple species models will be extensively discussed.

Part II of these lecture notes, consisting of Chapters 5 to 8, is devoted to applications in ecological dynamics. In Chapter 5, we present some general theory for semilinear equations modeling a single population, including the monotone iteration method due to Sattinger, and the relationship of linear and nonlinear stability of equilibria. Then we specialize to the case of diffusive logistic equations, and discuss the convergence to equilibria, the asymptotic behavior of equilibria as the diffusion rate tends to zero or infinity, as well as the concept of the critical domain size. In Chapter 6, we discuss the Fisher–KPP equation on the real line. We introduce the notion of spreading speed, and give an elementary proof of the classical result for the homogeneous case, based on constructing generalized super/subsolutions. We also discuss some recent results concerning the shifting habitat, and the possibility of nonlocal selection of the spreading speed. Chapter 7 is devoted to the diffusive Lotka–Volterra model of two competing species in bounded domains. We first cast these equations into the setting of the strongly monotone dynamical systems and derive some general conclusions. Next, we investigate the large time dynamics by way of constructing Lyapunov functions and applying LaSalle's invariance principle. Particular attention is given to the problem of evolution of dispersal, where the selection of slow dispersal due to Hastings is discussed in the broader context of possibly weak competition. In the latter case, the full dynamics can still be characterized by a brilliant result of He and Ni which demonstrates the linear stability and uniqueness of the positive equilibrium, whenever one exists. The outcome of the competition dynamics may be reversed with the addition of advection. We illustrate this by showing that faster dispersal becomes advantageous in some circumstances. In Chapter 8, we discuss the dynamics of phytoplankton populations in a water column. Here the individuals are engaged in a nonlocal competition for light via shading. Surprisingly, in the two-species case the nonlocal PDE model is in fact order-preserving, albeit with respect to a nonstandard cone. This result facilitates the classification of phytoplankton dynamics. When the number of species is greater than two, the system is no longer order-preserving. Here we present a method to analyze the population dynamics for a large number of interacting species, based on the theory of normalized principal Floquet bundles.

Evolutionary dynamics is the subject of Part III, which consists of Chapters 9 to 10. In Chapter 9, we discuss the adaptive dynamics framework, which is a conceptual framework enabling the exploration of evolutionary questions using ecological models. We discuss the framework of adaptive dynamics in the context of a river population model, hoping to offer to the readers a PDE viewpoint of the theory. Specifically, we introduce in concrete terms the key notions of invasion exponent, selection gradient, singular strategy, convergence stable strategy, evolutionary stable strategy (ESS), continuously stable strategy, neighborhood invader strategy, dimorphism (coexistence of strategies) and evolutionary branching point. In Chapter 10, we discuss the so-called selection-mutation models, which describes population structured by a continuous trait. They can be regarded as the analogue of N-species competition models, as N tends to infinity. We start with a model without spatial structure, due to Magal and Webb, and show the convergence to equilibrium, then we move to discuss the continuous-trait model with spatial structure, due to Perthame and Souganidis, on the evolution of dispersal in spatially varying but temporally constant environment. These results suggest the relationship of selection-mutation models with the framework of adaptive dynamics.

Part IV consists of Appendices A to E. We introduce to the readers several useful tools from nonlinear functional analysis and dynamical systems. These analytical tools are well known but their proofs are developed in different separate sources. Here, we give a relatively self-contained treatment of these topics, by combining materials from selected research papers and monographs in these areas, with some of our own modifications. In Appendix A, we derive the fixed point index from the well known Leray Schauder degree. In Appendix B, we introduce the concept of a positive cone, and that of an ordered Banach space. We present a self-contained proof of the Krein–Rutman Theorem for positively homogeneous maps of degree one $f : K \to K$ that are monotone with respect to K, and then derive from that the classical Krein–Rutman Theorem for compact positive linear operators. In Appendix C, we discuss dynamical systems in ordered Banach spaces that are generated by monotone and subhomogeneous maps. We show that such a system has a globally attracting fixed point. We will also prove an analogous result for continuous-time semiflows. In Appendix D, we consider general monotone dynamical systems and prove the Dancer–Hess Lemma concerning the existence of a connecting orbit between two ordered fixed points, and prove the limit set trichotomy. We then extend the result to continuous semiflows. In Appendix E, we present the theory of abstract competitive systems, and develop the trichotomy result due to Hsu, Smith and Waltman. In the case when the mapping (or semiflow) is continuously differentiable, we present a new condition to achieve a stronger trichotomy result.

We have benefited tremendously from communications and collaborations with many colleagues and friends during the preparation of this set of lecture notes and we thank all of them. First of all, we thank our common thesis advisor, Wei-Ming Ni for his support through these many years. We are grateful to Benoît Perthame for suggesting the publication of this work. We learned a great deal from the works of our friends and our collaborators, Robert Stephen Cantrell, Xinfu Chen, Chris Cosner, Yihong Du, Avner Friedman, Alan Hastings, Sze-Bi Hsu, Vivian Hutson, Mark Lewis, Suzanne Lenhart, Frithjof Lutscher, Konstantin Mischaikow, Peter Poláčik, Wenxian Shen, Xuefeng Wang, Michael Winkler, Yaping Wu, Youshan Tao, Eiji Yanagida, Shoji Yotsutani and Qixiao Ye, to whom we are very grateful. We would like to acknowledge Zhucheng Jin, Alexis Leculier, Shuang Liu, and Xiao Yu for reading parts of the manuscript, and the US National Science Foundation for its support via the grant DMS-1853561. KYL is also supported by the European Union's

Horizon 2020 research and innovation programme via grant agreement No. 740623, and he would like to thank the Institut Henri Poincaré for providing an excellent environment where part of this manuscript was completed. YL is also supported by the research funds from Shanghai Jiao Tong University, the Shanghai Frontier Research Center on Modern Analysis (CMA-Shanghai), the National Science Foundation of China and MOE-LSC.

Last but not the least, KYL thanks his wife Wendy Xu and daughter Hazel, YL thanks his wife Jianling Huang and son Ankai, for their unconditional support and understanding throughout this endeavor.

Columbus, Ohio, USA *King-Yeung Lam*
Shanghai, China *Yuan Lou*
August 2022

Contents

Part I
Linear Theory

Chapter 1
The Maximum Principle and the Principal Eigenvalues for Single Equations

Abstract In this chapter we will discuss the maximum principle for single parabolic and elliptic equations, where we introduce the useful concept of super- and subsolutions in a generalized sense which enables the gluing of families super- or subsolutions. Next, we derive the existence of principal eigenvalue by applying the Krein–Rutman theorem. We will also determine the limit of principal eigenvalue as the diffusion rate tends to zero or infinity. Finally, we will give a characterization of the maximum principle based on the existence of a strict positive supersolution.

1.1 The Maximum Principle for Single Parabolic Equations

Let Ω be a bounded domain in $\mathbb{R}^N$ with smooth boundary $\partial\Omega$, and let $\mathbf{n} = (n_i)_{i=1}^N$ be the unit outward normal vector on $\partial\Omega$. For $T > 0$, define

$$\Omega_T = \Omega \times (0,T], \quad \overline{\Omega}_T = \overline{\Omega} \times [0,T], \quad S\Omega_T = \partial\Omega \times (0,T], \quad P\Omega_T = \overline{\Omega}_T \setminus \Omega_T.$$

Consider the linear parabolic operator in non-divergence form with continuous coefficients

$$u_t + \mathcal{L}u \equiv u_t - a^{ij}D_{ij}u - b^i D_i u - cu = f \quad \text{in } \Omega_T, \tag{1.1}$$

endowed with the oblique boundary condition

$$\mathcal{B}u \equiv p^i D_i u + p^0 u = g \quad \text{on } S\Omega_T, \tag{1.2}$$

where we adopt the convention to sum over repeated indices. For each i, p^i satisfies

$$p^i(x,t)n_i(x) > 0 \quad \text{and} \quad p^0 \geq 0 \quad \text{on } \partial\Omega \times [0,T].$$

Setting $(p_i)_{i=1}^N$ to be the outward unit normal vector $\mathbf{n}$ on $\partial\Omega$, (1.2) reduces to

$$\mathbf{n} \cdot \nabla u + p^0 u = g \quad \text{on } S\Omega_T,$$

K. -Y. Lam, Y. Lou, *Introduction to Reaction-Diffusion Equations*, Lecture Notes on Mathematical Modelling in the Life Sciences, https://doi.org/10.1007/978-3-031-20422-7_1

and we obtain the Neumann boundary condition when $p^0 = 0$, or the Robin boundary condition when $p^0 \geq 0$ is nontrivial. We note that most of the results in this chapter continue to hold for the Dirichlet boundary condition.

In this chapter, we assume $a^{ij}, b^i, c \in C^0(\overline{\Omega}_T)$ and the uniform ellipticity condition; i.e. there exists some constant $\lambda_0 > 0$ such that

$$\lambda_0|\xi|^2 \leq a^{ij}(x,t)\xi_j\xi_j \quad \text{for all } \xi \in \mathbb{R}^N, \quad (x,t) \in \Omega_T. \tag{1.3}$$

We define the notions of classical and generalized super- and subsolutions for the initial-boundary-value problem

$$u_t + \mathcal{L}u = f(x,t,u,Du) \quad \text{in } \Omega_T, \quad \text{and} \quad \mathcal{B}u = g(x,t) \quad \text{on } S\Omega_T. \tag{1.4}$$

Definition 1.1.1 1. We say that $u \in C^{2,1}(\Omega_T)$ satisfies

$$u_t + \mathcal{L}u \leq f(x,t,u,Du) \quad \text{in } \Omega_T \tag{1.5}$$

in the classical sense if the inequality is satisfied everywhere in Ω_T. In this case, we call u a classical subsolution of $u_t + \mathcal{L}u = f$ in Ω_T. We similarly define the notion of classical supersolution by reversing the inequality (1.5).

2. We say that $u \in C^{1,0}(\overline{\Omega}_T)$ satisfies

$$\mathcal{B}u(x,t) \leq g(x,t) \quad (\text{resp. } \mathcal{B}u \geq g(x,t)) \quad \text{on } S\Omega_T,$$

in the classical sense if the inequality is satisfied everywhere on $S\Omega_T$.

3. We say that $u \in C(\overline{\Omega}_T)$ satisfies (1.5) in the generalized sense if, for each $(x_0,t_0) \in \Omega_T$, there exist a neighborhood U of (x_0,t_0) in Ω_T, and $\tilde{u} \in C^{2,1}(\overline{U})$ such that $\tilde{u} \leq u$ in U, $\tilde{u}(x_0,t_0) = u(x_0,t_0)$ and

$$\tilde{u}_t(x_0,t_0) + \mathcal{L}\tilde{u}(x_0,t_0) \leq f(x_0,t_0,\tilde{u}(x_0,t_0),D\tilde{u}(x_0,t_0)). \tag{1.6}$$

In such a case, we say that u is a generalized subsolution of $u_t + \mathcal{L}u = f$ in Ω_T.

4. We say that $u \in C(\overline{\Omega}_T)$ satisfies

$$u_t + \mathcal{L}u \geq f(x,t,u,Du) \quad \text{in } \Omega_T$$

in the generalized sense if, for each $(x_0,t_0) \in \Omega_T$, there exist a neighborhood U of (x_0,t_0) in Ω_T, and $\tilde{u} \in C^{2,1}(\overline{U})$ such that $\tilde{u} \geq u$ in U, $\tilde{u}(x_0,t_0) = u(x_0,t_0)$ and

$$\tilde{u}_t(x_0,t_0) + \mathcal{L}\tilde{u}(x_0,t_0) \geq f(x_0,t_0,\tilde{u}(x_0,t_0),D\tilde{u}(x_0,t_0)). \tag{1.7}$$

In such a case, we say that u is a generalized supersolution of $u_t + \mathcal{L}u = f$ in Ω_T.

5. We say that $u \in C(\overline{\Omega}_T)$ satisfies

$$u_t + \mathcal{L}u \leq f(x,t,u,Du) \quad \text{in } \Omega_T \quad \text{and} \quad \mathcal{B}u \leq g(x,t) \quad \text{on } S\Omega_T \tag{1.8}$$

in the generalized sense if for each $(x_0,t_0) \in \overline{\Omega} \times (0,T]$, there exist a neighborhood U of (x_0,t_0) in $\overline{\Omega}_T$ and a function $\tilde{u} \in C^{2,1}(\overline{U})$ such that $\tilde{u} \leq u$ in U, $\tilde{u}(x_0,t_0) =$

$u(x_0, t_0)$, and $\tilde{u}$ satisfies (1.6) if $(x_0, t_0) \in \Omega_T$, or

$$\mathcal{B}\tilde{u}(x_0, t_0) \le g(x_0, t_0) \quad \text{if } (x_0, t_0) \in S\Omega_T.$$

In such a case, we say that u is a generalized subsolution of (1.4).

6. We say that $u \in C(\overline{\Omega}_T)$ satisfies

$$u_t + \mathcal{L}u \ge f(x, t, u, Du) \quad \text{in } \Omega_T \quad \text{and} \quad \mathcal{B}u \ge g(x, t) \quad \text{on } S\Omega_T \tag{1.9}$$

in the generalized sense if for each $(x_0, t_0) \in \overline{\Omega} \times (0, T]$, there exist a neighborhood U of (x_0, t_0) in $\overline{\Omega}_T$ and a function $\tilde{u} \in C^{2,1}(\overline{\Omega}_T)$ such that $\tilde{u} \ge u$ in U, $\tilde{u}(x_0, t_0) = u(x_0, t_0)$, and $\tilde{u}$ satisfies (1.7) if $(x_0, t_0) \in \Omega_T$, or

$$\mathcal{B}\tilde{u}(x_0, t_0) \ge g(x_0, t_0) \quad \text{if } (x_0, t_0) \in S\Omega_T.$$

In such a case, we say that u is a generalized supersolution of (1.4).

Remark 1.1.2 1. We define $u_t + \mathcal{L}u < f(x, t, u, Du)$ in Ω_T in the generalized sense if statement 3 in the above definition holds with the inequality in (1.6) being replaced by a strict inequality. We also define $u_t + \mathcal{L}u > f(x, t, u, Du)$ analogously.

2. If $u_t + \mathcal{L}u \le f(x, t, u, Du)$ in Ω_T (resp. $u_t + \mathcal{L}u < f(x, t, u, Du)$ in Ω_T or $\mathcal{B}u \le 0$ in $S\Omega_T$) in the classical sense, then it automatically holds in the generalized sense.
3. Let u_i $(i = 1, 2)$ be two classical subsolutions of $u_t + \mathcal{L}u = f(x, t, u, Du)$ in Ω_T, then $u = \max\{u_1, u_2\}$ is a generalized subsolution of the same equation in Ω_T.
4. Let u_i $(i = 1, 2)$ be two classical supersolutions of $u_t + \mathcal{L}u = f(x, t, u, Du)$ in Ω_T, then $u = \min\{u_1, u_2\}$ is a generalized supersolution of the same equation in Ω_T.
5. Let A_1, A_2 be relatively open subsets in $\overline{\Omega}_T$ such that $\overline{\Omega}_T \subset A_1 \cup A_2$. Suppose

   a. u_i $(i = 1, 2)$ are generalized subsolutions of $u_t + \mathcal{L}u = f(x, t, u, Du)$ in $A_i \cap \Omega_T$, and
   b. $u_1 < u_2$ on $(\partial A_1) \cap A_2 \cap \Omega_T$ and $u_1 > u_2$ on $A_1 \cap (\partial A_2) \cap \Omega_T$,

   then $u = \max\{u_1, u_2\}$ belongs to $C(\Omega_T)$ and is a generalized subsolution of the same equation in Ω_T. A similar statement holds if u_i are subsolutions of $u_t + \mathcal{L}u = f(x, t, u, Du)$ in $A_i \cap \Omega_T$ and $\mathcal{B} = g$ on $A_i \cap S\Omega_T$. A similar statement holds for generalized supersolutions.
6. For operators in divergence form with measurable coefficients, the notion of weak super- and subsolutions can be defined by integration by parts with nonnegative test functions. See [2] and [11] (and also Problem 1.4.4) for gluing of the weak super- and subsolutions in the H^1 framework. For operators in non-divergence form with continuous coefficients, a general theory has been developed for super- and subsolutions in the viscosity sense [10]. Here we simply observe that the classical notions of super- and subsolutions can be somewhat generalized (with no change to the proofs) without using more technical machinery. This covers many common constructions [8, 26].

Theorem 1.1.3 (Weak Maximum Principle) *Let $u \in C(\overline{\Omega}_T)$ be given.*

1. *If $u_t + \mathcal{L}u \le 0$ holds in the generalized sense in Ω_T and $u \le 0$ in $P\Omega_T$, then*

$$u \le 0 \quad \text{in } \Omega_T.$$

2. *If $u_t + \mathcal{L}u \le 0$ holds in the generalized sense in Ω_T and $c \le 0$ in Ω_T, then*

$$\max_{\overline{\Omega}_T} u \le \max_{P\Omega_T} u^+,$$

where $u^+ := \max\{u, 0\}$.

Remark 1.1.4 Let $u \in C(\overline{\Omega}_T)$ be given. Suppose $u_t + \mathcal{L}u \le 0$ and $u_t + \mathcal{L}u \ge 0$ in Ω_T in the generalized sense, and that $c \le 0$. Then

$$\max_{\overline{\Omega}_T} |u| \le \max_{P\Omega_T} |u| \quad \text{in } \Omega_T.$$

See Problem 1.4.1.

Remark 1.1.5 For the weak maximum principle when Ω is an unbounded domain, or the entire space, an additional growth condition at infinity has to be imposed. See Section 6.2 in Chapter 6 for the counterpart of Theorem 1.1.3 in that setting.

Proof First, we prove the second assertion. It is enough to prove that

$$\max_{\overline{\Omega}_T}(u - \epsilon t) \le \max_{P\Omega_T}(u - \epsilon t)^+ \quad \text{for each } \epsilon > 0.$$

Suppose to the contrary that there exist $\epsilon > 0$ and $(x_0, t_0) \in \Omega_T$ such that $\max_{\overline{\Omega}_T}(u - \epsilon t) = u(x_0, t_0) - \epsilon t_0 > \max_{P\Omega_T}(u - \epsilon t)^+ \ge 0$. For the point (x_0, t_0), take the smooth function $\tilde{u}$ according to the definition of generalized subsolution, so that

$$\begin{cases} \tilde{u} \le u & \text{in a neighborhood of } (x_0, t_0), \\ \tilde{u}(x_0, t_0) = u(x_0, t_0) > 0 & \text{and} \quad \tilde{u}_t + \mathcal{L}\tilde{u} \le 0 \quad \text{at } (x_0, t_0). \end{cases} \tag{1.10}$$

Since $u(x,t) - \epsilon t$ has a local maximum at (x_0, t_0), it follows that $\tilde{u}(x,t) - \epsilon t$ also has a local maximum at (x_0, t_0). Hence,

$$\tilde{u}_t \ge \epsilon, \quad a^{ij} D_{ij}\tilde{u} + b^j D_j \tilde{u} \le 0, \quad c\tilde{u} \le 0, \quad \text{at } (x_0, t_0),$$

so that

$$\tilde{u}_t + \mathcal{L}\tilde{u} \ge \epsilon > 0 \quad \text{at } (x_0, t_0),$$

which is in contradiction with (1.10). This proves the second assertion.

For the first assertion, let $v(x,t) = \mathrm{e}^{-t \sup c} u(x,t)$, then v satisfies

$$v_t - a^{ij} D_{ij} v - b^j D_j v - \tilde{c} v \le 0 \quad \text{in } \Omega_T$$

in the generalized sense, where $\tilde{c} = c - \sup_{\Omega_T} c \leq 0$ in Ω_T. We can then repeat the previous step to deduce that

$$\max_{\overline{\Omega}_T} v(x,t) \leq \max_{P\Omega_T} v(x,t) = \max_{P\Omega_T} e^{t \sup_{\Omega_T} c} u(x,t) \leq 0.$$

This is equivalent to $u(x,t) \leq 0$ in Ω_T. This proves the first assertion. □

We also prove that, when $c < 0$ everywhere, then a generalized subsolution u cannot attain a positive local maximum in the interior.

Lemma 1.1.6 *Let Ω be a domain in $\mathbb{R}^N$ which is possibly unbounded. Suppose u satisfies $u_t + \mathcal{L}u \leq 0$ in the generalized sense in Ω_T, where $a^{ij}, b^i, c \in C_{loc}(\Omega_T)$ such that $a^{ij}\xi_i\xi_j > 0$ for all $\xi = (\xi_i) \in \mathbb{R}^N$. If $c < 0$ in Ω_T (but not necessarily bounded from below), then u cannot attain a positive local maximum at an interior point.*

Proof Suppose to the contrary that u has a local maximum point $(x_0, t_0) \in \Omega_T$ such that $u(x_0, t_0) > 0$. Then by the definition of generalized subsolution, there exists a smooth function $\tilde{u}$ such that $\tilde{u} \leq u$ in a neighborhood of (x_0, t_0) and equality holds at (x_0, t_0). Hence, $\tilde{u}$ also attains a positive local maximum value at (x_0, t_0), so that

$$\tilde{u}_t \geq 0, \quad a^{ij} D_{ij}\tilde{u} + b^i D_i \tilde{u} \leq 0, \quad c\tilde{u} < 0 \quad \text{at } (x_0, t_0).$$

This contradicts the fact that $\tilde{u}_t + \mathcal{L}u \leq 0$. □

Having proved the weak maximum principle for generalized super- and subsolutions, we can derive the Boundary Point Lemma and Strong Maximum Principle, and Growth Lemma, since they are all predicated on being able to compare u with a suitable classical barrier function via the weak maximum principle.

Theorem 1.1.7 (Boundary Point Lemma) *Let $R > 0$ and $Y = (y, s) \in \mathbb{R}^{N+1}$ be given. Consider the lower paraboloid*

$$P_R = P_R(Y) = \{X = (x,t) : |x - y|^2 + (s - t) < R^2, \text{ and } t < s\}.$$

Suppose $u \in C(\overline{P}_R)$ satisfies $u_t + \mathcal{L}u \leq 0$ in P_R in the generalized sense, and that there is an $X_1 = (x_1, s)$ with $|x_1 - y| = R$ such that

$$u(X_1) > u(X) \quad \text{and} \quad c(X)u(X_1) \leq 0 \quad \text{for all } X \in P_R. \tag{1.11}$$

Then $\beta \cdot \nabla_x u(X_1) > 0$ for any vector β such that $\beta \cdot (x_1 - y) > 0$, in the sense that

$$\liminf_{h \to 0+} \frac{u(x_1, s) - u(x_1 - \beta h, s)}{h} > 0.$$

Proof This is originally due to Nirenberg [32], who used an ellipsoid instead of a paraboloid. See [28, Chap. II, Lemma 2.8] for the proof. Here we observe that the proof works not only for classical subsolutions, but also generalized subsolutions. This is based on observing that $u(X_1) - u$ and the auxiliary function $h_R(r) =$

$e^{-\alpha r^2} - e^{-\alpha R^2}$ form a super- and sub-solution pair in the domain $P_R \setminus \overline{P_{R/2}}$, in the generalized sense. Here

$$\begin{cases} X = (x_1, ..., x_N, t) \quad \text{and} \quad X_1 = (x_{1,1}, ..., x_{N,1}, t_1), \\ r^2 = |t - t_1| + \sum_{i=1}^N |x_i - x_{i,1}|^2. \end{cases}$$

Therefore, fixing $\epsilon > 0$ so that

$$u(x_1, s) - u(x, t) \geq \epsilon h_R(r) \quad \text{on } \{X \in \partial P_{R/2},\ t < s\}$$

and recalling that $u(x_1, s) - u(x, t) \geq 0 = \epsilon h_R(r)$ on $\{X \in \partial P_R,\ t < s\}$, one may apply the weak maximum principle to obtain

$$u(x_1, s) - u(x, s) \geq \epsilon h_R(r)$$

for all $(x, s) \in P_R \setminus P_{R/2}$. □

Lemma 1.1.8 (Growth Lemma due to Krylov–Safonov) *Let $R > 0$ and $\alpha > 0$ be positive constants and set*

$$Q = \{(x, t) : |x| < R, -\alpha R^2 < t < 0\}.$$

Then there exists a uniform positive constant κ such that for each

$$u \in \left\{\tilde{u} \in C(\overline{\Omega}_T) : \tilde{u} \geq 0 \text{ in } Q, \text{ and } \tilde{u}_t + \mathcal{L}\tilde{u} \leq 0 \text{ in } Q \text{ in the generalized sense}\right\}$$

if

$$u \geq h \quad \text{in } \{|x| < \epsilon R,\ t = -\alpha R^2\}, \quad \text{for some } h > 0,\ 0 < \epsilon < 1,$$

then

$$u \geq \frac{\epsilon^\kappa h}{2} \quad \text{in } \{|x| < R/2,\ t = 0\}.$$

Proof See [28, Lemma 2.6]. □

Theorem 1.1.9 (Strong Maximum Principle) *Assume*

$$u_t + \mathcal{L}u \leq 0 \quad \text{in } \Omega_T, \quad \mathcal{B}u \leq 0 \quad \text{on } S\Omega_T, \quad \text{in the generalized sense.}$$

1. *If $u(x, 0) \leq 0$ in Ω, then $u \leq 0$ in Ω_T.*
2. *If, in addition, $u(x_0, t_0) = 0$ for some $(x_0, t_0) \in \overline{\Omega} \times (0, T]$, then*

$$u \equiv 0 \quad \text{in } \overline{\Omega} \times (0, t_0].$$

Proof We first show part 1. By replacing $u(x, t)$ with $e^{-t \sup c} u(x, t)$ if necessary, we may assume $c \leq 0$ in Ω_T. Suppose $u(x, 0) \leq 0$ in Ω, we need to show that $u \leq 0$ in Ω_T. It suffices to show that $v(x, t) = u(x, t) - \epsilon t$ is negative in Ω_T for each $\epsilon > 0$.

Suppose to the contrary that there exists an $\epsilon > 0$ such that $\sup_{\Omega_T} v = v(x_0, t_0) > 0$ for some $(x_0, t_0) \in \overline{\Omega}_T$, then necessarily $t_0 > 0$.
Step 1. We first discuss the case $x_0 \in \operatorname{Int}\Omega$. In this case, there exist a neighborhood U of (x_0, t_0) and a smooth function $\tilde{u}$ satisfying

$$\tilde{u}_t + \mathcal{L}\tilde{u} \leq 0 \quad \text{at } (x_0, t_0) \tag{1.12}$$

and that

$$v(x,t) = u(x,t) - \epsilon t \geq \tilde{u}(x,t) - \epsilon t, \quad \text{such that equality holds at } (x_0, t_0).$$

Since v attains a positive local maximum point at (x_0, t_0), the smooth function $\tilde{u}(x,t) - \epsilon t$ also attains a positive maximum value at (x_0, t_0), i.e.

$$\tilde{u}_t \geq \epsilon, \quad a^{ij} D_{ij}\tilde{u} + b^j D_j \tilde{u} \leq 0, \quad c\tilde{u} \leq 0, \quad \text{at } (x_0, t_0),$$

so that

$$\tilde{u}_t + \mathcal{L}\tilde{u} \geq \epsilon > 0 \quad \text{at } (x_0, t_0),$$

which is in contradiction with (1.12).
Step 2. Suppose $\sup_{\Omega_T} v > 0$ and $v < \sup_{\Omega_T} v$ in Ω_T, i.e. v attains its positive maximum at some $(x_0, t_0) \in \partial\Omega \times (0, T]$. By applying the Boundary Point Lemma (Theorem 1.1.7) we deduce that

$$p^j D_j \tilde{u} \geq p^j D_j v > 0 \quad \text{at } (x_0, t_0).$$

But this is impossible since, according to the definition of generalized subsolutions,

$$p^0 \tilde{u} \geq 0 \quad \text{and} \quad p^j D_j \tilde{u} + p^0 \tilde{u} \leq 0 \quad \text{at } (x_0, t_0).$$

Step 3. By Steps 1 and 2, $v(x,t) = u(x,t) - \epsilon t \leq 0$ in Ω_T for all $\epsilon > 0$. Letting $\epsilon \to 0$, we conclude that $u \leq 0$ in Ω_T. In summary, $u(x, 0) \leq 0$ in Ω implies $u \leq 0$ in Ω_T. This proves assertion 1.
Step 4. Suppose that $u(x_0, t_0) = 0$ for some $(x_0, t_0) \in \Omega \times (0, T]$. Then the Growth Lemma implies $u(x,t) \equiv 0$ in $\Omega \times [0, t_0)$. By continuity, we have $u(x,t) \equiv 0$ in $\overline{\Omega} \times [0, t_0]$.
Step 5. Suppose that $u(x_0, t_0) = 0$ for some $(x_0, t_0) \in S\Omega_T$ and $u < 0$ in $\Omega \times [0, t_0]$. We may repeat the arguments in Step 2 to obtain a contradiction. This completes the proof. □

1.2 The Comparison Principle for Semilinear Equations

Consider the semilinear parabolic equation:

$$u_t + \mathcal{L}u = f(x, t, u, Du) \quad \text{in } \Omega_T, \quad \mathcal{B}u = 0 \quad \text{on } S\Omega_T, \tag{1.13}$$

where $\mathcal{L}$ is a non-divergence form operator with continuous coefficients, $f(x,t,s,p)$ is C^α in x, $C^{\alpha/2}$ in t with $\alpha \in (0,1)$, and C^1 in (s,p); $\mathcal{B}u$ is the oblique boundary operator.

Corollary 1.2.1 *Suppose $u, v \in C(\overline{\Omega}_T)$ satisfies*

$$\begin{cases} u_t + \mathcal{L}u \le f(x,t,u,Du) \quad \textit{and} \quad v_t + Lv \ge f(x,t,v,Dv) & \textit{in } \Omega_T, \\ \mathcal{B}(u-v) \le 0 & \textit{in } S\Omega_T \end{cases} \tag{1.14}$$

in the generalized sense. If $u(x,0) \le v(x,0)$ in Ω, then $u \le v$ in Ω_T.

Proof Suppose $u, v \in C^{2,1}(\overline{\Omega}_T)$ and satisfies (1.14) in the classical sense. It suffices to observe that $w = u - v$ satisfies

$$w_t + \mathcal{L}w \le b(x,t)\cdot Dw + c(x,t)w \quad \text{in } \Omega_T, \quad \mathcal{B}w \le 0 \quad \text{on } S\Omega_T$$

in the generalized sense, where

$$\begin{cases} b(x,t) = \int_0^1 D_4 f(x,t,u(x,t),\xi Du(x,t) + (1-\xi)Dv(x,t))\,\mathrm{d}\xi, \\ c(x,t) = \int_0^1 D_3 f(x,t,\xi u(x,t) + (1-\xi)v(x,t), Dv(x,t))\,\mathrm{d}\xi, \end{cases}$$

and $D_i f$ is the partial derivative of f with respect to the i-th variable. The conclusion follows from part 1 of Theorem 1.1.9. We leave the general case as an exercise (see Problem 1.4.2). □

Definition 1.2.2 Suppose f is independent of t, i.e. $f = f(x,u,Du)$. We say that $\underline{\theta} \in C^2(\Omega) \cap C^1(\overline{\Omega})$ is a subequilibrium of (1.13) in the classical sense if it satisfies in the classical sense the following:

$$\mathcal{L}\underline{\theta} \le f(x,\underline{\theta},D\underline{\theta}) \quad \text{in } \Omega, \qquad \mathcal{B}\underline{\theta} \le 0 \quad \text{on } \partial\Omega. \tag{1.15}$$

And we say that $\underline{\theta} \in C(\overline{\Omega})$ is a subequilibrium of (1.13) in the generalized sense if for each $x_0 \in \overline{\Omega}$, there exist a neighborhood U of x_0 in $\overline{\Omega}$ and a function $\tilde{\theta} \in C^{2,1}(\overline{\Omega})$ such that $\underline{\theta} \ge \tilde{\theta}$ in U, $\underline{\theta}(x_0) = \tilde{\theta}(x_0)$, and $\tilde{\theta}$ satisfies

$$\mathcal{L}\tilde{\theta}(x_0) \le f(x_0,\tilde{\theta}(x_0),D\tilde{\theta}(x_0)) \quad \text{if } x_0 \in \Omega,$$

or

$$\mathcal{B}\tilde{\theta}(x_0) \le 0 \quad \text{if } x_0 \in \partial\Omega.$$

Analogously, we define the notion of superequilibrium in the classical and generalized sense by reversing the above inequalities.

Remark 1.2.3 Suppose f is independent of t. If $\underline{\theta}$ is a subequilibrium of (1.13) in the classical sense (resp. generalized sense), then it is automatically a subsolution of (1.13) in the classical sense (resp. generalized sense). A similar statement holds for superequilibrium.

Corollary 1.2.4 (Monotone iteration method due to D. Sattinger [34]) *Assume that $f = f(x, u, p)$ is independent of t. Suppose $u \in C^{2,1}(\Omega_T) \cap C^{1,0}(\overline{\Omega} \times (0, T]) \cap C(\overline{\Omega}_T)$ is a classical solution of $u_t + \mathcal{L}u = f(x, u, Du)$ and $u_0(x)$ is a subequilibrium (resp. superequilibrium) in the generalized sense, then $t \mapsto u(x, t)$ is nondecreasing (resp. nonincreasing) for every $x \in \Omega$. Moreover, if $T = \infty$, and*

$$\sup_{t \geq 0} \|u(\cdot, t)\|_{L^\infty(\Omega)} < +\infty, \qquad \limsup_{t \to \infty} \|f(\cdot, u(\cdot, t), Du(\cdot, t))\|_{L^\infty(\Omega)} < +\infty,$$

then $u_\infty(x) := \lim_{t \to \infty} u(x, t)$ converges in $C(\overline{\Omega})$, such that $u_\infty \in W^{2,p}(\Omega)$ satisfies $\mathcal{B}u_\infty = 0$ in the classical sense, and is a strong solution of

$$\mathcal{L}u_\infty = f(x, u_\infty, Du_\infty) \quad \text{in } \Omega. \tag{1.16}$$

Proof By Corollary 1.2.1, we have

$$u(x, t_0) \geq u_0(x) \quad \text{for any } (x, t_0) \in \Omega \times (0, T).$$

Next, fix $t_0 \in (0, T)$ and apply the comparison principle to the solutions to $u_t + \mathcal{L}u = f(x, u)$ with initial conditions $u_0(x)$ and $u(x, t_0)$. We deduce that

$$u(x, t + t_0) \geq u(x, t) \quad \text{for any } x \in \Omega,\ t \in (0, T - t_0].$$

This proves the first part.

For the second part, observe first that since $u(x, t)$ is monotone and bounded as $t \to \infty$, the limit $u_\infty(x) := \lim\limits_{t \to \infty} u(x, t)$ exists in the pointwise sense and belongs to $L^\infty(\Omega)$. Next, by the boundedness of u and $f(x, t, u, Du)$ in $L^\infty(\Omega \times [0, \infty))$, we can apply the L^p estimates to deduce that

$$\sup_{t \geq 1} \|u\|_{W^{2,1,p}(\Omega \times [t, t+1])} < \infty.$$

This means $u(x, t + k) - u_\infty(x) \to 0$ as $k \to \infty$ weakly in $W^{2,1,p}(\Omega \times [0, 1])$ and (thanks to Sobolev embedding) strongly in $C^{1+\alpha,(1+\alpha)/2}(\overline{\Omega} \times [0, 1])$. It is standard to check that u_∞ is a strong solution of (1.16), and classical solution of $\mathcal{B}u_\infty = 0$ on $\partial\Omega$. □

Theorem 1.2.5 (Interior Harnack's inequality) *Let $u \in W^{2,1}_{N+1,loc}(\Omega \times (0, T)) \cap C(\overline{\Omega}_T)$ be a strong solution of $u_t + \mathcal{L}u = 0$ in $\Omega \times (0, T)$ on Ω_T. Suppose u is nonnegative on Ω_T. For each compact set $K \subset \Omega$ and $\delta > 0$, there exists a $C_1 = C_1(K)$ such that*

$$\sup_K u(\cdot, s) \leq C_1 \inf_K u(\cdot, t) \quad \textit{for every } s, t \in [0, T] \textit{ satisfying } s \geq \delta,\ t - s \geq \delta. \tag{1.17}$$

If we further assume that $u \in C^{1,0}(\overline{\Omega}_T)$ and $\mathcal{B}u = 0$ on $S\Omega_T$ in the classical sense, then C_1 can be chosen independent of K, i.e.

$$\sup_{\Omega} u(\cdot, s) \le C_1 \inf_{\Omega} u(\cdot, t) \quad \textit{for every } s, t \in [0, T] \textit{ satisfying } s \ge \delta,\ t - s \ge \delta. \tag{1.18}$$

Proof For the proof of (1.17) see, e.g. [13, Theorem 2.2]. The estimate (1.18) is contained in the proof of [22, Theorem 2.5] (which is, in turn, based on the weak Harnack's inequalities [22, Lemma 3.5] and [29, Theorem 7.10]). Here, we shall give a quick proof of (1.18) for the special case of u being a nonnegative solution of the Neumann problem

$$\begin{cases} u_t - \Delta u = c(x,t)u & \text{in } \Omega_T, \\ \partial_n u := n_i(x) D_i u = 0 & \text{on } S\Omega_T, \\ u(x,0) = u_0(x) & \text{in } \Omega, \end{cases} \tag{1.19}$$

where Δ is the Laplace operator; Ω is a bounded domain with smooth boundary $\partial\Omega$; $\mathbf{n}(x) = (n_i(x))$ is the outward unit normal vector on $\partial\Omega$; and $c \in L^\infty(\Omega_T)$. In this case, let $\tilde{u}$ be an extension of u obtained via reflecting across the spatial boundary $S\Omega_T$. It is a standard fact that $\tilde{u}$ satisfies a uniformly parabolic equation in $\tilde{\Omega} \times [0, T]$, for some bounded open set $\tilde{\Omega}$ containing the closure of Ω. Now, we may deduce (1.18) from (1.17). □

Next, we give a useful result due to J. Húska [22, Theorem 2.5], for nonnegative strong solutions. Note that classical solutions are automatically strong solutions.

Theorem 1.2.6 (Harnack principle) *Let $c \in L^\infty(\Omega_T)$, $T \ge \delta_0$. Then there exists a C (depending on δ_0 among other things, but independent of T) such that for every nonnegative strong solution $u \in W^{2,1}_{N+1,loc}(\Omega \times (0, T)) \cap C^{0,1}(\overline{\Omega}_T)$ of $u_t + \mathcal{L}u = 0$ in $\Omega \times (0, T)$, which satisfies $\mathcal{B}u = 0$ on $S\Omega_T$ in the classical sense, we have*

$$\sup_{x \in \Omega} u(x,t) \le C \inf_{x \in \Omega} u(x,t) \quad \textit{for each } t \in [\delta_0, T].$$

Proof By comparing the solution u with the supersolution $\overline{u} = e^{\|c\|_\infty (t - t_1)}$ over the set $\Omega \times [t_1, t_2]$, we deduce

$$\sup_{x \in \Omega} u(x, t_2) \le e^{\|c\|_\infty (t_2 - t_1)} \sup_{x \in \Omega} u(x, t_1) \quad \text{for every } 0 \le t_1 < t_2 \le T. \tag{1.20}$$

Next, we apply (1.18) of Theorem 1.2.5 and obtain

$$\sup_{\Omega} u(\cdot, t - \delta_0) \le C \inf_{\Omega} u(\cdot, t). \tag{1.21}$$

Finally, we combine (1.20) and (1.21) to deduce that there exists $C' = Ce^{\|c\|_\infty \delta_0}$ such that

$$\sup_{\Omega} u(\cdot, t) \le e^{\|c\|_\infty \delta_0} \sup_{\Omega} u(\cdot, t - \delta_0) \le C' \inf_{\Omega} u(\cdot, t). \tag{1.22}$$

This completes the proof. □

Fix $X_0 = (x_0, t_0)$. Define

$$Q(X_0, R) = \{(x, t) : |x - x_0| < R, t_0 - R^2 < t < t_0\}.$$

Theorem 1.2.7 (Local maximum principle) *If $u \in W^{2,1}_{N+1}(Q(X_0, R))$ satisfies $u_t + \mathcal{L}u \le f$ in $Q(X_0, R)$ in the strong sense, then for any $p > 0$ and $r \in (0, 1)$, there is a constant C determined only by p, r and bounds of the coefficients such that*

$$\sup_{Q(X_0, rR)} u \le C \left[\left(R^{-N-2} \int_{Q(X_0,R)} (u^+)^p \, \mathrm{d}X \right)^{1/p} + R^{N/(N+1)} \|f\|_{L^{N+1}(Q(X_0,R))} \right].$$

Proof See [28, Theorem 7.36]. □

Corollary 1.2.8 *Suppose $u \in W^{2,1}_{N+1}(\Omega_T)$ satisfies $u_t - \Delta u + b^i D_i u + cu \le f$ in Ω_T in the strong sense, and satisfies the Neumann boundary condition $n_i D_i u = 0$ on $S\Omega_T$ in the classical sense. Then for any $p > 0$ and $t_0 \in (0, T)$, there is a constant C determined by p, the domain Ω_T, and bounds of the coefficients such that*

$$\sup_{\Omega \times (t_0, T]} u \le C \left[\|u^+\|_{L^p(\Omega_T)} + \|f\|_{L^{N+1}(\Omega_T)} \right].$$

Proof This follows by first reflecting across the spatial boundary, to extend u such that it is a strong solution to $u_t + \Delta u \le f$ in a larger cylinder domain Ω'_T, and such that

$$\sup_{\Omega'_T} u^+ \le C \sup_{\Omega_T} u^+.$$

The conclusion follows from the local maximum principle (Theorem 1.2.7). We omit the details. □

1.3 The Principal Eigenvalue for Linear Elliptic Operators

We recall the classical Krein–Rutman theorem for positive compact linear operators.

Definition 1.3.1 Let K be a subset of a Banach space X.

1. The set K is a *cone* if (i) K is closed and convex, (ii) $\mu K \subset K$ for all $\mu \ge 0$, and (iii) $K \cap (-K) = \{0\}$.
2. K is a *total cone* if it is a cone and $K - K = \{x - y : x, y \in K\}$ is dense in X.
3. K is a *solid cone* if it is a cone with nonempty interior.

Note that a cone K is total if it is solid.

Definition 1.3.2 The cone K in X induces a partial ordering of X. For $x, y \in X$, we write

$$x \le y \qquad (\text{resp.} \qquad x < y, \qquad x \ll y)$$

if

$$y - x \in K \quad (\text{resp. } y - x \in K \setminus \{0\}, \quad y - x \in \operatorname{Int} K).$$

In such a case, we say that X is an ordered Banach space with order induced by the cone K.

We state the classical theorems due to Krein and Rutman [25]. The proofs can be found in Appendix B.

Theorem 1.3.3 (Krein and Rutman, weak form) *Suppose that X is a real ordered Banach space with order induced by a total cone K, and $T : X \to X$ is a compact linear operator with spectral radius $r(T)$. If $T(K) \subset K$ and $r(T) > 0$, then there exist $x_0 \in K \setminus \{0\}$ and $f_0 \in K^* \setminus \{0\}$ such that*

$$Tx_0 = rx_0 \quad \text{and} \quad T^* f_0 = r f_0,$$

where $r = r(T)$, T^* is the adjoint of T, and K^* is the adjoint cone

$$K^* = \{f \in X^* : f(x) \ge 0 \quad \text{for all } x \in K\}.$$

If K is a solid cone in X and T is strongly monotone, then a stronger version of the theorem holds.

Theorem 1.3.4 (Krein and Rutman, strong form) *Suppose that X is a real ordered Banach space with order induced by the cone K. If K is a solid cone and $T : X \to X$ is a compact linear operator which is strongly monotone, i.e. $T(K \setminus \{0\}) \subset \operatorname{Int} K$, then the following statements hold.*

1. *The spectral radius $r(T)$ is positive and is a simple eigenvalue with an eigenvector $v \in \operatorname{Int} K$. Moreover, there is no other eigenvalue with an eigenvector in $K \setminus \{0\}$.*
2. *There exists an $\epsilon > 0$ such that $|\lambda| < r(T) - \epsilon$ for all eigenvalues $\lambda \in \mathbb{C} \setminus \{r(T)\}$.*

In the following section, we will choose to work with eigenfunctions in the strong sense. This is consistent with our applying the Krein–Rutman theorem to the inverse of the elliptic operator $\mathcal{L}$ in the space $C(\bar{\Omega})$ of continuous function on $\bar{\Omega}$. Precisely, the eigenfunctions for the elliptic problem thus belong to $\cap_{p>1} W^{2,p}(\Omega)$. We remark that if all the coefficients are assumed to be Hölder continuous, then the eigenfunctions will be classical by virtue of the Schauder theory.

We will need the following touching lemma for strong solutions to elliptic problems with oblique derivative conditions.

Lemma 1.3.5 *Let a^{ij}, b^i, c, p^i, p^0 be independent of time, and let $p > N$. Suppose $w \in W^{2,p}(\Omega)$ is a nonnegative, strong solution to*

$$\begin{cases} \mathcal{L}w \geq 0 & \text{in } \Omega, \\ \mathcal{B}w \geq 0 & \text{on } \partial\Omega. \end{cases}$$

If $\inf_\Omega w = 0$, then $w \equiv 0$ in $\overline{\Omega}$.

Note that by Sobolev imbedding, w satisfies $\mathcal{B}w \geq 0$ in the classical sense on $\partial\Omega$.

Proof By the interior Harnack's inequality for nonnegative strong supersolutions [16, Theorem 9.22] asserts that, for each x_0 and $R > 0$ such that $B_{2R}(x_0) \subset \Omega$,

$$\left(\frac{1}{|B_R|}\int_{B_R} w^p\right)^{1/p} \leq C \inf_{B_R} w,$$

where p and C are positive constants independent of w. It follows that the set $A = \{x \in \Omega : w(x) = 0\}$ is open in Ω. Since $w \in C(\bar{\Omega})$, we see that A is also closed. Hence $A = \Omega$ or A is empty. In the first case, $w \equiv 0$ and we are done. In the latter case $w > 0$ in Ω and $w(x_0) = 0$ for some $x_0 \in \partial\Omega$. By the Hopf boundary lemma[1], this implies that $\mathcal{B}w = p^i D_i w > 0$ at x_0, which is a contradiction. Hence it must hold that $w \equiv 0$. □

Theorem 1.3.6 *If a^{ij}, b^i, c, p^i, p^0 are independent of time, then the elliptic eigenvalue problem*

$$\begin{cases} \mathcal{L}\phi = \mu\phi & \text{in } \Omega, \\ \mathcal{B}\phi = 0 & \text{on } \partial\Omega \end{cases} \tag{1.23}$$

has a principal eigenvalue μ_1, in the sense that

1. *$\mu_1 \in \mathbb{R}$ is a simple eigenvalue of* (1.23).
2. *$\mu_1 < \inf \operatorname{Re} \mu$, where the infimum is taken over all eigenvalues μ of* (1.23) *which are distinct from μ_1.*
3. *The eigenfunction $\phi_1(x)$ corresponding to μ_1 can be chosen such that $\phi_1 > 0$ in $\overline{\Omega}$.*
4. *If μ is an eigenvalue of* (2.1) *with a nonnegative eigenfunction, then $\mu = \mu_1$.*

Remark 1.3.7 It follows from Theorem 1.3.6 that if the boundary value problem

$$\mathcal{L}w = \mu_0 w \ \text{ in } \Omega, \quad \mathcal{B}w = 0 \ \text{ on } \partial\Omega$$

has a positive solution for some $\mu_0 \in \mathbb{R}$, then μ_0 is necessarily the principal eigenvalue of (1.23).

[1] By the Alexandrov–Bakelman–Pucci maximum principle [16, Theorem 9.1], one can repeat the proof of [16, Lemma 3.4] or [30, Lemma 1.24] to compare the supersolution w (in the strong sense) with the barrier function.

Remark 1.3.8 In fact, the same conclusion holds for the weighted eigenvalue problem:

$$\begin{cases} \mathcal{L}\phi = \mu\beta\phi & \text{in } \Omega, \\ \mathcal{B}\phi = 0 & \text{on } \partial\Omega, \end{cases} \tag{1.24}$$

where $\beta \in C(\overline{\Omega})$ is strictly positive on $\overline{\Omega}$. See Problem 1.4.3.

Proof In the following, we present a proof based on applying the Krein–Rutman theorem to the inverse of elliptic operator $\mathcal{L}$. A shorter proof is possible by considering the semigroup operator of the parabolic problem; this will be presented in Chapter 2. If the elliptic problem is not linear (e.g. $\mathcal{L}$ is only positively homogeneous of degree one and is well-defined only in a cone), then it may be necessary to use the proof based on the semigroup operator; see [19] for an example in this situation.

Recall that $\mathcal{L} = -a^{ij}D_{ij} - b^i D_i - c$, where $a^{ij}, b^i, c \in C^0(\overline{\Omega})$ satisfy the ellipticity condition (1.3). Replacing μ by $\mu - \|c\|_{C(\overline{\Omega})} - 1$ in (1.23), we may assume without loss of generality that $c \leq -1$ in Ω.

Consider the inhomogeneous linear problem

$$\begin{cases} \mathcal{L}v = f & \text{in } \Omega, \\ \mathcal{B}v = 0 & \text{on } \partial\Omega. \end{cases} \tag{1.25}$$

Given $f \in C(\overline{\Omega})$, it is well-known that (1.25) has a unique solution $v \in \cap_{p>1} W^{2,p}(\Omega)$ (see, e.g. [30, Theorem 6.30]). Hence, we can define the linear operator $T : C(\overline{\Omega}) \to C(\overline{\Omega})$ such that $Tf = v$. Moreover, for each $p > 1$, there exists a C such that

$$\|v\|_{W^{2,p}(\Omega)} \leq C\|f\|_{L^p(\Omega)} \leq C'\|f\|_{C(\overline{\Omega})}.$$

The Sobolev embedding theorem (see [16, Chapter 7]) asserts that $W^{2,p}(\Omega)$ is compactly embedded in $C^1(\overline{\Omega})$ for $p > N$. We can therefore take $p > N$ in the above inequality and deduce that the linear operator T is compact.

Next, we remark that T is strongly monotone, by the classical maximum principle for elliptic equations. Indeed, this can be derived from the maximum principle for parabolic equations by observing that, by Duhamel's principle, we have

$$Tf = \int_0^\infty \Phi_t[f]\,\mathrm{d}t,$$

where $\Phi_t : C(\overline{\Omega}) \to C(\overline{\Omega})$ is the semigroup operator so that $\Phi_t[u_0] = u(\cdot, t)$, with $u(x, t)$ being the unique solution to the initial boundary value problem

$$\begin{cases} u_t + \mathcal{L}u = 0 & \text{in } \Omega_T, \\ Bu = 0 & \text{on } S\Omega_T, \\ u(x,0) = u_0(x) & \text{in } \Omega. \end{cases}$$

Since Φ_t is strongly monotone for each $t > 0$ (thanks to the Strong Maximum Principle for linear parabolic equations), so is T.

Finally, we invoke the strong version of the Krein–Rutman theorem (Theorem 1.3.4) to deduce that T has a principal eigenvalue Λ_1, in the sense that (i) $\Lambda_1 \in \mathbb{R}$ is a simple eigenvalue of T; (ii) $\Lambda_1 > \sup|\Lambda|$, where the supremum is taken over all other eigenvalues of T; (iii) the eigenfunction ϕ_1 corresponding to Λ_1 can be chosen so that $\phi_1 > 0$ in $\overline{\Omega}$; (iv) if Λ is an eigenvalue of T with a nonnegative eigenfunction, then $\Lambda = \Lambda_1$. By the definition of T, we deduce that the problem (1.23) has the principal eigenvalue $\mu_1 = \frac{1}{\Lambda_1}$ with the desired properties, except for assertion 2.

We prove assertion 2 by adapting an argument from [12, Chapter 6]. Given any other eigenvalue $\mu \in \mathbb{C}$ of (1.23) such that $\operatorname{Re}\mu \leq \mu_1$, we will show that $\mu = \mu_1$.

Indeed, let $\mathcal{L}u = \mu u$ for some $u \not\equiv 0$, set

$$v = \frac{u}{\phi_1},$$

then direct computation yields

$$\mu v = \frac{1}{\phi_1}\mathcal{L}(v\phi_1) = \mathcal{L}v + cv - \frac{2}{\phi_1}a^{ij}D_j\phi_1 D_i v + \frac{v}{\phi_1}\mathcal{L}\phi_1. \tag{1.26}$$

Writing $\tilde{b}^i = b^i + \frac{1}{\phi_1}a^{ij}D_j\phi_1$ and

$$Kv := -a^{ij}D_{ij}v - \tilde{b}^i D_i v,$$

we deduce from (1.26) that

$$Kv = (\mu - \mu_1)v \quad \text{in } \Omega. \tag{1.27}$$

Taking the complex conjugate, we also derive

$$K\bar{v} = (\bar{\mu} - \mu_1)\bar{v} \quad \text{in } \Omega. \tag{1.28}$$

Next, we compute

$$K(|v|^2) = K(v\bar{v}) = \bar{v}Kv + vK\bar{v} - 2a^{ij}D_i v D_j v \leq \bar{v}Kv + vK\bar{v} - 2\lambda_0|Dv|^2, \tag{1.29}$$

where we used

$$a^{ij}\xi_i\bar{\xi}_j = a^{ij}[\operatorname{Re}\xi_i \operatorname{Re}\xi_j + \operatorname{Im}\xi_i \operatorname{Im}\xi_j] \geq \lambda_0|\xi|^2.$$

Substituting (1.27) and (1.28) into (1.29), we obtain

$$K(|v|^2) \leq 2(\operatorname{Re}\mu - \mu_1)|v|^2 - 2\lambda_0|Dv|^2 \quad \text{in } \Omega.$$

If $\operatorname{Re}\mu \leq \mu_1$, we claim that v is constant, i.e. $u \in \operatorname{span}\{\phi_1\}$. Indeed, let $w = (\sup_\Omega |v|^2) - |v|^2$, then w satisfies

$$\begin{cases} Kw \geq 2\lambda_0 |Dv|^2 \geq 0 \quad \text{and} \quad w \geq, \not\equiv 0 & \text{in } \Omega, \\ p^j D_j w = 0 & \text{on } \partial\Omega. \end{cases} \tag{1.30}$$

By construction, $\inf_\Omega w = 0$. By the touching lemma (see Lemma 1.3.5), it follows that $w \equiv 0$ in Ω. Substituting this back into (1.30), we have that $|Dv| \equiv 0$ in Ω. This implies that $u \in \text{span}\{\phi_1\}$. Hence, $\mu = \mu_1$ and the proof is complete. □

The following lemma gives a characterization of the maximum principle by the existence of a positive strict supersolution. The reader may skip the proof on first reading.

Lemma 1.3.9 *Suppose that the assumptions of Theorem* 1.3.6 *hold. Let* Ω *be a bounded domain with smooth boundary* $\partial\Omega$*, and let* $\Omega' \subset \Omega$ *be a proper subdomain. Suppose there exists a* $w \in C(\overline{\Omega}')$ *such that* $w > 0$ *in* $\overline{\Omega}'$ *and*

$$\begin{cases} \mathcal{L}w \geq 0 & \text{in } \Omega', \\ \mathcal{B}w \geq 0 & \text{in } (\partial\Omega') \cap (\partial\Omega) \end{cases} \tag{1.31}$$

in the generalized sense. Then for any $u \in C(\overline{\Omega})$ *such that*

$$\begin{cases} \mathcal{L}u \leq 0 & \text{in } \Omega', \\ \mathcal{B}u \leq 0 & \text{in } (\partial\Omega') \cap (\partial\Omega), \\ u \leq 0 & \text{on } (\partial\Omega') \cap \Omega \end{cases} \tag{1.32}$$

in the generalized sense, we have

$$u \equiv 0 \quad \text{in } \Omega', \quad \text{or} \quad u < 0 \quad \text{in } \Omega' \cup [(\partial\Omega') \cap (\partial\Omega)].$$

Proof Let

$$\partial\Omega' = \Gamma_1 \cup \Gamma_2, \quad \text{where } \Gamma_1 = (\partial\Omega') \cap (\partial\Omega), \quad \Gamma_2 = (\partial\Omega') \cap \Omega.$$

Since Ω' is a proper subset of Ω, and both are connected, Γ_2 is a nonempty set on which $w > 0 \geq u$. Hence $u \neq kw$ for all $k > 0$. By the strong maximum principle, it suffices to show that $u \leq 0$ in Ω'. Suppose not, then there exists a constant $k > 0$ such that $v = u - kw$ satisfies

$$v \leq 0 \quad \text{in } \Omega', \quad \text{and } v(x_0) = 0 \quad \text{for some } x_0 \in \overline{\Omega}',$$

and also satisfies (1.32) in the generalized sense. On the one hand, the strong maximum principle implies $x_0 \notin \Omega'$. On the other hand, $w > 0$ on $\overline{\Gamma}_2$, so that $v < 0$ on $\overline{\Gamma}_2$ and hence $x_0 \notin \overline{\Gamma}_2$. Hence, $v < 0$ in Ω' and $v(x_0) = 0$ for some $x_0 \in \partial\Omega' \setminus \overline{\Gamma}_2$. Therefore, x_0 lies on the interior of Γ_1, which is the smooth part of $\partial\Omega'$. The Hopf Boundary Lemma implies

$$\mathcal{B}v = p^i D_i v + p^0 v = p^i D_i v > 0 \quad \text{at } x_0.$$

This is a contradiction, since $\mathcal{B}v = \mathcal{B}u - k\mathcal{B}w \leq 0$. □

Corollary 1.3.10 *Suppose that the assumptions of Theorem* 1.3.6 *hold. Let Ω be a bounded domain with smooth boundary $\partial\Omega$. Suppose there exists a $w \in C(\overline{\Omega})$ such that $w > 0$ in $\overline{\Omega}$ and*

$$\begin{cases} \mathcal{L}w \geq 0 & \text{in } \Omega, \\ \mathcal{B}w \geq 0 & \text{in } \partial\Omega \end{cases} \tag{1.33}$$

in the generalized sense. Then for any $u \in C(\overline{\Omega})$ such that

$$\begin{cases} \mathcal{L}u \leq 0 & \text{in } \Omega, \\ \mathcal{B}u \leq 0 & \text{in } \partial\Omega \end{cases} \tag{1.34}$$

in the generalized sense, one of the following holds.

- *$u < 0$ in $\overline{\Omega}$;*
- *$u \equiv 0$ in $\overline{\Omega}$;*
- *$u \equiv kw$ in Ω, and equalities hold in* (1.33) *in the classical sense.*

Remark 1.3.11 If w is a strict positive supersolution, then it must hold that $u \leq 0$, i.e. the maximum principle holds. If there are no strict positive supersolutions, then it must hold that $\mu_1 \leq 0$, where μ_1 is the principal eigenvalue of (1.23) (otherwise the corresponding eigenfunction $\phi_1 > 0$ is a strict positive supersolution). In such a case the maximum principle fails as

$$\mathcal{L}\phi_1 \leq 0 \quad \text{in } \Omega, \quad \mathcal{B}\phi_1 = 0$$

but $\phi_1 > 0$ in $\overline{\Omega}$.

Proof If $u \leq 0$ in Ω, then the strong maximum principle asserts that either $u \equiv 0$ or $u < 0$ in $\overline{\Omega}$. If $u > 0$ somewhere in Ω, repeat the proof of Lemma 1.3.9 with $\Omega' = \Omega$, then there exists a $k > 0$ such that $v = u - kw$ satisfies either $v \equiv 0$ or $v < 0$ in $\overline{\Omega}$. By the choice of k, we must have $v \equiv 0$. This implies that $u \in \operatorname{span}\{w\}$. □

One can estimate the principal eigenvalue by constructing appropriate super/subsolutions, as the following two lemmas illustrate. For simplicity, we assume in addition that $a^{ij}, b^i, c \in C^\alpha(\overline{\Omega})$ and $p^i \in C^{1+\alpha}(\overline{\Omega})$, so that the principal eigenfunction $\phi \in C^{2+\alpha}(\overline{\Omega})$ thanks to the Schauder theory (see [30, Chapter 2]).

Lemma 1.3.12 *Suppose that the assumptions of Theorem* 1.3.6 *hold and let μ_1 be the principal eigenvalue of* (1.23). *Suppose, in addition, that there exist a function $w \in C(\overline{\Omega})$ and a constant $\underline{\mu}$ such that $w > 0$ in $\overline{\Omega}$ is a supersolution in the sense that*

$$\mathcal{L}w \geq \underline{\mu} w \quad \text{in } \Omega, \quad \text{and} \quad \mathcal{B}w \geq 0 \quad \text{on } \partial\Omega \tag{1.35}$$

in the generalized sense. Then $\mu_1 \geq \underline{\mu}$ and equality holds if and only if w is the corresponding eigenfunction, i.e. equalities hold in (1.35).

Proof Let μ_1 and $\phi > 0$ be the principal eigenvalue and positive eigenfunction of (1.23). Define $u = \phi - kw$, where $k > 0$ is the least real number such that $\phi - kw \le 0$.

Suppose that $\mu_1 \le \underline{\mu}$. It remains to show that $\mu_1 = \underline{\mu}$ and w satisfies (1.23) in the classical sense. Indeed, Corollary 1.3.10 implies that $u = \phi - kw \in \text{span}\{\phi\}$. By the choice of k, we must have $u = 0$. Hence $w \in \text{span}\{\phi\}$. □

Lemma 1.3.13 *Suppose that the assumptions of Theorem* 1.3.6 *hold and let* μ_1 *be the principal eigenvalue of* (1.23). *Suppose, in addition, that there exist a nonnegative and nontrivial function* $w \in C(\overline{\Omega})$ *and a constant* $\overline{\mu}$ *such that* w *is a subsolution of* (1.23), *i.e.*

$$\mathcal{L}w \le \overline{\mu}w \quad \text{in } \Omega, \quad \text{and} \quad \mathcal{B}w \le 0 \quad \text{on } \partial\Omega, \tag{1.36}$$

in the generalized sense. Then $\mu_1 \le \overline{\mu}$, *and equality holds if and only if* w *is the corresponding eigenfunction, i.e. equality holds in* (1.36).

Proof Let μ_1 and $\phi > 0$ be the principal eigenvalue and positive eigenfunction of (1.23). Define $u = w - k\phi$, where $k > 0$ is the least real number such that $\phi - kw \le 0$.

Suppose that $\mu_1 \ge \overline{\mu}$, it remains to show that $\mu_1 = \overline{\mu}$ and w satisfies (1.23) in the classical sense. Indeed, Corollary 1.3.10 implies that $u = w - k\phi \in \text{span}\{\phi\}$. By the choice of k, we must have $u = 0$. Hence $w \in \text{span}\{\phi\}$. □

We end this chapter with a classical result: the monotonicity dependence of eigenvalue on the growth rate $c(\cdot)$ and the diffusion rate d. Consider the following elliptic eigenvalue problem.

$$\begin{cases} -d\Delta\phi - c\phi = \mu\phi & \text{in } \Omega, \\ \mathbf{n} \cdot \nabla\phi + p^0\phi = 0 & \text{on } \partial\Omega, \end{cases} \tag{1.37}$$

where $d > 0$ and $c \in C^{\alpha}(\overline{\Omega})$ and $0 \le p^0 \in C^{1+\alpha}(\overline{\Omega})$ for some $\alpha \in (0, 1)$. It follows that the principal eigenfunction $\phi_1 \in C^{2+\alpha}(\overline{\Omega})$ satisfies (1.37) in the classical sense.

Lemma 1.3.14 *Let* $\mu_1(c)$ *be the principal eigenvalue of* (1.37), *and let* $\mu_1(c')$ *be the principal eigenvalue of* (1.37) *with* c *replaced by* c'. *If* $c \le c'$ *in* Ω, *then* $\mu_1(c) \ge \mu_1(c')$.

Proof This follows from Lemma 1.3.12. □

We prove the well-known results concerning the monotonicity of eigenvalue with respect to the diffusion rate. Note that this monotonicity fails in general when the drift is nonzero.

Proposition 1.3.15 *Let* μ_1 *be the principal eigenvalue of* (1.37). *Then* $\mu_1 = \mu_1(d)$ *is smooth and nondecreasing in the diffusion rate* $d > 0$. *If, in addition,* c *is nonconstant or* $p^0 \not\equiv 0$, *then* $\frac{\partial}{\partial d}\mu_1 > 0$ *for* $d > 0$.

Proof By scaling, we may assume without loss of generality that $|\Omega| = 1$. Fix a positive eigenfunction ϕ_1 corresponding to μ_1 and normalize it by $\int_\Omega |\phi_1|^2 = 1$.

Assuming smoothness, we denote by $'$ the partial derivative with respect to the diffusion rate d, and differentiate (1.37) to obtain

$$\begin{cases} d\Delta\phi_1' + c\phi_1' + \mu_1\phi_1' + \Delta\phi_1 = -\mu_1'\phi_1 & \text{in } \Omega, \\ \mathbf{n}\cdot\nabla\phi_1' + p^0\phi_1' = 0 & \text{on } \partial\Omega. \end{cases} \tag{1.38}$$

We claim that

$$\int_\Omega \phi_1\Delta\phi_1' = \int_\Omega \phi_1'\Delta\phi_1. \tag{1.39}$$

Indeed,

$$\begin{aligned} \int_\Omega \phi_1\Delta\phi_1' &= -\int_\Omega \nabla\phi_1\cdot\nabla\phi_1' + \int_{\partial\Omega}\phi_1(\mathbf{n}\cdot\nabla\phi_1') \\ &= \int_\Omega \phi_1'\Delta\phi_1 + \int_{\partial\Omega}\left[\phi_1(\mathbf{n}\cdot\nabla\phi_1') - \phi_1'(\mathbf{n}\cdot\nabla\phi_1)\right] \\ &= \int_\Omega \phi_1'\Delta\phi_1 + \int_{\partial\Omega}\left[\phi_1(-p^0\phi_1') - \phi_1'(-p^0\phi_1)\right] = \int_\Omega \phi_1'\Delta\phi_1, \end{aligned}$$

where we used the homogeneous Robin boundary condition.

Multiplying (1.38) by ϕ_1 and using (1.39), we obtain

$$\begin{aligned} -\mu_1'\int_\Omega |\phi_1|^2 &= \int_\Omega \phi_1(d\Delta\phi_1' + c\phi_1' + \mu_1\phi_1') + \int_\Omega \phi_1\Delta\phi_1 \\ &= \int_\Omega \phi_1(d\Delta\phi_1' + c\phi_1' + \mu_1\phi_1') - \int_\Omega |\nabla\phi_1|^2 + \int_{\partial\Omega}\phi_1(\mathbf{n}\cdot\nabla\phi_1) \\ &= \int_\Omega \phi_1'(d\Delta\phi_1 + c\phi_1 + \mu_1\phi_1) - \int_\Omega |\nabla\phi_1|^2 - \int_{\partial\Omega} p^0|\phi_1|^2 \\ &= -\int_\Omega |\nabla\phi_1|^2 - \int_{\partial\Omega} p^0|\phi_1|^2 \le 0. \end{aligned}$$

This proves that $\mu_1' \ge 0$ for all $d > 0$, i.e. $d \mapsto \mu_1$ is nondecreasing. If, in addition, either c is nonconstant, or $p^0 \not\equiv 0$, then ϕ_1 is nonconstant, so that $\mu_1' > 0$ for all $d > 0$.

It remains to prove the smooth dependence of (μ_1, ϕ_1) on d. We use an idea due to A. Lazer (see [4, Lemma 1.2]) via the Implicit Function Theorem by regarding (μ_1, ϕ_1) as the unique solution of the mapping $\mathcal{F} : \mathbb{R}\times Y\times(0,\infty) \to C^\alpha(\overline{\Omega})$, given by

$$\mathcal{F}\begin{pmatrix}\mu\\ \phi\\ d\end{pmatrix} = \begin{pmatrix} d\Delta\phi + c\phi + \mu\phi \\ \frac{1}{2}\int_\Omega |\phi|^2\,\mathrm{d}x \end{pmatrix},$$

where

$$Y = \{\phi \in C^{2+\alpha}(\overline{\Omega}) : \mathbf{n}\cdot\nabla\phi + p^0\phi = 0 \text{ on } \partial\Omega\}.$$

For each $d > 0$, Theorem 1.3.6 asserts the existence of $(\mu_1, \phi_1) = (\mu_1(d), \phi_1(x; d))$ such that

$$\mathcal{F}(\mu_1(d), \phi_1(\cdot; d), d) = 0.$$

To prove the smooth dependence of (μ_1, ϕ_1) on d, it suffices to show that for each fixed $d > 0$, the linear mapping

$$D_{(\mu,\phi)}\mathcal{F}(\mu_1(d), \phi_1(\cdot; d), d) : \mathbb{R} \times Y \to C^\alpha(\overline{\Omega}) \times \mathbb{R}$$

is invertible. To this end, given $(f, \bar{g}) \in C^\alpha(\overline{\Omega}) \times \mathbb{R}$, we need to prove the existence and uniqueness of $(\bar{h}, w) \in \mathbb{R} \times Y$ such that

$$\begin{cases} d\Delta w + cw + \mu_1 w + \bar{h}\phi_1 = f & \text{in } \Omega, \\ \mathbf{n} \cdot \nabla w + p^0 w = 0 & \text{on } \partial\Omega, \\ \int_\Omega \phi_1 w = \bar{g}. \end{cases} \tag{1.40}$$

First we show the existence. To this end, we choose $\bar{h} = \int_\Omega f\phi_1$, so that

$$\int_\Omega (f - \bar{h}\phi_1)\phi_1 = \left(\int_\Omega f\phi_1\right) - \bar{h} = 0. \tag{1.41}$$

Next, we define $\hat{w} \in W^{1,2}(\Omega)$ to be the unique weak solution to

$$\begin{cases} d\Delta \hat{w} + c\hat{w} + \mu_1 \hat{w} = f - \bar{h}\phi_1 & \text{in } \Omega, \\ \mathbf{n} \cdot \nabla \hat{w} + p^0 \hat{w} = 0 & \text{on } \partial\Omega, \\ \int_\Omega \hat{w}\phi_1 = 0. \end{cases}$$

Such a choice of $\hat{w}$ is well-defined in view of the Fredholm alternative [16, Theorem 5.11], thanks to (1.41) and the fact that μ_1 is a simple eigenvalue with eigenfunction ϕ_1. Furthermore, $\hat{w} \in C^{2+\alpha}(\overline{\Omega})$ according to elliptic regularity theory [30, Theorem 4.40]. Finally, define $w = \hat{w} + \bar{g}\phi_1$, and observe that $(\bar{h}, w)$ satisfies (1.40). This proves the existence.

To show uniqueness, we set $(f, \bar{g}) = (0, 0)$ in (1.40) and proceed to show $(\bar{h}, w) = (0, 0)$. Indeed, multiplying the first equation of (1.40) by ϕ_1, and integrating, we have

$$-\bar{h}\int_\Omega |\phi_1|^2 = \int_\Omega \phi_1(d\Delta w + cw + \mu_1 w) = \int_\Omega w(d\Delta\phi_1 + c\phi_1 + \mu_1\phi_1) = 0,$$

where we used the Robin boundary condition of w and ϕ_1, as we integrate by parts twice for the second equality. Thus $\bar{h} = 0$ and (1.40) becomes

$$\begin{cases} d\Delta w + cw + \mu_1 w = 0 & \text{in } \Omega, \\ \mathbf{n} \cdot \nabla w + p^0 w = 0 & \text{on } \partial\Omega, \\ \int_\Omega w\phi_1 = 0. \end{cases}$$

Since μ_1 is a simple eigenvalue, it follows that $w = k\phi_1$ for some $k \in \mathbb{R}$, so that the integral condition becomes $\int_\Omega k|\phi_1|^2 = 0$. Since $\phi_1 \not\equiv 0$, we have $k = 0$. i.e. $w = 0$. This completes the proof of the smooth dependence of (μ_1, ϕ_1) on d. □

Proposition 1.3.16 *Let μ_1 be the principal eigenvalue of* (1.37), *where $p^0 \geq 0$. Then*

$$\mu_1 \geq -\max_{\overline{\Omega}} c. \tag{1.42}$$

Furthermore, we have

$$\lim_{d\to 0+} \mu_1 = -\max_{\overline{\Omega}} c. \tag{1.43}$$

Proof First, we integrate the first equation of

$$\begin{cases} -d\Delta\phi_1 - c\phi_1 = \mu_1\phi_1 & \text{in } \Omega, \\ \mathbf{n}\cdot\nabla\phi_1 + p^0\phi_1 = 0 & \text{on } \partial\Omega, \end{cases} \tag{1.44}$$

over Ω to obtain

$$-\int_\Omega c\phi_1 \leq \mu_1 \int_\Omega \phi_1. \tag{1.45}$$

In particular, (1.42) holds. Note that (1.42) can also be established by applying Lemma 1.3.12, which is the comparison principle for the principal eigenvalues; see Problem 1.4.9.

Next, we show (1.43). For this purpose, fix $x_0 \in \Omega$ and $0 < r < \mathrm{dist}(x_0, \partial\Omega)$, and let $\tilde{\lambda}$ and $\tilde{\phi}$ be the principal eigenvalue and the positive eigenfunction of

$$-\Delta\tilde{\phi} = \tilde{\lambda}\tilde{\phi} \quad \text{in } B_r(x_0), \quad \tilde{\phi} = 0 \quad \text{on } \partial B_r(x_0).$$

Since $\phi_1 > 0$ in $\overline{B_r(x_0)}$ and $\tilde{\phi} = 0$ on $\partial B_r(x_0)$, up to multiplication of $\tilde{\phi}$ by a positive constant, we may assume that

$$\phi_1 \geq \tilde{\phi} \quad \text{in } B_r(x_0) \quad \text{and} \quad \phi_1(x_0') = \tilde{\phi}(x_0') > 0 \quad \text{for some } x_0' \in B_r(x_0).$$

Therefore, $\Delta\phi_1(x_0') \geq \Delta\tilde{\phi}(x_0')$. Evaluating (1.44) at the point x_0', we have

$$d\tilde{\lambda}\tilde{\phi}(x_0') = -d\Delta\tilde{\phi}(x_0') \geq (c(x_0') + \mu_1)\tilde{\phi}(x_0').$$

Dividing by $\tilde{\phi}(x_0') > 0$, we have

$$\mu_1 \leq -c(x_0') + d\tilde{\lambda} \leq -\inf_{B_r(x_0)} c + d\tilde{\lambda}.$$

Taking lim sup as $d \to 0$, and then sending $r \to 0$, we have

$$\limsup_{d\to 0} \mu_1 \leq -c(x_0) \quad \text{for each } x_0 \in \Omega.$$

Since $x_0 \in \Omega$ is arbitrary, we obtain

$$\limsup_{d\to 0} \mu_1 \leq -\max_{\overline{\Omega}} c. \tag{1.46}$$

Combining with the first inequality in (1.42), we obtain (1.43). □

Next, we prove an elementary result of semi-classical analysis.

Proposition 1.3.17 *Let μ_1 be the principal eigenvalue of* (1.37), *where $p^0 \geq 0$. Let ϕ_1 be a positive eigenfunction associated with μ_1, which is normalized by* $\sup_\Omega \phi_1 = 1$, *then as $d \to 0$,*

$$\phi_1 \to 0 \quad \textit{locally uniformly in } \Omega_0,$$

where $\Omega_0 = \{x \in \overline{\Omega} : c(x) < \sup_\Omega c\}$.

Remark 1.3.18 In particular, the principal eigenfunction ϕ_1 (normalized by $\sup_\Omega \phi_1 = 1$) concentrates on the set where c attains its global maximum value.

Proof Set $d = \epsilon^2$, then ϕ_1 and μ_1 satisfies

$$-\epsilon^2 \Delta\phi_1 = (c(x) + \mu_1)\phi_1 \quad \text{in } \Omega.$$

Next, we introduce the WKB-ansatz (see [14])

$$w_\epsilon = -\epsilon \log \phi_1.$$

Then $w_\epsilon \geq 0$, and satisfies

$$\begin{cases} -\epsilon \Delta w_\epsilon + |\nabla_x w_\epsilon|^2 + c(x) + \mu_1 = 0 & \text{in } \Omega, \\ \mathbf{n} \cdot \nabla w_\epsilon \geq 0 & \text{on } \partial\Omega, \\ \inf_\Omega w_\epsilon = 0. \end{cases} \tag{1.47}$$

Define the semi-relaxed limit

$$w_*(x) := \liminf_{\substack{\epsilon \to 0 \\ x' \to x}} w_\epsilon(x').$$

Since w_ϵ are nonnegative, we deduce that w_* is a well-defined nonnegative function in $\overline{\Omega}$, if we allow the possibility that $w_* = +\infty$ at certain points. The following claim uses the concept of viscosity solution implicitly.

Claim $w_* : \overline{\Omega} \to [0, \infty]$ is lower semicontinuous[2] in $\overline{\Omega}$, $\inf_{\overline{\Omega}} w_* = 0$, and satisfies

$$w_*(x) > 0 \quad \text{in } \Omega_0 = \{x \in \overline{\Omega} : c(x) < \sup_\Omega c\}. \tag{1.48}$$

The lower semicontinuity follows by construction. Next, choose a point $x_\epsilon \in \overline{\Omega}$ such that $w_\epsilon(x_\epsilon) = 0$, we can then pass to a sequence such that $x_\epsilon \to x_1$ for some

[2] A function $w : \overline{\Omega} \to [-\infty, \infty]$ is said to be lower (resp. upper) semicontinuous at the point x_0 if

$$\liminf_{x \to x_0} \geq f(x_0) \qquad (\text{resp.} \quad \limsup_{x \to x_0} \leq f(x_0)).$$

It is lower (resp. upper) semicontinuous in a set $A \subset \overline{\Omega}$ if it is lower (resp. upper) semicontinuous at every point of A.

$x_1 \in \bar{\Omega}$. It then follows that $w_*(x_1) \leq 0$. Since w_* is nonnegative, this proves that $\inf_{\overline{\Omega}} w_* = 0$.

Next, suppose to the contrary that there exists an $x_0 \in \overline{\Omega}$ such that $c(x_0) < \sup_\Omega c$ and $w_*(x_0) = 0$. Then $w_* + |x - x_0|^2$ has a strict local minimum at x_0. By construction, there exists a $y_\epsilon \in \overline{\Omega}$ such that $y_\epsilon \to x_0$ and such that $w_\epsilon + |x - x_0|^2$ attains a local minimum at y_ϵ. We first consider the case when x_0 is an interior point, wherein y_ϵ are interior points as well. Then

$$\nabla(w_\epsilon + |x - x_0|^2) = 0 \quad \text{and} \quad \Delta(w_\epsilon + |x - x_0|^2) \geq 0 \quad \text{at } y_\epsilon,$$

so that

$$c(y_\epsilon) + \mu_1 = -|\nabla_x w_\epsilon(y_\epsilon)|^2 + \epsilon \Delta w_\epsilon(y_\epsilon) \geq -|2(y_\epsilon - x_0)|^2 - 2N\epsilon,$$

where N is the dimension of Ω. Letting $\epsilon \to 0$, then $\mu_1 \to -\sup_\Omega c$ (in view of Proposition 1.3.16) and we obtain

$$c(x_0) - \sup_\Omega c \geq 0.$$

This contradicts $c(x_0) < \sup_\Omega c$. Hence, (1.48) holds if $x_0 \in \Omega$. If $x_0 \in \partial\Omega$, we consider instead the local minimum point y_ϵ of $w_\epsilon(x) + |x - x_0 + \epsilon n_0|^2$, where n_0 is the outer unit normal vector at x_0. Using the boundary condition of w_ϵ, there exists an $r > 0$ such that for all ϵ small,

$$\mathbf{n} \cdot \nabla(w_\epsilon(x) + |x - x_0 + \epsilon n_0|^2) > 0 \quad \text{on } \partial\Omega \cap B_r(x_0).$$

It follows that the local minimum y_ϵ is attained in the interior, and one can argue similarly as in the case $x_0 \in \Omega$. Hence, (1.48) holds and the claim is proved.

Next, define

$$\Omega_0^\delta = \{x \in \Omega : c(x) < \sup_\Omega c - \delta\},$$

where $\delta > 0$ is the same as in the beginning of the proof. By the lower semicontinuity of w_*, the following positive number is well-defined.

$$\eta := \frac{1}{2} \inf_{\Omega_0^\delta} w_* > 0.$$

Hence, there exists a $\delta' \in (0, \delta)$ such that for $\epsilon \in (0, \delta')$, we have $\inf_{\Omega_0^\delta} w_\epsilon \geq \eta$. Hence, we deduce that

$$\sup_{\{x\in\Omega:\ c(x)<\sup c-\delta\}} \phi_1 \leq e^{-\frac{\eta}{\epsilon}} \quad \text{for all } 0 < \epsilon \ll 1.$$

Since δ is arbitrary, this proves that $\phi_1 \to 0$ in $C_{loc}\left(\{x \in \overline{\Omega} : c(x) < \max_{\overline{\Omega}} c\}\right)$. □

Proposition 1.3.19 *Let μ_1 and ϕ_1 be the principal eigenvalue and positive eigenfunction of* (1.37). *If $p^0 \equiv 0$, then*

$$-\max_{\overline{\Omega}} c \le \mu_1 \le -\frac{1}{|\Omega|}\int_\Omega c. \tag{1.49}$$

Furthermore, we have

$$\lim_{d\to 0+} \mu_1 = -\max_{\overline{\Omega}} c, \tag{1.50}$$

$$\lim_{d\to\infty} \mu_1 = -\frac{1}{|\Omega|}\int_\Omega c, \tag{1.51}$$

and that, as $d \to \infty$, $\phi_1 \to 1$ weakly in $W^{2,p}(\Omega)$ and strongly in $C^1(\overline{\Omega})$.

Proof First, we integrate

$$\begin{cases} -d\Delta\phi_1 - c\phi_1 = \mu_1\phi_1 & \text{in } \Omega, \\ \mathbf{n}\cdot\nabla\phi_1 = 0 & \text{on } \partial\Omega \end{cases} \tag{1.52}$$

over Ω to obtain

$$-\int_\Omega c\phi_1 = \mu_1 \int_\Omega \phi_1. \tag{1.53}$$

In particular,

$$-\max_{\overline{\Omega}} c \le \mu_1 \le -\min_{\overline{\Omega}} c. \tag{1.54}$$

Next, divide (1.52) by ϕ_1 (note that $\phi_1 > 0$ in $\overline{\Omega}$) and integrate by parts,

$$|\Omega|\mu_1 = -d\int_\Omega \frac{|\nabla\phi_1|^2}{|\phi_1|^2} - \int_\Omega c \le -\int_\Omega c. \tag{1.55}$$

Combining (1.54) and (1.55), we derive (1.49). We also note that (1.50) is proved in Proposition 1.3.19.

To show (1.51), we normalize the principal eigenfunction ϕ_1 by $\sup_\Omega \phi_1 = 1$. Dividing (1.52) by d, we have

$$-\Delta\phi_1 = \frac{1}{d}(c + \mu_1)\phi_1. \tag{1.56}$$

By using elliptic L^p estimates, we can pass to a subsequence to get $\phi_1 \to \phi_\infty$ weakly in $W^{2,p}(\Omega)$ and strongly in $C^1(\overline{\Omega})$, where ϕ_∞ is a strong solution of

$$-\Delta\phi_\infty = 0 \quad \text{in } \Omega, \qquad \text{and} \qquad \mathbf{n}\cdot\nabla\phi_\infty = 0 \quad \text{on } \partial\Omega.$$

i.e. ϕ_∞ is a constant. Since $\phi_1 \to \phi_\infty$ uniformly and $\sup_\Omega \phi = 1$, it follows that $\phi_\infty \equiv 1$. In conclusion, $\phi_1 \to 1$ in $C(\overline{\Omega})$ as $d \to \infty$. Letting $d \to \infty$ in (1.53), we obtain the desired result. □

To close the chapter, we consider the one-dimensional domain $\Omega = (0, L)$, and state a result concerning the dependence on drift.

Lemma 1.3.20 *Let μ_1 be the principal eigenvalue of*

$$d\psi'' + \alpha\psi' + c\psi + \mu_1\psi = 0 \quad \text{in } (0, L), \quad \text{and} \quad \psi'(0) = \psi'(L) = 0.$$

If $c(x)$ is nonconstant and nonincreasing, then $\frac{\partial}{\partial\alpha}\mu_1 > 0$ for $\alpha \in \mathbb{R}$.

Proof Refer to [20, Lemma 5.2] or Lemma 8.3.7 in Chapter 8. □

1.4 Further Reading

In this text, we assume the knowledge of the regularity theory for elliptic and parabolic equations. A nice overview of the basic theory can be found in [21] with complete references to find specific proofs. The classical reference for regularity of elliptic equations is [16], and a more recent account of the oblique derivative problem can be found in [30]; see also [9]. For the regularity theory of parabolic equations, see [15, 28]. An alternative to construct solutions to parabolic equations is via semigroup theory, which is based on the variation of constants formulation. This latter approach relies only on elliptic regularity estimates; see [17, 31].

An influential paper on monotone methods for elliptic and parabolic equations is [34], which contains in particular Corollary 1.2.4.

The classical reference for maximum principles is [33], in which the maximum principle in more general, noncylindrical domains, are treated for both elliptic and parabolic equations.

We applied the Krein–Rutman theorem 1.3.3, for compact linear operator preserving a cone, to derive the principal eigenvalue for elliptic problems; see Appendix B for the proof of various versions of the Krein–Rutman theorem. Our estimate for the spectral gap (assertion 2 of Theorem 1.3.6) is taken from [12, Chapter 6]. The characterization of maximum principle by principal eigenvalue is a classical result; see, e.g. [3] for the Dirichlet case in general bounded domains. A more practical "rule of thumb" can be stated as follows: for a given linear elliptic boundary value problem, the validity of the maximum principle is equivalent to the existence of a strict positive supersolution; see Corollary 1.3.10. Eigenvalue problems with integral operators arise when studying persistence questions in nonlocal diffusion models; see, e.g. [5, 27] for the spectral theory.

We also mention here the recent work by Chang et al. [6], where the authors prove in an elementary way the strong version of the Krein–Rutman theorem for semi-strongly positive operators (including strongly positive operators), and study the relationship between the semi-strong positivity and the ideal-irreducibility in a Banach lattice, as well as the upper and lower spectral radii for reducible linear positive operators. For reducible operators, they prove that the lower spectral radius always serves as the least upper bound of the set of eigenvalues pertaining to positive eigenvectors, and the upper spectral radius the greatest lower bound of the set. Moreover, they apply these abstract results on some PDE examples.

The smooth dependence of the principal eigenvalue, proved in Proposition 1.3.15, is due to an idea of Lazer, which appeared in [4]. See also Chapter 4 for the recent generalization to the principal Floquet bundle.

In the absence of advection, we proved the monotonicity of the principal eigenvalue in diffusion rate in Proposition 1.3.15. This is connected to the so-called *reduction principle* due to L. Altenberg [1] concerning the monotonicity of the spectral bound $s(\rho A + Q)$ in $\rho \in \mathbb{R}_+$, where A is a compact linear operator in a Banach space with positive resolvents, and Q is a multiplicative operator. This, in turn, is a generalization of a theorem due to Karlin [23] for the finite-dimensional case; see also [7, 24].

In Proposition 1.3.17, we use the WKB-ansatz to show that the eigenfunction of (1.37), when normalized in L^1, tends to a measure μ supported on the set of global maximum points of the zero-th order coefficient. The determination of this measure is known as the semi-classical limits; see [18] and the references therein. In general, if the zero-th order coefficient is a Morse function, then the set of global maximum points is finite and the weight of μ at each of these point is determined by the associated Hessian.

Problems

1.4.1 Let $\mathcal{L} = -a^{ij}D_{ij} - b^j D_j - c$ be uniformly elliptic and have continuous coefficients. Suppose $c \le 0$ and $u \in C(\overline{\Omega}_T)$ satisfies $u_t + \mathcal{L}u \le 0$ and $u_t + \mathcal{L}u \ge 0$ in Ω_T in the generalized sense, then

$$\max_{\overline{\Omega}_T} |u| \le \max_{P\Omega_T} |u| \quad \text{in } \Omega_T.$$

1.4.2 Prove Corollary 1.2.1 by adopting the following steps.

(a) By the transformation $\tilde{u}(x,t) = \mathrm{e}^{-Mt}u(x,t)$, one may assume without loss of generality that $u \mapsto f(x,t,u,p)$ is strictly decreasing.

(b) Suppose there exists an $\epsilon > 0$ such that $u - v - \epsilon t$ attains a positive local maximum at some $(x_0, t_0) \in \Omega \times (0, T]$, then there exist smooth functions $\tilde{u}, \tilde{v}$ such that $\tilde{w} = \tilde{u} - \tilde{v}$ attains a local maximum at (x_0, t_0), whence

$$0 > f(x_0, t_0, \tilde{u}, \nabla u) - f(x_0, t_0, \tilde{v}, \nabla v) = \tilde{w}_t - \mathcal{L}\tilde{w} \ge \epsilon,$$

which is impossible.

(c) Suppose $u - v$ attains a positive local maximum at some $(x_0, t_0) \in \partial\Omega \times (0, T]$, and $(u-v)(x,t) < (u-v)(x_0, t_0)$ for all $(x,t) \in \Omega \times (0,T]$, then one can similarly choose smooth functions $\tilde{u}, \tilde{v}$ such that $\tilde{w} = \tilde{u} - \tilde{v}$ satisfies $\tilde{w}_t - \mathcal{L}\tilde{w} \ge \epsilon$ at (x_0, t_0). Apply the Boundary Point Lemma to obtain a contradiction.

1.4.3 [Generalization of Theorem 1.3.6 to the elliptic eigenvalue problem with a positive weight.] Suppose $a^{ij}, b^i, c, p^i, p^0, \beta$ are independent of time, and $\inf_\Omega \beta > 0$. Show that the elliptic eigenvalue problem (1.24) has a principal eigenvalue μ_1, in the sense that the conclusions of Theorem 1.3.6 hold.

1.4.4 [This problem provides a justification of constructing a global subsolution by gluing local subsolutions, in the H^1 framework.] Consider the elliptic equation in divergence form

$$\begin{cases} -D_i(a^{ij}D_ju + bu) + c^iD_iu + du = f & \text{in } \Omega, \\ n_i \cdot (a^{ij}D_ju + bu) = g & \text{on } \partial\Omega, \end{cases}$$

where $\mathbf{n} = (n_i)$ is the unit outer normal vector on $\partial\Omega$. Prove that if Ω can be partitioned such that $\Omega = \cup\overline{A}_i$, and $\overline{A}_i \subset B_i$, where A_i, B_i are open sets, if u_i is a weak subsolution in B_i for each i, and if $u := \max\{u_i\}$ satisfies $u = u_i$ in A_i, then u is a weak subsolution in Ω. [Hint: First, by approximation one can assume without loss of generality that u_i is C^1 for all i. In this case, the conclusion holds if zero is a regular value of $u_i - u_j$. By approximating u_i by $u_i + \epsilon_i$, we can always assume that zero is a regular value of $u_i - u_j$, for all $i \neq j$.]

1.4.5 Let μ_1 be the principal eigenvalue of

$$\begin{cases} -d\partial_i(a^{ij}(x)D_j\phi) - c(x)\phi = \mu\phi & \text{in } \Omega, \\ n_i(x)a^{ij}(x)D_j\phi + p^0\phi = 0 & \text{on } \partial\Omega, \end{cases} \tag{1.57}$$

where $\mathbf{n}(x) = (n_i(x))$ is the outward unit normal vector on $\partial\Omega$, d is a positive constant, $a^{ij}, c, p^0 \in C(\overline{\Omega})$ are such that a^{ij} is uniformly elliptic and $p^0 \geq 0$. Show that μ_1 is nondecreasing in d. Also, show that μ_1 is strictly increasing in d if either c is nonconstant or p^0 is nontrivial.

1.4.6 Let Ω be a bounded domain in $\mathbb{R}^N$ with smooth boundary, and let $\mathcal{L}$ be an elliptic operator satisfying the condition in the beginning of the chapter, with $a^{ij}, b^j, c \in C^\alpha(\overline{\Omega})$. Let λ_N be the principal eigenvalue of the Laplacian

$$\mathcal{L}\varphi + \lambda\varphi = 0 \quad \text{in } \Omega, \quad \mathbf{n} \cdot \nabla\varphi = 0 \text{ on } \partial\Omega,$$

and λ_D be the principal eigenvalue of the same equation with the zero Dirichlet boundary condition $\varphi = 0$ on $\partial\Omega$. Show that $\lambda_D > \lambda_N$.

1.4.7 Under the assumptions of Theorem 1.2.6, show that for each $\delta_0, M > 0$, there exists a C depending on δ_0, M such that

$$\sup_\Omega u(\cdot, t_1) \leq C \inf_\Omega u(\cdot, t_2) \quad \text{for any } t_1, t_2 \in [\delta_0, T], \text{ such that } |t_1 - t_2| \leq M.$$

(Note that the constant C is independent of T, and also that t_1 can be greater than, equal to, or less than t_2.)

1.4.8 Let $\theta(x)$ be a positive equilibrium of

$$\begin{cases} \partial_t u + \mathcal{L}u = uF(x,u) & \text{in } \Omega \times (0,\infty), \\ \mathcal{B}u = 0 & \text{on } \partial\Omega \times (0,\infty), \\ u(x,0) = u_0(x) & \text{in } \Omega. \end{cases} \tag{1.58}$$

Suppose $\overline{\theta}$ is a positive superequilibrium, i.e.

$$\mathcal{L}\overline{\theta} \geq \overline{\theta}F(x,\overline{\theta}) \text{ in } \Omega, \quad \text{and} \quad \mathcal{B}\overline{\theta} \geq 0 \text{ on } \partial\Omega.$$

If $s \mapsto F(x,s)$ is strictly decreasing in s, show that

$$\theta(x) \leq \overline{\theta}(x) \quad \text{in } \Omega.$$

Similarly, one can also show $\theta(x) \geq \underline{\theta}(x)$ in Ω for each nonnegative subequilibrium $\underline{\theta}$. [Hint: Apply Corollary 1.3.10 and Remark 1.3.11 by choosing w and u in the statement of the corollary as $w = \theta$ and $u = \theta - \overline{\theta}$.]

1.4.9 Let μ_1 be the principal eigenvalue of (1.37), where $d > 0$ and $c \in C^\alpha(\overline{\Omega})$ and $0 \leq p^0 \in C^{1+\alpha}(\overline{\Omega})$ for some $\alpha \in (0,1)$. Show that $\mu_1 \geq -\sup_\Omega c$ by applying Lemma 1.3.12.

References

1. L. Altenberg, *Resolvent positive linear operators exhibit the reduction phenomenon*, Proc. Natl. Acad. Sci. USA, 109 (2012), pp. 3705–3710.
2. H. Berestycki and P.-L. Lions, *Some applications of the method of super and subsolutions*, in Bifurcation and nonlinear eigenvalue problems (Proc., Session, Univ. Paris XIII, Villetaneuse, 1978), vol. 782 of Lecture Notes in Math., Springer, Berlin, 1980, pp. 16–41.
3. H. Berestycki, L. Nirenberg, and S. R. S. Varadhan, *The principal eigenvalue and maximum principle for second-order elliptic operators in general domains*, Comm. Pure Appl. Math., 47 (1994), pp. 47–92.
4. R. S. Cantrell and C. Cosner, *On the steady-state problem for the Volterra–Lotka competition model with diffusion*, Houston J. Math., 13 (1987), pp. 337–352.
5. R. S. Cantrell, C. Cosner, and K.-Y. Lam, *Resident-invader dynamics in infinite dimensional systems*, J. Differential Equations, 263 (2017), pp. 4565–4616.
6. K.-C. Chang, X. Wang, and X. Wu, *On the spectral theory of positive operators and PDE applications*, Discrete Contin. Dyn. Syst., 40 (2020), pp. 3171–3200.
7. S. Chen, J. Shi, Z. Shuai, and Y. Wu, *Two novel proofs of spectral monotonicity of perturbed essentially nonnegative matrices with applications in population dynamics*, SIAM Journal on Applied Mathematics, 82 (2022), pp. 654–676.
8. X. Chen, K.-Y. Lam, and Y. Lou, *Dynamics of a reaction-diffusion-advection model for two competing species*, Discrete Contin. Dyn. Syst., 32 (2012), pp. 3841–3859.
9. Y.-Z. Chen and L.-C. Wu, *Second order elliptic equations and elliptic systems*, vol. 174 of Translations of Mathematical Monographs, American Mathematical Society, Providence, RI, 1998. Translated from the 1991 Chinese original by Bei Hu.
10. M. G. Crandall, H. Ishii, and P.-L. Lions, *User's guide to viscosity solutions of second order partial differential equations*, Bull. Amer. Math. Soc. (N.S.), 27 (1992), pp. 1–67.

11. Y. Du, *Order structure and topological methods in nonlinear partial differential equations. Vol. 1: Maximum principles and applications*, vol. 2 of Series in Partial Differential Equations and Applications, World Scientific Publishing Co. Pte. Ltd., Hackensack, NJ, 2006.
12. L. C. Evans, *Partial differential equations*, vol. 19 of Graduate Studies in Mathematics, American Mathematical Society, Providence, RI, second ed., 2010.
13. E. B. Fabes, M. V. Safonov, and Y. Yuan, *Behavior near the boundary of positive solutions of second order parabolic equations. II*, Trans. Amer. Math. Soc., 351 (1999), pp. 4947–4961.
14. M. I. Freidlin and A. D. Wentzell, *Random perturbations of dynamical systems*, vol. 260 of Grundlehren der mathematischen Wissenschaften [Fundamental Principles of Mathematical Sciences], Springer, Heidelberg, third ed., 2012. Translated from the 1979 Russian original by Joseph Szücs.
15. A. Friedman, *Partial differential equations of parabolic type*, Prentice-Hall, Inc., Englewood Cliffs, N.J., 1964.
16. D. Gilbarg and N. S. Trudinger, *Elliptic partial differential equations of second order*, Classics in Mathematics, Springer-Verlag, Berlin, 2001. Reprint of the 1998 edition.
17. D. Henry, *Geometric theory of semilinear parabolic equations*, vol. 840 of Lecture Notes in Mathematics, Springer-Verlag, Berlin-New York, 1981.
18. D. Holcman and I. Kupka, *Singular perturbation for the first eigenfunction and blow-up analysis*, Forum Math., 18 (2006), pp. 445–518.
19. S.-B. Hsu, K.-Y. Lam, and F.-B. Wang, *Single species growth consuming inorganic carbon with internal storage in a poorly mixed habitat*, J. Math. Biol., 75 (2017), pp. 1775–1825.
20. S.-B. Hsu and Y. Lou, *Single phytoplankton species growth with light and advection in a water column*, SIAM J. Appl. Math., 70 (2010), pp. 2942–2974.
21. B. Hu, *Blow-up theories for semilinear parabolic equations*, vol. 2018 of Lecture Notes in Mathematics, Springer, Heidelberg, 2011.
22. J. Húska, *Harnack inequality and exponential separation for oblique derivative problems on Lipschitz domains*, J. Differential Equations, 226 (2006), pp. 541–557.
23. S. Karlin, *Classifications of selection-migration structures and conditions for a protected polymorphism*, Evol. Biol, 14 (1982), pp. 61–204.
24. S. Kirkland, C.-K. Li, and S. J. Schreiber, *On the evolution of dispersal in patchy landscapes*, SIAM J. Appl. Math., 66 (2006), pp. 1366–1382.
25. M. G. Kreĭn and M. A. Rutman, *Linear operators leaving invariant a cone in a Banach space*, Uspehi Matem. Nauk (N. S.), 3 (1948), pp. 3–95.
26. K.-Y. Lam, Y. Lou, and F. Lutscher, *The emergence of range limits in advective environments*, SIAM J. Appl. Math., 76 (2016), pp. 641–662.
27. F. Li, J. Coville, and X. Wang, *On eigenvalue problems arising from nonlocal diffusion models*, Discrete Contin. Dyn. Syst., 37 (2017), pp. 879–903.
28. G. M. Lieberman, *Second order parabolic differential equations*, World Scientific Publishing Co., Inc., River Edge, NJ, 1996.
29. ———, *Pointwise estimates for oblique derivative problems in nonsmooth domains*, J. Differential Equations, 173 (2001), pp. 178–211.
30. ———, *Oblique derivative problems for elliptic equations*, World Scientific Publishing Co. Pte. Ltd., Hackensack, NJ, 2013.
31. A. Lunardi, *Analytic semigroups and optimal regularity in parabolic problems*, Modern Birkhäuser Classics, Birkhäuser/Springer Basel AG, Basel, 1995. [2013 reprint of the 1995 original] [MR1329547].
32. L. Nirenberg, *A strong maximum principle for parabolic equations*, Comm. Pure Appl. Math., 6 (1953), pp. 167–177.
33. M. H. Protter and H. F. Weinberger, *Maximum principles in differential equations*, Springer-Verlag, New York, 1984. Corrected reprint of the 1967 original.
34. D. H. Sattinger, *Monotone methods in nonlinear elliptic and parabolic boundary value problems*, Indiana Univ. Math. J., 21 (1971/72), pp. 979–1000.

Chapter 2
The Principal Eigenvalue for Periodic-Parabolic Problems

Abstract In this chapter, we establish the existence of the principal eigenvalue for periodic-parabolic operators. We study some qualitative properties of the principal eigenvalues, including their small and large diffusion limits and monotonicity. As an application of the principal eigenvalue theory, we investigate the evolution of dispersal in spatially and temporally varying environments. Finally we present some references for the principal eigenvalue theory and other applications.

2.1 Existence of the Principal Eigenvalue for Periodic-Parabolic Problems

In the following section, we will choose to work with eigenfunctions that satisfy the PDE in the strong sense. This is consistent with our applying the Krein–Rutman Theorem to the semigroup operator $\Phi_t = \mathrm{e}^{-t\mathcal{L}}$ in the space $C(\overline{\Omega})$ of continuous functions on $\overline{\Omega}$. In such a case, the eigenfunctions for the parabolic problem (resp. elliptic problem) belong to $\cap_{p>1} W_p^{2,1}(\Omega_T)$ (resp. $\cap_{p>1} W^{2,p}(\Omega)$). We remark that if all the coefficients are assumed to be Hölder continuous, then the eigenfunctions will be classical by virtue of the Schauder theory [22, Chapter IV].

Theorem 2.1.1 *If a^{ij}, b^i, c, p^i, p^0 are T-periodic and continuous, then the periodic-parabolic eigenvalue problem*

$$\begin{cases} \varphi_t + \mathcal{L}\varphi = \lambda\varphi & \text{in } \Omega_T, \\ \mathcal{B}\varphi = 0 & \text{on } S\Omega_T, \\ \varphi(x,0) = \varphi(x,T) & \text{for } x \in \Omega \end{cases} \tag{2.1}$$

has a principal eigenvalue λ_1, in the sense that

1. *$\lambda_1 \in \mathbb{R}$ is a simple eigenvalue of* (2.1).
2. *$\lambda_1 < \inf_{\lambda \neq \lambda_1} \operatorname{Re}\lambda$, where the infimum is taken over all eigenvalues λ of* (2.1) *which are distinct from λ_1.*

K. -Y. Lam, Y. Lou, *Introduction to Reaction-Diffusion Equations*, Lecture Notes on Mathematical Modelling in the Life Sciences, https://doi.org/10.1007/978-3-031-20422-7_2

3. *The eigenfunction corresponding to* λ_1 *can be chosen strictly positive in* $\overline{\Omega}_T$.
4. *If* λ *is an eigenvalue of* (2.1) *with a nonnegative eigenfunction, then* $\lambda = \lambda_1$.

Proof For each $t > 0$, consider the evolution operator $\Phi_t : C(\overline{\Omega}) \to C(\overline{\Omega})$, which for any solution $u(x, t)$ to

$$\begin{cases} u_t + \mathcal{L}u = 0 & \text{in } \Omega_T, \\ \mathcal{B}u = 0 & \text{on } S\Omega_T, \\ u(x, 0) = u_0(x) & \text{in } \Omega, \end{cases}$$

satisfies

$$\Phi_t[u_0] = u(\cdot, t) \quad \text{for } t > 0.$$

We claim that λ is an eigenvalue of (2.1) with eigenfunction $\varphi(x, t)$ if and only if $e^{\lambda T}$ is an eigenvalue of Φ_T with eigenfunction $\varphi(x, 0)$. Indeed, suppose $(\lambda, \varphi(x, t))$ is an eigenpair of (2.1), then it follows that

$$\Phi_t[\varphi(\cdot, 0)] = e^{-\lambda t}\varphi(\cdot, t) \quad \text{for all } t \in [0, T].$$

In particular,

$$\Phi_T[\varphi(\cdot, 0)] = e^{-\lambda T}\varphi(\cdot, T) = e^{-\lambda T}\varphi(\cdot, 0),$$

i.e. $\varphi(\cdot, 0)$ is an eigenfunction of Φ_T associated with the eigenvalue $e^{-\lambda T}$. On the other hand, suppose $\Phi_T[\phi] = \Lambda\phi$, then

$$\lambda = -\frac{\log \Lambda}{T} \quad \text{and} \quad \varphi(\cdot, t) = e^{\lambda t}\Phi_t[\phi] \tag{2.2}$$

defines an eigenpair of (2.1).

Next, observe that the operator $\Phi_T : C(\overline{\Omega}) \to C(\overline{\Omega})$ is compact, and strongly monotone. The compactness follows from the parabolic L^p estimate, whereas the strong monotonicity follows from the strong maximum principle for linear parabolic equations with oblique boundary conditions (Theorem 1.1.9).

Finally, we invoke Theorem 1.3.4 to deduce that Φ_T has a principal eigenvalue Λ_1, in the sense that (i) $\Lambda_1 \in \mathbb{R}$ is a simple eigenvalue of Φ_T; (ii) $\Lambda_1 > \sup |\Lambda|$, where the supremum is taken over all other eigenvalues of Φ_T; (iii) the eigenfunction ϕ_1 corresponding to Λ_1 can be chosen strictly positive in $\overline{\Omega}$; and (iv) if Λ is an eigenvalue of Φ_T with a nonnegative eigenfunction, then $\Lambda = \Lambda_1$. By virtue of the correspondence (2.2), we deduce that the problem (2.1) has the principal eigenvalue $\lambda_1 = \frac{1}{T}\log \Lambda_1$ with the desired properties. □

An alternative proof of the existence of principal eigenvalue for the elliptic problem can be given via the semigroup operator Φ_t.

Proof (Alternative proof of Theorem 1.3.6) Let $\Phi_t : C(\overline{\Omega}) \to C(\overline{\Omega})$ be the semigroup operator which for any solution $u(x, t)$ to

$$\begin{cases} u_t + \mathcal{L}u = 0 & \text{in } \Omega_T, \\ \mathcal{B}u = 0 & \text{on } S\Omega_T, \\ u(x,0) = u_0(x) & \text{in } \Omega, \end{cases}$$

where all coefficients are independent of t, satisfies

$$\Phi_t[u_0] = u(\cdot, t) \quad \text{for } t > 0.$$

Then Φ_t is strongly monotone, by the strong maximum principle for linear parabolic equations.

Step 1. Suppose (1.23) has a real eigenvalue μ_1 with a nonnegative, nontrivial eigenfunction ϕ_1, then it is easy to see that

$$\Phi_t[\phi_1] = \mathrm{e}^{-\mu_1 t}\phi_1 \quad \text{for each } t > 0.$$

Since ϕ_1 is nonnegative and nontrivial, it follows from the Krein–Rutman Theorem (Theorem 1.3.4 applied to the linear operator Φ_t) that $\phi_1 > 0$ in $\overline{\Omega}$, $r(\Phi_t) = \mathrm{e}^{-\mu_1 t}$ for each $t > 0$, and μ_1 satisfies the properties 1 to 4 as stated in the theorem.

Step 2. It remains to show the existence of an eigenvalue μ_1 corresponding to a strictly positive eigenfunction ϕ_1. For this purpose, let $\Phi_1 = \Phi_t\big|_{t=1}$. By the Krein–Rutman Theorem (Theorem 1.3.4), $r(\Phi_1) > 0$ and has a principal eigenfunction ϕ_1 such that $\phi_1 > 0$ in $\overline{\Omega}$. Define $\mu_1 = -\log r_1$.

Step 3. Let $u(x,t) = \Phi_t[\phi_1]$. It remains to show that

$$u(x,t) = \mathrm{e}^{-\mu_1 t}\phi_1(x) \quad \text{for } t > 0. \tag{2.3}$$

Note that (2.3) implies μ_1 is an eigenvalue of (1.23) with an eigenfunction $\phi_1 > 0$ in $\overline{\Omega}$, which finishes the proof.

To show (2.3), we first observe that, by construction, we have $u(x,n) = \mathrm{e}^{-\mu_1 n}\phi_1(x)$ for $n \in \mathbb{N}$.

Next, consider the positive compact operator $\Phi_{1/2}$, then the Krein–Rutman Theorem (Theorem 1.3.4) implies $r(\Phi_{1/2}) > 0$ and $\Phi_{1/2}(\hat{\phi}) = r(\Phi_{1/2})\hat{\phi}$ for some positive continuous function $\hat{\phi}$. This implies

$$\Phi_1(\hat{\phi}) = \Phi_{1/2} \circ \Phi_{1/2}(\hat{\phi}) = r(\Phi_{1/2})^2\hat{\phi}.$$

Hence $r(\Phi_{1/2})^2$ is an eigenvalue with a positive eigenfunction $\hat{\phi}$. By the characterization of the principal eigenvalue in Theorem 1.3.4, we must have

$$r(\Phi_{1/2})^2 = r(\Phi_1) = \mathrm{e}^{-\mu_1} \quad \text{and} \quad \hat{\phi} \in \mathrm{span}\{\phi_1\}.$$

Hence, (2.3) holds for $t = n/2$ for $n \in \mathbb{N}$.

By induction, it follows that (2.3) holds for $t = \frac{n}{2^m}$ for $n, m \in \mathbb{N}$. Since the set of such t's are dense in $\mathbb{R}_+$, it then follows by continuity that (2.3) holds for $t > 0$. This proves the claim. We leave the verification of other properties as an exercise for the readers. □

2.2 Qualitative Properties of Periodic Principal Eigenvalues

In this section, we discuss various asymptotic behaviors of the principal eigenvalues for periodic-parabolic operators. To avoid mathematical technicalities we will limit most of our discussions to the following problem, leaving more general situations for discussion and further reading:

$$\begin{cases} \partial_t \varphi - d\Delta\varphi - c(x,t)\varphi = \lambda\varphi, & x \in \Omega, t \in [0,1], \\ \mathbf{n} \cdot \nabla\varphi = 0, & x \in \partial\Omega, t \in [0,1], \\ \varphi(x,0) = \varphi(x,1), & x \in \Omega. \end{cases} \tag{2.4}$$

Here $c(x,t)$ is assumed to be Hölder continuous, time-periodic with the unit period. As shown by Theorem 2.1.1, the principal eigenvalue of (2.4), which we will denote by λ_1, is real and it has the smallest real part among all eigenvalues. Denote its eigenfunction by φ_1, which can be assumed to be strictly positive in $\overline{\Omega} \times [0,1]$.

Given any 1-periodic function $p(x,t)$, denote the temporal average by

$$\hat{p}(x) := \int_0^1 p(x,s)\mathrm{d}s.$$

In the following, we give some examples of principal eigenvalues and principal eigenfunctions for various special functions c:

- For $c = c(t)$ a 1-periodic function,

$$\lambda_1 = -\int_0^1 c(t)\,\mathrm{d}t \quad \text{and} \quad \varphi_1 = \varphi_1(t) = \exp\left(-\lambda_1 t + \int_0^t c(s)\,\mathrm{d}s\right)$$

solve

$$\varphi_1' - c(t)\varphi_1 = \lambda_1\varphi_1, \qquad \varphi_1(t+1) = \varphi_1(t).$$

- For $c = c(x)$, λ_1 and $\varphi_1 = \varphi_1(x) > 0$ solve

$$\begin{cases} -d\Delta\varphi - c(x)\varphi = \lambda\varphi & \text{in } \Omega, \\ \mathbf{n} \cdot \nabla\varphi = 0 & \text{on } \partial\Omega. \end{cases} \tag{2.5}$$

That is, λ_1 is the smallest eigenvalue of the linear elliptic eigenvalue problem (2.5) with φ_1 as the corresponding eigenfunction.

- For $c(x,t) = a(x) + b(t)$ with 1-periodic function b,

$$\lambda_1 = \mu_1 - \int_0^1 b(t)\,\mathrm{d}t, \qquad \varphi_1(x,t) = w(x)\mathrm{e}^{-\int_0^t b(s)\,\mathrm{d}s + t\left(\int_0^1 b(s)\,\mathrm{d}s\right)},$$

where μ_1 and $w > 0$ solve

$$\begin{cases} -d\Delta w - a(x)w = \mu_1 w & \text{in } \Omega, \\ \mathbf{n}\cdot\nabla w = 0 & \text{on } \partial\Omega. \end{cases} \tag{2.6}$$

The following result provides the upper and lower bounds for the principal eigenvalue of (2.4), which are uniform for all diffusion rates.

Lemma 2.2.1 *The following estimates hold for any $d > 0$:*

$$-\int_0^1 \max_{x\in\overline{\Omega}} c(x,t)\,\mathrm{d}t \le \lambda_1 \le -\frac{1}{|\Omega|}\int_\Omega\int_0^1 c(x,t)\,\mathrm{d}t\mathrm{d}x. \tag{2.7}$$

Furthermore, equalities hold if and only if c is independent of x.

Proof Let λ_1 denote the principal eigenvalue of (2.4) with corresponding eigenfunction φ_1, which is uniquely determined by $\int_0^1\int_\Omega \varphi_1\,\mathrm{d}x\mathrm{d}t = 1$. Integrating the equation of φ_1 in Ω,

$$\frac{\mathrm{d}}{\mathrm{d}t}\int_\Omega \varphi_1 - \int_\Omega c\varphi_1 = \lambda_1\int_\Omega \varphi_1, \qquad \forall t > 0.$$

It follows that

$$\frac{\mathrm{d}}{\mathrm{d}t}\int_\Omega \varphi_1 - \left(\max_{x\in\overline{\Omega}} c(x,t)\right)\int_\Omega \varphi_1 \le \lambda_1\int_\Omega \varphi_1, \qquad \forall t > 0.$$

Dividing the above equation by $\int_\Omega \varphi_1(x,t)$ and integrating in $t \in (0,1)$, by the time-periodicity of φ_1 we obtain the first inequality of (2.7). Notice from the proof that the first equality in (2.7) holds if and only if

$$\int_0^1\int_\Omega [c(x,t) - \max_{x\in\overline{\Omega}} c(x,t)]\varphi_1\,\mathrm{d}x\,\mathrm{d}t = 0.$$

This implies $c(x,t) = \max_{\overline{\Omega}} c(\cdot,t)$ holds for any t, i.e. c is independent of x.

To prove the second inequality, we divide the equation of φ_1 by φ_1 to obtain

$$\partial_t(\ln\varphi_1) - \frac{\Delta\varphi_1}{\varphi_1} - c = \lambda_1.$$

Integrating the above equation in $\Omega\times(0,1)$, we get

$$-\int_0^1\int_\Omega \frac{|\nabla_x\varphi_1|^2}{\varphi_1^2} - \int_0^1\int_\Omega c(x,t) = \lambda_1,$$

from which the second inequality of (2.7) follows. From the proof we see that the second equality in (2.7) holds if and only if $\nabla_x \varphi_1 \equiv 0$, i.e. φ_1 is independent of x. This implies that c is also independent of x and vice versa. □

Remark 2.2.2 Another way to prove the first inequality of (2.7) is to apply the comparison principle for the principal eigenvalues to (2.7) and

$$\begin{cases} \partial_t \varphi - d\Delta\varphi - \left(\max\limits_{x\in\overline{\Omega}} c(x,t)\right)\varphi = \lambda\varphi, & x \in \Omega, t \in [0,1], \\ \mathbf{n}\cdot\nabla\varphi = 0, & x \in \partial\Omega, t \in [0,1], \\ \varphi(x,0) = \varphi(x,1), & x \in \Omega, \end{cases}$$

for which the principal eigenvalue is given by $-\int_0^1 \max_{x\in\overline{\Omega}} c(x,t)\,dt$. Hence, the following estimates could hold for some more general operator $\mathcal{L}$ and oblique derivative boundary operators $\mathcal{B} = p^i D_i$ (for which $p^0 \equiv 0$):

$$-\int_0^1 \max_{x\in\overline{\Omega}} c(x,t)\,dt \le \lambda_1 \le -\int_0^1 \min_{x\in\overline{\Omega}} c(x,t)\,dt \qquad \forall d > 0. \tag{2.8}$$

Remark 2.2.3 The following characterization is given in [5]:

$$\int_0^1 \max_{x\in\overline{\Omega}} c(x,t)\,dt = \sup_{x(t)\in\mathcal{S}} \int_0^1 c(x(t),t)\,dt, \tag{2.9}$$

where $\mathcal{S} = \{x(t) \in C[0,\infty) : x(t) = x(t+1)\}$.

Remark 2.2.4 Another interesting and important quantity is

$$\max_{x\in\overline{\Omega}} \hat{c}(x) = \max_{x\in\overline{\Omega}} \int_0^1 c(x,t)\,dt.$$

It is easy to see that

$$\int_0^1 \max_{x\in\overline{\Omega}} c(x,t)\,dt \ge \max_{x\in\overline{\Omega}} \int_0^1 c(x,t)\,dt \ge \frac{1}{|\Omega|}\int_\Omega\int_0^1 c(x,t)\,dxdt. \tag{2.10}$$

The first equality holds if and only if there exists some x_0 such that $c(x_0,t) = \max\limits_{x\in\overline{\Omega}} c(x,t)$ for all $t \in [0,1]$. The second equality holds iff $\hat{c}(x) = \int_0^1 c(x,t)\,dt$ is independent of x. These quantities seem to reflect the spatial and temporal heterogeneity of c from different aspects, and they will play critical roles in understanding the asymptotic behavior of λ_1, as will be seen clearly in the rest of this chapter.

2.2.1 The Hutson–Shen–Vickers Lemma

To improve the upper bound of the principal eigenvalue of (2.4), as given in (2.7), we consider the following linear elliptic eigenvalue problem:

$$\begin{cases} -d\Delta\psi - \left(\int_0^1 c(x,t)\,\mathrm{d}t\right)\psi = \mu\psi, & x \in \Omega, \\ \mathbf{n}\cdot\nabla\psi = 0, & x \in \partial\Omega. \end{cases} \tag{2.11}$$

Let μ_1 denote the smallest eigenvalue of (2.11).

Lemma 2.2.5 *Let λ_1 be the principal eigenvalue of* (2.4). *Then*

$$\lambda_1 \le \mu_1 \quad \textit{holds for all } d > 0.$$

Furthermore, equality holds if and only if $c(x,t) - \hat{c}(x)$ is independent of x.

Proof Let λ_1 and φ_1 be the principal eigenvalue and the associated positive eigenfunction of (2.4), respectively. For simplicity let $d = 1$. Set

$$w(x) = \mathrm{e}^{\int_0^1 \log\varphi_1(x,t)\mathrm{d}t}.$$

By direct calculations,

$$w_{x_i} = w\int_0^1 \frac{\varphi_{1,x_i}}{\varphi_1}\,\mathrm{d}t,$$

$$w_{x_ix_i} = w\left(\int_0^1 \frac{\varphi_{1,x_i}}{\varphi_1}\,\mathrm{d}t\right)^2 + w\int_0^1 \frac{\varphi_{1,x_ix_i}\varphi_1 - (\varphi_{1,x_i})^2}{\varphi_1^2}\,\mathrm{d}t \le w\int_0^1 \frac{\varphi_{1,x_ix_i}}{\varphi_1}\,\mathrm{d}t,$$

where the inequality follows from the Cauchy–Schwarz inequality.

It follows that $\mathbf{n}\cdot\nabla w = 0$ on $\partial\Omega$ and

$$\Delta w \le w\int_0^1 \frac{\Delta\varphi_1}{\varphi_1}\,\mathrm{d}t \quad \text{in } \Omega.$$

By the equation of φ_1, it follows that

$$\Delta w \le w\int_0^1 \frac{\varphi_{1,t} - (c+\lambda_1)\varphi_1}{\varphi_1}\,\mathrm{d}t = -w\left(\int_0^1 c(x,t)\,\mathrm{d}t + \lambda_1\right). \tag{2.12}$$

Let $\mu_1 \in \mathbb{R}$ and $\psi_1 > 0$ denote the principal eigenvalue and eigenfunction of (2.11). Multiplying (2.11) by w, the inequality (2.12) of w by ψ_1, subtracting and integrating the results in Ω, we obtain $\lambda_1 \le \mu_1$.

From the above argument we see that $\mu_1 = \lambda_1$ if and only if $\nabla(\log\varphi_1)$ is independent of t, i.e. $\varphi_1 = u(x)v(t)$ for some positive functions $u(x)$ and $v(t)$, such that v is 1-periodic in t.

It remains to show that $\varphi_1(x,t) = u(x)v(t)$ if and only if $c(x,t) - \hat{c}(x)$ is independent of x.

On the one hand, if $\varphi_1 = u(x)v(t)$, then (2.4) implies

$$v'(t)u(x) = v(t)\left[\Delta u(x) + c(x,t)u(x) + \lambda_1 u(x)\right]. \tag{2.13}$$

Dividing the above by $v(t)$, which is positive, and integrating in t, we obtain

$$-\Delta u - \hat{c}(x)u = \lambda_1 u \quad \text{in } \Omega.$$

Substituting back into (2.13), we deduce that

$$v'(t)u(x) = v(t)[c(x,t) - \hat{c}(x)]u(x).$$

Canceling $u(x)$ from both sides, we have

$$c(x,t) - \hat{c}(x) = \frac{v'(t)}{v(t)}.$$

This proves the necessity.

On the other hand, suppose $c(x,t) = \hat{c}(x) + b(t)$ for some 1-periodic function $b(t)$. Then $\int_0^1 b(t)\,\mathrm{d}t = 0$ by construction.

Let μ_1 and $\psi_1(x)$ be the principal eigenvalue and eigenfunction of (2.11), and define $\varphi_1(x,t) = \psi_1(x)\exp(\int_0^t b(s)\,\mathrm{d}s)$. One can then verify that μ_1 and φ_1 form an eigenvalue and eigenfunction pair of (2.4). This proves sufficiency. □

Remark 2.2.6 Lemma 2.2.5 gives a better upper bound than (2.7). In fact, Hutson et al. proved a more general result; see [19] for the details.

Remark 2.2.7 Lemma 2.2.5 says that the addition of temporal variation tends to destabilize an equilibrium under all circumstances. Biologically interpreted, this means that spatio-temporal heterogeneity of resources always favors persistence.

2.2.2 Small diffusion limit

Determining the asymptotic behavior of λ_1 for sufficiently small d is an interesting and nontrivial question. A well-known result in this connection can be stated as follows:

Lemma 2.2.8 *The following limit holds:*

$$\lim_{d\to 0} \lambda_1 = -\max_{x\in\overline{\Omega}} \int_0^1 c(x,t)\mathrm{d}t. \tag{2.14}$$

Proof A consequence of the Hutson–Shen–Vickers Lemma is

$$\limsup_{d\to 0+} \lambda_1 \le \lim_{d\to 0} \mu_1 = -\max_{x\in\overline{\Omega}} \int_0^1 c(x,t)\,dt,$$

where the last limit is a consequence of Proposition 1.3.16.

It suffices to show that

$$\liminf_{d\to 0+} \lambda_1 \ge -\max_{x\in\overline{\Omega}} \int_0^1 c(x,t)\,dt.$$

Since λ_1 is uniformly bounded (Lemma 2.2.1), we may pass to a sequence and assume that $\lambda_1 \to \lambda_*$ for some λ_*.

Next, we argue by contradiction and suppose that

$$\lambda_* < -\max_{x\in\overline{\Omega}} \int_0^1 c(x,t)\,dt. \tag{2.15}$$

By suitable normalization, we may further assume that

$$\max_{\overline{\Omega}\times\mathbb{R}} \varphi_d = 1 \quad \text{and} \quad \varphi_d(x_d, t_d) = 1 \quad \text{for some } x_d \in \overline{\Omega},\ t_d \ge 0.$$

Set $y = \frac{1}{\sqrt{d}}(x - x_d)$ and define

$$\psi_d(y,t) = \varphi_d(x_d + \sqrt{d}y, t).$$

Then ψ_d satisfies

$$0 \le \psi_d \le 1 \quad \text{in } \Omega_d \times \mathbb{R}, \quad \psi_d(0, t_d) = 1$$

and

$$\begin{cases} \partial_t \psi_d - \Delta\psi_d - c_d(y,t)\psi_d = \lambda_1 \psi_d, & y \in \Omega_d,\ t \in [0,1], \\ \nabla\psi_d \cdot \mathbf{n} = 0, & y \in \partial\Omega_d,\ t \in [0,1], \\ \psi_d(x,0) = \psi_d(x,1), & y \in \Omega_d, \end{cases} \tag{2.16}$$

where

$$c_d(y,t) := c(x_d + \sqrt{d}y, t) \quad \text{and} \quad \Omega_d := \{y : x_d + \sqrt{d}y \in \Omega\}.$$

Passing to a subsequence if necessary, we may assume that $x_d \to x_* \in \overline{\Omega}$, $t_d \to t_* \in [0,1]$ and $\lambda_1 \to \lambda_* \in \mathbb{R}$. For the case $x_* \in \Omega$, $\Omega_d \to \mathbb{R}^N$ as $d \to 0$. By parabolic regularity theory, we can pass to a sequence, so that $\psi_d \to \psi_*$ uniformly in any compact subset of $\mathbb{R}^N \times \mathbb{R}$ for some $\psi_* \in C^{2,1}(\mathbb{R}^N \times \mathbb{R})$ which satisfies $0 \le \psi_* \le 1$ in $\mathbb{R}^N \times \mathbb{R}$, $\psi_*(0, t^*) = 1$, and

$$\begin{cases} \partial_t \psi_* - \Delta\psi_* - c(x_*, t)\psi_* = \lambda_* \psi_*, \ y \in \mathbb{R}^N, t \in [0, 1], \\ \psi_*(y, 0) = \psi_*(y, 1), \qquad\qquad\qquad y \in \mathbb{R}^N. \end{cases} \tag{2.17}$$

Next, we construct the supersolution

$$p(t) = \exp\left(\int_0^t [c(x_*, s) + \lambda_*]\, ds\right).$$

Since $p(0) = 1 = \sup_\Omega \psi_*(\cdot, 0)$, we deduce that $p(1) \geq \sup_\Omega \psi_*(\cdot, 1)$. However, ψ_* is 1-periodic, thus we have $p(1) \geq 1$, i.e.

$$\lambda_* + \int_0^1 c(x_*, t)\, dt \geq 0.$$

This is in contradiction with (2.15).

Next, suppose $x_* \in \partial\Omega$. If $\mathrm{dist}(x_d, \partial\Omega)/\sqrt{d} \to \infty$, ψ_* satisfies again (2.17) in $\mathbb{R}^N \times \mathbb{R}$, and we can argue similarly as above. Otherwise, ψ_* satisfies (2.17) in $H \times \mathbb{R}$, where ∂H is a hyperplane. One can then use the Neumann boundary condition to extend the function ψ_* by reflection across ∂H. The resulting function is again a 1-periodic in time solution of an equation in the form of (2.17). Hence, a contradiction can be similarly reached. For details, see the proof of Proposition 17.3 of [13]. □

Our previous proof is based upon the blowup argument from [13]. Another possible approach is to use the WKB ansatz and the limiting Hamilton–Jacobi equation to derive the small diffusion limit of λ_1 [23].

The high order expansion of λ_1 for a small diffusion rate seems to be of independent interest. For instance, if we assume

$$-\max_{x \in \overline{\Omega}} \int_0^1 c(x, t)\, dt < -\frac{1}{|\Omega|} \int_\Omega \int_0^1 c(x, t)\, dt dx, \tag{2.18}$$

it is unknown whether λ_1 is monotone increasing for a small diffusion rate. Note that such conclusion holds true when c is independent of t.

Remark 2.2.9 We conjecture that if

$$\int_0^1 \max_{x \in \overline{\Omega}} c(x, t)\, dt = \max_{x \in \overline{\Omega}} \int_0^1 c(x, t)\, dt \tag{2.19}$$

holds, then λ_1 is monotone non-decreasing in d. Note that the conjecture holds when c is independent of t; (2.19) automatically holds in this case. See also Prob. 2.4.6 for another example.

2.2.3 Large diffusion limit

The large diffusion limit of the principal eigenvalue for time-periodic operators is well known, and the result is similar to that for principal eigenvalues of elliptic operators. The proof of the following result is due to Hutson et al. [18, Lemma 2.4].

Lemma 2.2.10 *The following limit holds:*

$$\lim_{d\to\infty} \lambda_1 = -\frac{1}{|\Omega|}\int_\Omega\int_0^1 c(x,t)\,\mathrm{d}t\mathrm{d}x. \tag{2.20}$$

Proof Assume, without loss of generality, that $|\Omega| = 1$. Let $\varphi_1 > 0$ denote the eigenfunction corresponding to λ_1, determined by $\int_\Omega\int_0^1 \varphi_1^2 = 1$. Multiplying the equation of φ_1 by φ_1 and integrating the results,

$$\int_\Omega\int_0^1 |\nabla_x\varphi_1|^2 = \frac{1}{d}\int_\Omega\int_0^1 (c(x,t)+\lambda_1)\varphi_1^2 \le \frac{C_1}{d},$$

where $C_1 = 2\sup c$, and we used the upper bound of λ_1 in (2.7).

Set $\Phi := \varphi_1 - \int_\Omega \varphi_1$. Then $\nabla_x\varphi_1 = \nabla_x\Phi$, i.e.

$$\int_\Omega\int_0^1 |\nabla_x\Phi|^2 \le \frac{C_1}{d}.$$

Now, Poincaré's inequality (see Problem 2.4.5) says that the constant

$$C_p = \inf_{\substack{w\in H^1(\Omega)\\ \int_\Omega w=0,\, w\not\equiv 0}} \frac{\int_\Omega |\nabla_x w(x)|^2\,\mathrm{d}x}{\int_\Omega |w(x)|^2\,\mathrm{d}x}$$

is positive. In particular, C_p depends on Ω but is independent of w. Since $\int_\Omega \Phi = 0$ for all t, it follows that

$$\int_\Omega\int_0^1 |\Phi|^2 \le \frac{C_2}{d}, \qquad \text{where } C_2 = C_p C_1. \tag{2.21}$$

Integrating the equation of φ_1 in Ω,

$$\partial_t\int_\Omega \varphi_1 = \int_\Omega (c(x,t)+\lambda_1)\varphi_1, \qquad \forall t>0.$$

Set $\bar\varphi(t) = \int_\Omega \varphi_1$ and write $\varphi_1 = \bar\varphi + \Phi$.

$$\bar\varphi_t = \bar\varphi(t)\int_\Omega (c(x,t)+\lambda_1) + \int_\Omega [(c(x,t)+\lambda_1)\Phi].$$

As λ_1 is bounded in d,

$$\int_0^1\int_\Omega |(c(x,t)+\lambda_1)\Phi| \le C_3\int_0^1\int_\Omega |\Phi| \le C_4\left(\int_0^1\int_\Omega \Phi^2\right)^{1/2} \le \frac{C_5}{d^{1/2}}.$$

By the equation of $\bar{\varphi}$,

$$e^{-\int_0^t\int_\Omega (c+\lambda_1)}\bar{\varphi}(t) = \bar{\varphi}(0) + o(1), \tag{2.22}$$

where the $o(1)$ term converges to zero uniformly in t as $d \to +\infty$.

Next, we claim that

$$\liminf_{d\to+\infty} \bar{\varphi}(0) > 0. \tag{2.23}$$

Suppose not, then (2.22) implies that $\sup_{t\in[0,1]} \bar{\varphi}(t) \to 0$ along a sequence $d = d_k \to +\infty$. This, together with (2.21) implies that

$$\int_\Omega\int_0^1 \varphi_1^2 \to 0,$$

which contradicts $\int_\Omega\int_0^1 \varphi_1^2 = 1$. This proves (2.23).

To proceed, recall that $\bar{\varphi}(0) = \bar{\varphi}(1)$ by periodicity, so (2.22) becomes

$$e^{-\int_0^1\int_\Omega (c+\lambda_1)}\bar{\varphi}(0) = \bar{\varphi}(0) + o(1).$$

Since $\bar{\varphi}(0) \not\to 0$ by (2.23), we have proved (2.20). □

Let $d > 0$ and $c \in C(\overline{\Omega}\times\mathbb{R})$ be given. Denote by $\lambda(d,c)$ the principal eigenvalue of (2.4).

Corollary 2.2.11 *Suppose $c(x,t)$ is nonconstant in x, and that $\hat{c}(x)$ is constant in x. Then the following two statements hold.*

(i) *There exists a $d_1 > 0$ such that $\partial_d\lambda(d_1,c) < 0$.*
(ii) *There exist $0 < d_1 < d_2$ such that $\lambda(d_1,c) = \lambda(d_2,c)$.*

Proof This is [18, Theorem 2.2]. By Lemmas 2.2.1 and 2.2.10,

$$\lambda(d,c) < -\frac{1}{|\Omega|}\int_\Omega \hat{c}(x)\,dx \quad \text{and} \quad \lim_{d\to\infty}\lambda(d,c) = -\frac{1}{|\Omega|}\int_\Omega \hat{c}(x)\,dx.$$

Moreover, since $\hat{c}$ is constant, Lemma 2.2.8 says that

$$\lim_{d\to 0}\lambda(d,c) = -\frac{1}{|\Omega|}\int_\Omega \hat{c}(x)\,dx.$$

It is then clear that (i) and (ii) hold. □

Remark 2.2.12 The high order expansion of λ_1 for a sufficiently large diffusion rate d is given in Bai et al. [2]. In particular, it is shown in [2] that λ_1 is monotone non-decreasing for a sufficiently large diffusion rate d. See also [21] for some related results on the principal Floquet bundle and applications to the evolution of dispersal in spatially heterogeneous and temporally varying environments.

2.2.4 Monotonicity in frequency

In this subsection we address an interesting question raised by Hutson et al. [18] on the dependence of the principal eigenvalue of time-periodic parabolic operators on the parameter frequency.

Consider the periodic-parabolic eigenvalue problem

$$\begin{cases} \partial_t u - \Delta u - c(x, \tau t)u = \lambda(\tau)u, \ x \in \Omega, t \in [0, 1/\tau], \\ \nabla u \cdot \mathbf{n} = 0, & x \in \partial\Omega, t \in [0, 1/\tau], \\ u(x, 0) = u(x, 1/\tau), & x \in \Omega, \end{cases}$$

where the parameter $\tau > 0$ measures the frequency of the temporal variations of the environment. By a change of variable, the above is equivalent to

$$\begin{cases} L_\tau u := \tau\partial_t u - \Delta u - c(x, t)u = \lambda(\tau)u, \ x \in \Omega, t \in [0, 1], \\ \nabla u \cdot \mathbf{n} = 0, & x \in \partial\Omega, t \in [0, 1], \\ u(x, 0) = u(x, 1), & x \in \Omega, \end{cases} \tag{2.24}$$

where we denote by $\lambda(\tau)$ the principal eigenvalue, and by u_τ a positive eigenfunction associated with $\lambda(\tau)$.

In [18], Hutson et al. asked whether the principal eigenvalue $\lambda(\tau)$ is monotone in τ, which suggests that increasing the temporal variation of the environment tends to favor the persistence of populations. The following result from [24] provided an affirmative answer:

Theorem 2.2.13 *$\lambda(\tau)$ is non-decreasing in $\tau > 0$. Furthermore, the following two alternatives hold:*

(i) *If $c = \hat{c}(x) + g(t)$ for some 1-periodic function $g(t)$, then $\lambda(\tau)$ is constant for $\tau > 0$;*
(ii) *Otherwise, $\lambda'(\tau) > 0$ for every $\tau > 0$.*

The idea of the proof is to consider the adjoint problem of (2.24), i.e.

$$\begin{cases} L_\tau^* v := -\tau\partial_t v - \Delta v - c(x, t)v = \lambda(\tau)v, \ x \in \Omega, t \in [0, 1], \\ \nabla v \cdot \mathbf{n} = 0, & x \in \partial\Omega, t \in [0, 1], \\ v(x, 0) = v(x, 1), & x \in \Omega. \end{cases} \tag{2.25}$$

Let v_τ denote a positive eigenfunction of (2.25) corresponding to $\lambda(\tau)$, and denote, for convenience, $C = \Omega \times [0, 1]$. We normalize u_τ and v_τ such that $\int_C u_\tau^2 = \int_C u_\tau v_\tau = 1$ for any $\tau > 0$.

Define the sets $\mathbb{S}$ and $\mathbb{S}_+$ by

$$\mathbb{S} = \left\{\zeta \in C^{2,1}(C) \cap C^{1,1}(\overline{C}) : \ \zeta(x, 0) = \zeta(x, 1) \text{ in } \Omega, \nabla\zeta \cdot \mathbf{n} = 0 \text{ on } \partial\Omega \times [0, 1]\right\}$$

and

$$\mathbb{S}_+ = \{\zeta \in \mathbb{S} : \ \zeta > 0 \quad \text{in } \overline{C}\}.$$

Then define the functional $J_\tau : \mathbb{S}_+ \to \mathbb{R}$ by

$$J_\tau(\zeta) = \int_C u_\tau v_\tau \left(\frac{L_\tau \zeta}{\zeta}\right) \mathrm{d}x\mathrm{d}t, \quad \zeta \in \mathbb{S}_+. \tag{2.26}$$

Clearly, for any $\tau > 0$, $u_\tau, v_\tau \in \mathbb{S}_+$ and J_τ is well defined on the cone $\mathbb{S}_+$. The following property of J_τ turns out to be crucial.

Lemma 2.2.14 *For any $\zeta \in \mathbb{S}_+$, we have*

$$J_\tau(u_\tau) - J_\tau(\zeta) = \int_C u_\tau v_\tau \left|\nabla \log\left(\frac{\zeta}{u_\tau}\right)\right|^2. \tag{2.27}$$

Proof By the definition of J_τ, we observe that, for every $\zeta \in \mathbb{S}_+$,

$$\begin{aligned} J_\tau(\zeta) &= \tau \int_C u_\tau v_\tau \left(\frac{\zeta_t}{\zeta}\right) - \int_C u_\tau v_\tau \left[\frac{\Delta\zeta}{\zeta}\right] - \int_C u_\tau v_\tau c \\ &= \tau \int_C u_\tau v_\tau \partial_t \log\zeta - \int_0^1 \int_{\partial\Omega} u_\tau v_\tau (\nabla \log\zeta) \cdot \mathbf{n} \\ &\quad + \int_C \nabla(u_\tau v_\tau) \cdot \nabla \log\zeta - \int_C u_\tau v_\tau \left|\nabla \log\zeta\right|^2 - \int_C u_\tau v_\tau c. \end{aligned} \tag{2.28}$$

For each $\tau > 0$, $\zeta \in \mathbb{S}_+$, and $\varphi \in \mathbb{S}$, recall that the Gateaux derivative of J_τ at the point ζ is given by

$$DJ_\tau(\zeta)\varphi = \lim_{h\to 0} \frac{1}{h} \left[J_\tau(\zeta + h\varphi) - J_\tau(\zeta)\right].$$

We claim that u_τ is a critical point of J_τ in the sense that

$$DJ_\tau(u_\tau)\varphi = 0 \ \text{ for all } \varphi \in \mathbb{S}. \tag{2.29}$$

To prove (2.29), we first notice that $\zeta \in \mathbb{S}_+$ implies

$$\int_0^1 \int_{\partial\Omega} u_\tau v_\tau (\nabla \log\zeta) \cdot \mathbf{n} = 0.$$

It follows from (2.28) that

$$\begin{aligned} DJ_\tau(\zeta)\varphi = \tau \int_C u_\tau v_\tau \partial_t \left(\frac{\varphi}{\zeta}\right) + \int_C \nabla(u_\tau v_\tau) \cdot \nabla\left(\frac{\varphi}{\zeta}\right) \\ - 2\int_C u_\tau v_\tau (\nabla \log \zeta) \cdot \nabla\left(\frac{\varphi}{\zeta}\right) \end{aligned} \tag{2.30}$$

for any $\varphi \in \mathbb{S}$. Through straightforward calculations, we further have

$$\begin{aligned}
& DJ_\tau(u_\tau)\varphi \\
=& \tau \int_C u_\tau v_\tau \partial_t \left(\frac{\varphi}{u_\tau}\right) - 2\int_C u_\tau v_\tau (\nabla \log u_\tau) \cdot \nabla\left(\frac{\varphi}{u_\tau}\right) + \int_C \nabla(u_\tau v_\tau) \cdot \nabla\left(\frac{\varphi}{u_\tau}\right) \\
=& -\tau \int_C \left(\frac{\varphi}{u_\tau}\right) \partial_t(u_\tau v_\tau) - 2\int_0^1\int_{\partial\Omega} \left(\frac{\varphi v_\tau}{u_\tau}\right) \nabla u_\tau \cdot \mathbf{n} \\
& + 2\int_C \left(\frac{\varphi}{u_\tau}\right) (\nabla v_\tau \cdot \nabla u_\tau + v_\tau \Delta u_\tau) \\
& + \int_0^1\int_{\partial\Omega} \left(\frac{\varphi}{u_\tau}\right) \nabla(u_\tau v_\tau) \cdot \mathbf{n} - \int_C \left(\frac{\varphi}{u_\tau}\right) \Delta(u_\tau v_\tau) \\
=& -\tau \int_C \left(\frac{\varphi}{u_\tau}\right) \partial_t(u_\tau v_\tau) + 2\int_C \left(\frac{\varphi}{u_\tau}\right) \left\{\nabla v_\tau \cdot \nabla u_\tau + v_\tau \Delta u_\tau\right\} \\
& - \int_C \left(\frac{\varphi}{u_\tau}\right) \left\{v_\tau \Delta u_\tau + 2\nabla v_\tau \cdot \nabla u_\tau + u_\tau \Delta v_\tau\right\} \\
=& -\tau \int_C \left(\frac{\varphi}{u_\tau}\right) \partial_t(u_\tau v_\tau) + \int_C \left(\frac{\varphi v_\tau}{u_\tau}\right) \Delta u_\tau - \int_C \varphi \Delta v_\tau \\
=& -\int_C \left(\frac{v_\tau}{u_\tau}\right) \left\{\tau \partial_t u_\tau - \Delta u_\tau\right\} \varphi + \int_C \left\{-\tau \partial_t v_\tau - \Delta v_\tau\right\} \varphi \\
=& -\int_C \frac{v_\tau}{u_\tau} (L_\tau u_\tau) \varphi + \int_C (L_\tau^* v_\tau) \varphi.
\end{aligned}$$

In the above applications of the divergence theorem, the boundary integrals vanish due to the boundary conditions of u_τ and v_τ.

Since $L_\tau u_\tau = \lambda(\tau) u_\tau$ and $L_\tau^* v_\tau = \lambda(\tau) v_\tau$, we obtain

$$DJ_\tau(u_\tau)\varphi = -\lambda(\tau)\int_C v_\tau \varphi + \lambda(\tau)\int_C v_\tau \varphi = 0,$$

and (2.29) thus follows.

We now proceed to prove formula (2.27) through some tedious manipulations. Taking (2.28) into account, direct calculation shows

$$
\begin{aligned}
&J_\tau(u_\tau) - J_\tau(\zeta) + \tau \int_C u_\tau v_\tau \partial_t \log\left(\frac{\zeta}{u_\tau}\right) + \int_C \nabla(u_\tau v_\tau)\cdot \nabla \log\left(\frac{\zeta}{u_\tau}\right)\\
=&-\int_C u_\tau v_\tau\, |\nabla \log u_\tau|^2 + \int_C u_\tau v_\tau |\nabla \log \zeta|^2\\
=&\int_C u_\tau v_\tau\, (\nabla \log \zeta + \nabla \log u_\tau)\cdot(\nabla \log \zeta - \nabla \log u_\tau)\\
=&\int_C u_\tau v_\tau \left[2\nabla \log u_\tau + \nabla \log\left(\frac{\zeta}{u_\tau}\right)\right]\cdot \nabla \log\left(\frac{\zeta}{u_\tau}\right)\\
=&2\int_C u_\tau v_\tau (\nabla \log u_\tau)\cdot \nabla \log\left(\frac{\zeta}{u_\tau}\right) + \int_C u_\tau v_\tau \left|\nabla \log\left(\frac{\zeta}{u_\tau}\right)\right|^2.
\end{aligned}
$$

Now, setting $\zeta = u_\tau$ in (2.30), we have

$$
\begin{aligned}
DJ_\tau(u_\tau)\varphi = \tau \int_C u_\tau v_\tau \partial_t \left(\frac{\varphi}{u_\tau}\right) + \int_C \nabla(u_\tau v_\tau)\cdot \nabla\left(\frac{\varphi}{u_\tau}\right)\\
-2\int_C u_\tau v_\tau (\nabla \log u_\tau)\cdot \nabla\left(\frac{\varphi}{u_\tau}\right)
\end{aligned}
\tag{2.31}
$$

for all $\varphi \in \mathbb{S}$. Setting $\varphi = u_\tau \log\left(\frac{\zeta}{u_\tau}\right)$, which belongs to $\mathbb{S}$, we obtain

$$
J_\tau(u_\tau) - J_\tau(\zeta) = -DJ_\tau(u_\tau)\varphi + \int_C u_\tau v_\tau \left|\nabla \log\left(\frac{\zeta}{u_\tau}\right)\right|^2.
$$

As $DJ_\tau(u_\tau)\varphi = 0$, the proof of Lemma 2.2.14 is complete. □

With the help of Lemma 2.2.14, we are in a position to prove Theorem 2.2.13.

Proof (of Theorem 2.2.13) We substitute $u = u_\tau$ in (2.24) and differentiate the resulting equation with respect to τ. Denoting $\frac{\partial}{\partial \tau} u_\tau$ by u'_τ and $\frac{\mathrm{d}}{\mathrm{d}\tau}\lambda(\tau)$ by $\lambda'(\tau)$ for brevity, we obtain

$$
\begin{cases}
\partial_t u_\tau + \tau \partial_t u'_\tau - \Delta u'_\tau - c u'_\tau = \lambda(\tau) u'_\tau + \lambda'(\tau) u_\tau, & x \in \Omega, t \in [0,1],\\
\nabla u'_\tau \cdot \mathbf{n} = 0, & x \in \partial\Omega, t \in [0,1],\\
u'_\tau(x,0) = u'_\tau(x,1), & x \in \Omega.
\end{cases}
$$

We multiply the above equation by v_τ and integrate the result over C. Together with the fact that $L^*_\tau v_\tau = \lambda(\tau) v_\tau$ and the normalization $\int_C u_\tau v_\tau = 1$, we find that

$$
\lambda'(\tau) = \int_C v_\tau \partial_t u_\tau.
$$

Recalling the definitions of L_τ, L^*_τ and J_τ, we further derive

$$
\begin{aligned}
\int_C v_\mathcal{T}\partial_t(u_\mathcal{T}) =& \frac{1}{2\mathcal{T}}\int_C v_\mathcal{T}(L_\mathcal{T}-L_\mathcal{T}^*)u_\mathcal{T} \\
=& \frac{1}{2\mathcal{T}}\left[\int_C v_\mathcal{T}L_\mathcal{T}u_\mathcal{T} - \int_C u_\mathcal{T}L_\mathcal{T}v_\mathcal{T}\right] \\
=& \frac{1}{2\mathcal{T}}\Big[J_\mathcal{T}(u_\mathcal{T}) - J_\mathcal{T}(v_\mathcal{T})\Big].
\end{aligned}
$$

In view of $v_\mathcal{T} \in \mathbb{S}_+$, Lemma 2.2.14 implies

$$
\lambda'(\mathcal{T}) = \frac{1}{2\mathcal{T}}\int_C u_\mathcal{T} v_\mathcal{T}\left|\nabla \log\left(\frac{v_\mathcal{T}}{u_\mathcal{T}}\right)\right|^2, \tag{2.32}
$$

which shows that $\lambda'(\mathcal{T}) \geq 0$ for all $\mathcal{T} > 0$.

It remains to prove parts (i) and (ii) in Theorem 2.2.13. Again, let $\lambda(\mathcal{T}), u_\mathcal{T}$ be the principal eigenvalue and positive eigenfunction of (2.24), for which $c(x,t)$ can be written as

$$
c(x,t) = \hat{c}(x) + g(t).
$$

Without loss of generality, we may assume that $\int_0^1 g(t)\,\mathrm{d}t = 0$. Set

$$
U_\mathcal{T}(x,t) = \exp\left(-\frac{1}{\mathcal{T}}\int_0^t g(s)\mathrm{d}s\right)u_\mathcal{T},
$$

and then observe that $U_\mathcal{T}$ is a positive eigenfunction of the following problem

$$
\begin{cases}
\mathcal{T}\partial_t U_\mathcal{T} - \Delta U_\mathcal{T} - \hat{c}(x)U_\mathcal{T} = \lambda(\mathcal{T})U_\mathcal{T}, & x \in \Omega, t \in [0,1], \\
\nabla U_\mathcal{T}\cdot \mathbf{n} = 0, & x \in \partial\Omega, t \in [0,1], \\
U_\mathcal{T}(x,0) = U_\mathcal{T}(x,1), & x \in \Omega.
\end{cases} \tag{2.33}
$$

Observe that all coefficients of (2.33) are independent of t. By the uniqueness of principal eigenfunction (up to scaling), $U_\mathcal{T}$ is independent of t. As a result, $\lambda(\mathcal{T})$ is constant for $\mathcal{T} > 0$.

Finally, we show that if $\frac{\partial\lambda}{\partial\mathcal{T}} = 0$ for some $\mathcal{T}_0 > 0$, then c can be written as $c = \hat{c}(x) + g(t)$. According to (2.32), we have $u_{\mathcal{T}_0} = h(t)v_{\mathcal{T}_0}$ for some 1-periodic function $h(t) > 0$. Substituting $u_{\mathcal{T}_0} = h(t)v_{\mathcal{T}_0}$ into $L_{\mathcal{T}_0}u_{\mathcal{T}_0} = \lambda(\mathcal{T}_0)u_{\mathcal{T}_0}$ and using $L_{\mathcal{T}_0}^* v_{\mathcal{T}_0} = \lambda(\mathcal{T}_0)v_{\mathcal{T}_0}$, we can deduce

$$
h'(t)v_{\mathcal{T}_0} + 2h(t)\partial_t v_{\mathcal{T}_0} = 0.
$$

It then follows that $\partial_t \log v_{\mathcal{T}_0} = -\frac{h'(t)}{2h(t)}$ in C, which depends only on t. Hence, $v_{\mathcal{T}_0}$ is of the form $v_{\mathcal{T}_0} = X_{\mathcal{T}_0}(x)T_{\mathcal{T}_0}(t)$ with some 1-periodic function $T_{\mathcal{T}_0}(t) > 0$ in $[0,T]$ and function $X_{\mathcal{T}_0}(x) > 0$ in Ω. Again using $L_\mathcal{T}^* v_{\mathcal{T}_0} = \lambda(\mathcal{T}_0)v_{\mathcal{T}_0}$, we arrive at

$$
-\mathcal{T}_0\frac{T_{\mathcal{T}_0}'(t)}{T_{\mathcal{T}_0}(t)} - \frac{\Delta X_{\mathcal{T}_0}(x)}{X_{\mathcal{T}_0}(x)} - c(x,t) = \lambda(\mathcal{T}_0) \text{ in } C
$$

and we have expressed $c(x,t)$ in the form of $c(x,t) = \hat{c}(x) + g(t)$. The proof of Theorem 2.2.13 is now complete. □

Some applications of Theorem 2.2.13 to two-species competition models in spatially heterogeneous and temporally varying environment will be given in the next section; see Theorem 2.3.2.

Remark 2.2.15 The limits of $\lambda(\tau)$ as τ tends to zero or infinity are given in [24]. In fact, these limits and Theorem 2.2.13 can also be extended to Dirichlet and Robin boundary conditions, as done in [24]. As an application of Theorem 2.2.13, consider the case of zero Neumann boundary conditions, it holds that

$$\lim_{\tau \to \infty} \lambda(\tau) = \mu_1,$$

where μ_1 denotes the smallest eigenvalue of (2.11). Hence it follows from Theorem 2.2.13 that $\lambda(\tau) \leq \mu_1$ holds for all $\tau > 0$; i.e. this gives another proof of Lemma 2.2.5.

Remark 2.2.16 We refer to [23] for the extensions of Theorem 2.2.13, where the joint dependence of the principal eigenvalues on the frequency and the diffusion rate is investigated, with applications to some infectious disease models. To be more specific, as another application of Theorem 2.2.13, consider the principal eigenvalue, denoted by $\lambda(\tau, d)$, of the linear eigenvalue problem

$$\begin{cases} \tau\partial_t u - d\Delta u - c(x,t)u = \lambda u, & x \in \Omega, t \in [0,1], \\ \nabla u \cdot \mathbf{n} = 0, & x \in \partial\Omega, t \in [0,1], \\ u(x,0) = u(x,1), & x \in \Omega. \end{cases} \tag{2.34}$$

Among other things, the topological structure of the level set of $\lambda(\tau, d)$ is classified in [23], in which the monotonicity result in Theorem 2.2.13 plays an important role. This in turn yields new understandings of the dependence of the principal eigenvalue $\lambda(\tau, d)$ on the diffusion rate d; see [23] for further details.

2.3 Applications to Two-Species Competition Models in a Spatially and Temporally Varying Environment

In this section we present some applications of the principal eigenvalue theory to two-species competition models in spatially heterogeneous and temporally varying environment. To this end, we consider the following competition model, in which two species are identical except possibly for their diffusion rates and relaxation time:

$$\begin{cases} \tau\partial_t u - d_1\Delta u = u(m(x,t) - u - v), & x \in \Omega,\ t > 0, \\ \partial_t v - d_2\Delta v = v(m(x,t) - u - v), & x \in \Omega,\ t > 0, \\ \mathbf{n}\cdot\nabla u = \mathbf{n}\cdot\nabla v = 0, & x \in \partial\Omega,\ t > 0, \\ u(x,0) = u_0(x), \quad v(x,0) = v_0(x), & x \in \Omega, \end{cases} \tag{2.35}$$

where $\tau > 0$ is the relaxation time which accounts for possibly different time scales for two species to respond to the temporal variations in the environment, the intrinsic growth rate $m(x,t)$ is positive, continuous in $\overline{\Omega} \times \mathbb{R}$, and 1-periodic in t, which measures the environmental inhomogeneity in both space and time. When m is non-constant in x and independent of t, it is shown in [8] that the slower diffusing species is always selected. The following result in [18] shows that the same principle no longer holds in general time-periodic environments, where the two competing species are identical except for their diffusion rates.

Theorem 2.3.1 *Assume $\tau = 1$. There exist a 1-periodic function $m(x,t)$, and $0 < d_1 < d_2$ such that the semi-trivial periodic solution $E_2 = (0,\tilde{v})$ exists, and it is globally asymptotically stable among all solutions (u,v) of (2.35) such that $u_0 \geq, \not\equiv 0$ and $v_0 \geq, \not\equiv 0$.*

Proof Let $c(x,t)$ and d_1 be as given in Corollary 2.2.11(i). Let $\lambda_1 = \lambda(d_1,c)$ be the principal eigenvalue of (2.4) with $d = d_1$, and $\varphi_1(x,t)$ be the corresponding positive eigenfunction. Set

$$m(x,t) = c(x,t) + \lambda(d_1,c) + \eta\varphi_1(x,t) \quad \text{and} \quad \tilde{u}(x,t) = \eta\varphi_1(x,t),$$

where $\eta > 0$ is large enough so that $m(x,t) > 0$. It is then clear that $\tilde{u}$ is a positive periodic solution of the dynamical system generated by

$$\begin{cases} \partial_t\theta - d_1\Delta\theta = \theta(m(x,t) - \theta), & x \in \Omega,\ t > 0, \\ \mathbf{n}\cdot\nabla\theta = 0, & x \in \partial\Omega,\ t > 0, \\ \theta(x,0) = \theta_0(x), & x \in \Omega. \end{cases} \tag{2.36}$$

In fact, $\tilde{u}$ is the unique positive periodic solution of (2.36), since the Poincaré map Φ_1 of the dynamical system generated by (2.36) generates a subhomogeneous dynamical system (see Appendix C for details). By continuity, if we replace d_1 by $d_2 = d_1 + \epsilon$ with $0 < \epsilon \ll 1$, there is again a unique positive periodic solution $\tilde{v}$. Therefore, the system (2.35) has at least three periodic solutions

$$E_0 = (0,0), \quad E_1 = (\tilde{u},0) \quad \text{and} \quad E_2 = (0,\tilde{v}).$$

By the theory of monotone dynamical systems (Theorem E.1.9), it is enough to show that (i) $E_1 = (\tilde{u},0)$ is linearly unstable, and (ii) the system (2.35) has no positive periodic solutions.

For (i), let λ_{E_1} be the principal eigenvalue of

$$\begin{cases} \partial_t \varphi - d_2 \Delta \varphi = (m - \tilde{u})\varphi + \lambda_{E_1} \varphi, & x \in \Omega,\ t \in [0,1], \\ \mathbf{n} \cdot \nabla \varphi = 0, & x \in \partial\Omega,\ t \in [0,1], \\ \varphi(x,0) = \varphi(x,1), & x \in \Omega. \end{cases} \tag{2.37}$$

We need to show that $\lambda_{E_1} < 0$. Indeed,

$$\lambda_{E_1} = \lambda(d_2, m - \tilde{u}) = \lambda(d_1 + \epsilon, c) < \lambda(d_1, c) = 0 \quad \text{for small } \epsilon > 0,$$

where we used $\partial_d \lambda(d_1, c) < 0$. This proves (i).

Next, we prove (ii). Suppose to the contrary that there are sequences

$$(d_1, d_2) = (d_1, d_1 + \epsilon_k) \quad \text{and} \quad (u_k, v_k)$$

where $\epsilon_k \to 0$ and (u_k, v_k) is a (componentwise) positive periodic solution of the system (2.35) with $(d_1, d_2) = (d_1, d_1 + \epsilon_k)$. By the maximum principle,

$$\|u_k\|_\infty + \|v_k\|_\infty \le 2\|m\|_\infty.$$

We may apply a parabolic L^p estimate to pass to a subsequence so that

$$(u_k, v_k) \to (u_\infty, v_\infty) \quad \text{weakly in } W^{2,1,p}(\Omega \times [0,1]).$$

By Sobolev embedding, it also converges strongly in $C^{1+\alpha,(1+\alpha)/2}(\overline{\Omega} \times [0,1])$ for all $0 < \alpha < 1$. Moreover, $u_\infty + v_\infty$ is a positive periodic solution of (2.36). By uniqueness, we have $u_\infty + v_\infty = \tilde{u}$. Hence,

$$m - u_k - v_k = m - \tilde{u} + o(1) = c + \lambda(d_1, c) + o(1).$$

Now, by interpreting the existence of (u_k, v_k), we have

$$\lambda(d_1, m - u_k - v_k) = \lambda(d_1 + \epsilon_k, m - u_k - v_k) = 0.$$

By the mean value theorem, there exists $\hat{d}_k \in (d_1, d_1 + \epsilon_k)$ such that

$$\partial_d \lambda(\hat{d}_k, m - u_k - v_k) = 0 \quad \text{for all } k.$$

Letting $k \to \infty$, we have $\hat{d}_k \to d_1$ and $m - u_k - v_k \to c + \lambda(d_1, c)$, so that

$$\partial_d \lambda(d_1, c + \lambda(d_1, c)) = \partial_d \lambda(d_1, c) = 0.$$

This is a contradiction to the construction of c. □

Next we present an application of Theorem 2.2.13 which says that for two competing species which are identical except for their relaxation time, the species with the shorter relaxation time will drive the other species to extinction.

Theorem 2.3.2 *Assume $d_1 = d_2$, $m \in C^2$, $\int_\Omega \int_0^1 m(x,t)\,\mathrm{d}x\,\mathrm{d}t > 0$, and $\partial_{xt} m(x,t) \not\equiv \phi(x)g(t)$ for any $\phi \in C(\mathbb{R};\mathbb{R})$ and $g \in C([0,1];\mathbb{R})$ which is 1-periodic. Then the following conclusions hold:*

(i) *If $\tau > 1$, then the semi-trivial periodic solution $E_2 = (0, \tilde{v})$ is globally asymptotically stable among all solutions (u, v) of (2.35) such that $u_0 \geq, \not\equiv 0$ and $v_0 \geq, \not\equiv 0$;*
(ii) *If $\tau < 1$, then the semi-trivial periodic solution $E_1 = (\tilde{u}, 0)$ is globally asymptotically stable among all solutions (u, v) of (2.35) such that $u_0 \geq, \not\equiv 0$ and $v_0 \geq, \not\equiv 0$.*

Proof It suffices to establish part (i) as the proof of (ii) is identical to that of (i). Thanks again to Theorem E.1.9, it suffices to show that (a) $E_1 = (\tilde{u}, 0)$ is linearly unstable, and (b) system (2.35) has no positive periodic solutions.

For (a), let $\lambda(\tau, d, c)$ denote the principal eigenvalue of the linear problem

$$\begin{cases} \tau\partial_t\varphi - d\Delta\varphi = c\varphi + \lambda\varphi, & x \in \Omega,\ t \in [0,1], \\ \mathbf{n} \cdot \nabla\varphi = 0, & x \in \partial\Omega,\ t \in [0,1], \\ \varphi(x,0) = \varphi(x,1), & x \in \Omega. \end{cases} \tag{2.38}$$

The stability of E_1 is determined by the sign of $\lambda_{E_1} := \lambda(1, d_2, m - \tilde{u})$. As $d_1 = d_2$, it follows from Theorem 2.2.13 that

$$\lambda_{E_1} = \lambda(1, d_1, m - \tilde{u}) \leq \lambda(\tau, d_2, m - \tilde{u}) = 0,$$

where the last equality follows from the equation of $\tilde{u}$. It suffices to further show that $\lambda_{E_1} \neq 0$: if not, by Theorem 2.2.13, $\lambda_{E_1} = 0$ implies that $m - \tilde{u} = a(x) + b(t)$ for some functions $a(x)$ and $b(t)$. Since $m - \tilde{u}$ is 1-periodic in t, we may replace $a(x)$ by $a(x) + \int_0^1 b(s)\,\mathrm{d}s$ and $b(t)$ by $b(t) - \int_0^1 b(s)\,\mathrm{d}s$, so that $b(t)$ is 1-periodic and with zero average in $[0,1]$. Substituting this into the equation of $\tilde{u}$ we find that $\tilde{u} = w(x)\exp\left(\frac{1}{\tau}\int_0^t b(s)\,\mathrm{d}s\right)$, where $w > 0$ is a positive solution of

$$-d_1\Delta w = a(x)w \quad \text{in } \Omega, \quad \mathbf{n} \cdot \nabla w = 0 \quad \text{on } \partial\Omega. \tag{2.39}$$

Hence, $m = w(x)\exp\left(\frac{1}{\tau}\int_0^t b(s)\,\mathrm{d}s\right) + a(x) + b(t)$, which is a contradiction to our assumption. Hence, $\lambda_{E_1} < 0$, i.e. $(\tilde{u}, 0)$ is unstable.

To prove (b), we argue by contradiction: Suppose that system (2.35) has positive periodic solutions, which we denote by (u, v). Therefore,

$$\lambda(\tau, d_1, m - u - v) = \lambda(1, d_2, m - u - v) = 0.$$

As $d_1 = d_2$ and $\tau \neq 1$, by Theorem 2.2.13 we see that $m - u - v = a(x) + b(t)$ for some functions $a(x)$ and $b(t)$. Similarly as before, we may assume that $\int_0^1 b(s)\,\mathrm{d}s = 0$, to deduce that

$$(u(x,t), v(x,t)) = \left(k_1 \exp\left(\frac{1}{\tau}\int_0^t b(s)\,\mathrm{d}s\right) w(x), k_2 \exp\left(\int_0^t b(s)\,\mathrm{d}s\right) w(x)\right)$$
$$= (z_1(t)w(x), z_2(t)w(x)), \tag{2.40}$$

where $k_i > 0$ are constants and $w(x)$ is a positive solution of (2.39). Hence, $m(x,t) = w(x)(z_1(t) + z_2(t)) + a(x) + b(t)$, which contradicts our assumption on m_{xt} being non-separable. □

Remark 2.3.3 For applications of the principal eigenvalues to the two species competition model (2.35) in a spatially heterogeneous and temporally environment we refer to the work of Bai et al. [2], in which the case $\tau = 1$ and $d_1 \neq d_2$ is further investigated. A spatially discrete and time continuous version of (2.35) with general τ and d_i $(i = 1, 2)$ has been considered in [27].

2.4 Further Reading

In contrast to autonomous problems, the principal eigenvalues of periodic-parabolic operators lack a variational characterization, and much less is known about their qualitative properties. Here we present some further references for applications of principal eigenvalues to ecology, evolution and epidemiology for the interested readers.

The theory of principal eigenvalues for periodic-parabolic operators appears frequently in the study of persistence, competition and predation of species in ecology. For the applications of principal eigenvalues to the derivation of the minimal traveling wave speeds in time-periodic shifting environments, see [3, 11, 42]. For its applications in spatio-temporally degenerate environments, see [7, 9, 10] for the case involving single equations and [1, 36, 37] for the case involving cooperative systems. For the influence of advection/flow on the principal eigenvalues of time-periodic operators with applications, see [20, 25, 26, 28, 32, 33, 34, 35, 39].

Principal eigenvalue theory also plays important roles in the study of evolution questions. The evolution of dispersal can be studied by the examination of the invasibility of phenotypes in two-species competition models. The latter is characterized in terms of the principal eigenvalue of the linearized system at boundary equilibria or periodic solutions. For time-periodic environments, it was pioneered by the work of Hutson et al. [18]; see [2] for more recent developments, and [21] for the extension to N-competing species. For the concept of ideal free distribution in time-periodic environments and evolutionarily stable dispersal strategies, see [4, 5] for recent developments. Principal eigenvalue theory is also applied to describe the asymptotic dynamics in phenotypically structured population models, see [6, 12, 31, 41].

The investigation of reaction-diffusion models in epidemiology, particularly the characterization of the basic reproduction number, motivated the mathematical analysis of the principal eigenvalues of periodic-parabolic problems with weight functions. See, e.g. [23, 38, 40, 44, 45].

When the coefficients of the parabolic operators are non-autonomous and aperiodic, the notion of the principal Floquet bundle is a natural generalization of the principal eigenvalue, which describes the exponential separation property. This will be discussed in Chapter 4. For more general results and recent developments, see [14, 15, 16, 17, 29, 30, 43].

Problems

2.4.1 Let λ_1 be the principal eigenvalue of

$$\begin{cases} \varphi_t + \mathcal{L}\varphi = \lambda\varphi & \text{in } \Omega_T, \\ \mathcal{B}\varphi = 0 & \text{on } S\Omega_T, \\ \varphi(x,0) = \varphi(x,T) & \text{for } x \in \Omega, \end{cases} \tag{2.41}$$

where $\mathcal{L} = -a^{ij}D_{ij} - b^iD_i - c$ is a non-divergence form operator and $\mathcal{B} = p^iD_i + p^0$ is an oblique boundary operator, such that the coefficients a^{ij}, b^i, p^i, p^0 are independent of time and continuous in $\overline{\Omega}$, and $c \in C(\overline{\Omega} \times [0,T])$. Show that

$$-\int_0^1 \max_{x\in\overline{\Omega}} c(x,t)\,\mathrm{d}t \le \lambda_1 \le -\int_0^1 \min_{x\in\overline{\Omega}} c(x,t)\,\mathrm{d}t \qquad \forall d > 0. \tag{2.42}$$

[Hint: Observe that $-\frac{1}{T}\int_0^T \max_{x\in\overline{\Omega}} c(x,t)\,\mathrm{d}t$ is the principal eigenvalue of

$$\begin{cases} \partial_t\varphi - a^{ij}D_{ij}\varphi - b^iD_i\varphi = \left(\max_{x\in\overline{\Omega}} c(x,t)\right)\varphi + \lambda\varphi, & x \in \Omega, t \in [0,T], \\ \mathcal{B}\varphi = 0, & x \in \partial\Omega, t \in [0,T], \\ \varphi(x,0) = \varphi(x,T), & x \in \Omega, \end{cases}$$

and use eigenvalue comparison lemmas similar to Lemmas 1.3.12 and 1.3.13.]

2.4.2 Prove the Hutson–Shen–Vickers Lemma for (2.41). Precisely, let λ_1 be the principal eigenvalue of (2.41), and let μ_1 denote the smallest eigenvalue of

$$\begin{cases} -a^{ij}D_{ij}\varphi - b^iD_i\varphi - \left(\frac{1}{T}\int_0^T c(x,t)\,\mathrm{d}t\right)\psi = \mu\psi, \; x \in \Omega, \\ \mathcal{B}\psi = 0, \qquad x \in \partial\Omega. \end{cases}$$

Suppose a^{ij} and b^i are independent of t, show that $\lambda_1 \le \mu_1$, and find a condition such that equality holds.

2.4.3 Assume $c(x,t)$ is continuous in $\overline{\Omega}\times[0,1]$ and 1-periodic in time. Show that

$$\int_0^1 \max_{x\in\overline{\Omega}} c(x,t)\,\mathrm{d}t = \sup_{x(t)\in\mathcal{S}} \int_0^1 c(x(t),t)\,\mathrm{d}t,$$

where $\mathcal{S}=\{x(t)\in C[0,\infty) : x(t)=x(t+1)\}$.

2.4.4 Let $\lambda(\tau)$ denote the principal eigenvalue of (2.24). Show that

$$\lim_{\tau\to\infty}\lambda(\tau)=\mu_1,$$

where μ_1 denotes the smallest eigenvalue of (2.11).

2.4.5 (Poincaré's inequality) Given $\beta\in C(\overline{\Omega})$ such that $\inf_\Omega \beta>0$, show that there is a positive constant C_β such that for any $u\in H^1(\Omega)$ such that $\int \beta(x)u(x)\,\mathrm{d}x=0$, we have

$$\int_\Omega |u(x)|^2\,\mathrm{d}x \le C_\beta \int_\Omega |\nabla u|^2\,\mathrm{d}x.$$

In particular, for any strictly positive continuous functions α,γ, we have

$$\int_\Omega \alpha(x)|u(x)|^2\,\mathrm{d}x \le \frac{C_\beta \sup_\Omega \alpha}{\inf_\Omega \gamma}\int_\Omega \gamma(x)|\nabla u(x)|^2\,\mathrm{d}x.$$

2.4.6 Let $\lambda(d)$ denote the principal eigenvalue of (2.4). Assume that Ω is an interval and either $c_x\ge 0$ for all x and t or $c_x\le 0$ for all x and t, then $\lambda(d)$ is monotone non-increasing in d.

References

1. P. Álvarez Caudevilla, Y. Du, and R. Peng, *Qualitative analysis of a cooperative reaction-diffusion system in a spatiotemporally degenerate environment*, SIAM J. Math. Anal., 46 (2014), pp. 499–531.
2. X. Bai, X. He, and W.-M. Ni, *Dynamics of a periodic-parabolic Lotka–Volterra competition-diffusion system in heterogeneous environments*, J. Eur. Math. Soc., (2022). in press.
3. H. Berestycki and G. Nadin, *Asymptotic spreading for general heterogeneous equations*, Mem. Amer. Math. Soc., (2022). in press.
4. R. S. Cantrell and C. Cosner, *Evolutionary stability of ideal free dispersal under spatial heterogeneity and time periodicity*, Math. Biosci., 305 (2018), pp. 71–76.
5. R. S. Cantrell, C. Cosner, and K.-Y. Lam, *Ideal free dispersal under general spatial heterogeneity and time periodicity*, SIAM J. Appl. Math., 81 (2021), pp. 789–813.
6. C. Carrère and G. Nadin, *Influence of mutations in phenotypically-structured populations in time periodic environment*, Discrete Contin. Dyn. Syst. Ser. B, 25 (2020), pp. 3609–3630.
7. D. Daners and C. Thornett, *Periodic-parabolic eigenvalue problems with a large parameter and degeneration*, J. Differential Equations, 261 (2016), pp. 273–295.
8. J. Dockery, V. Hutson, K. Mischaikow, and M. Pernarowski, *The evolution of slow dispersal rates: a reaction diffusion model*, J. Math. Biol., 37 (1998), pp. 61–83.

9. Y. Du and R. Peng, *The periodic logistic equation with spatial and temporal degeneracies*, Trans. Amer. Math. Soc., 364 (2012), pp. 6039–6070.
10. ———, *Sharp spatiotemporal patterns in the diffusive time-periodic logistic equation*, J. Differential Equations, 254 (2013), pp. 3794–3816.
11. J. Fang, R. Peng, and X.-Q. Zhao, *Propagation dynamics of a reaction-diffusion equation in a time-periodic shifting environment*, J. Math. Pures Appl. (9), 147 (2021), pp. 1–28.
12. S. Figueroa Iglesias and S. Mirrahimi, *Long time evolutionary dynamics of phenotypically structured populations in time-periodic environments*, SIAM J. Math. Anal., 50 (2018), pp. 5537–5568.
13. P. Hess, *Periodic-parabolic boundary value problems and positivity*, vol. 247 of Pitman Research Notes in Mathematics Series, Longman Scientific & Technical, Harlow; copublished in the United States with John Wiley & Sons, Inc., New York, 1991.
14. J. Húska, *Exponential separation and principal Floquet bundles for linear parabolic equations on general bounded domains: nondivergence case*, Trans. Amer. Math. Soc., 360 (2008), pp. 4639–4679.
15. J. Húska and P. Poláčik, *The principal Floquet bundle and exponential separation for linear parabolic equations*, J. Dynam. Differential Equations, 16 (2004), pp. 347–375.
16. J. Húska, P. Poláčik, and M. V. Safonov, *Principal eigenvalues, spectral gaps and exponential separation between positive and sign-changing solutions of parabolic equations*, Discrete Contin. Dyn. Syst., (2005), pp. 427–435.
17. J. Húska, P. Poláčik, and M. V. Safonov, *Harnack inequalities, exponential separation, and perturbations of principal Floquet bundles for linear parabolic equations*, Ann. Inst. H. Poincaré C Anal. Non Linéaire, 24 (2007), pp. 711–739.
18. V. Hutson, K. Mischaikow, and P. Poláčik, *The evolution of dispersal rates in a heterogeneous time-periodic environment*, J. Math. Biol., 43 (2001), pp. 501–533.
19. V. Hutson, W. Shen, and G. T. Vickers, *Estimates for the principal spectrum point for certain time-dependent parabolic operators*, Proc. Amer. Math. Soc., 129 (2001), pp. 1669–1679.
20. D. Jiang, K.-Y. Lam, Y. Lou, and Z.-C. Wang, *Monotonicity and global dynamics of a nonlocal two-species phytoplankton model*, SIAM J. Appl. Math., 79 (2019), pp. 716–742.
21. K.-Y. Lam and Y. Lou, The principal Floquet bundle and the dynamics of fast diffusing communities. Manuscript submitted for publication, 2022.
22. G. M. Lieberman, *Second order parabolic differential equations*, World Scientific Publishing Co., Inc., River Edge, NJ, 1996.
23. S. Liu and Y. Lou, *Classifying the level set of principal eigenvalue for time-periodic parabolic operators and applications*, J. Funct. Anal., 282 (2022), pp. Paper No. 109338, 43.
24. S. Liu, Y. Lou, R. Peng, and M. Zhou, *Monotonicity of the principal eigenvalue for a linear time-periodic parabolic operator*, Proc. Amer. Math. Soc., 147 (2019), pp. 5291-5302.
25. ———, *Asymptotics of the principal eigenvalue for a linear time-periodic parabolic operator I: large advection*, SIAM J. Math. Anal., 53 (2021), pp. 5243–5277.
26. ———, *Asymptotics of the principal eigenvalue for a linear time-periodic parabolic operator II: Small diffusion*, Trans. Amer. Math. Soc., 374 (2021), pp. 4895–4930.
27. S. Liu, Y. Lou, and P. Song, *A new monotonicity for principal eigenvalues with applications to time-periodic patch models*, SIAM J. Appl. Math, 82 (2022). in press.
28. M. Ma and C. Ou, *Existence, uniqueness, stability and bifurcation of periodic patterns for a seasonal single phytoplankton model with self-shading effect*, J. Differential Equations, 263 (2017), pp. 5630–5655.
29. J. Mierczyński and W. Shen, *Exponential separation and principal Lyapunov exponent/spectrum for random/nonautonomous parabolic equations*, J. Differential Equations, 191 (2003), pp. 175–205.
30. ———, *Time averaging for nonautonomous/random linear parabolic equations*, Discrete Contin. Dyn. Syst. Ser. B, 9 (2008), pp. 661–699.
31. S. Mirrahimi, B. Perthame, and P. E. Souganidis, *Time fluctuations in a population model of adaptive dynamics*, Ann. Inst. H. Poincaré C Anal. Non Linéaire, 32 (2015), pp. 41–58.
32. G. Nadin, *The principal eigenvalue of a space-time periodic parabolic operator*, Ann. Mat. Pura Appl. (4), 188 (2009), pp. 269–295.

33. G. Nadin, *Some dependence results between the spreading speed and the coefficients of the space-time periodic Fisher-KPP equation*, European J. Appl. Math., 22 (2011), pp. 169–185.
34. J. Nolen, M. Rudd, and J. Xin, *Existence of KPP fronts in spatially-temporally periodic advection and variational principle for propagation speeds*, Dyn. Partial Differ. Equ., 2 (2005), pp. 1–24.
35. J. Nolen and J. Xin, *Existence of KPP type fronts in space-time periodic shear flows and a study of minimal speeds based on variational principle*, Discrete Contin. Dyn. Syst., 13 (2005), pp. 1217–1234.
36. R. Peng, *Long-time behavior of a cooperative periodic-parabolic system in a spatiotemporally degenerate environment*, J. Differential Equations, 259 (2015), pp. 2903–2947.
37. ———, *Long-time behavior of a cooperative periodic-parabolic system: temporal degeneracy versus spatial degeneracy*, Calc. Var. Partial Differential Equations, 53 (2015), pp. 179–219.
38. R. Peng and X.-Q. Zhao, *A reaction-diffusion SIS epidemic model in a time-periodic environment*, Nonlinearity, 25 (2012), pp. 1451–1471.
39. ———, *A nonlocal and periodic reaction-diffusion-advection model of a single phytoplankton species*, J. Math. Biol., 72 (2016), pp. 755–791.
40. L. Pu and Z. Lin, *A diffusive SIS epidemic model in a heterogeneous and periodically evolving environment*, Math. Biosci. Eng., 16 (2019), pp. 3094–3110.
41. L. Roques, F. Patout, O. Bonnefon, and G. Martin, *Adaptation in general temporally changing environments*, SIAM J. Appl. Math., 80 (2020), pp. 2420–2447.
42. W. Shen, Z. Shen, S. Xue, and D. Zhou, *Population dynamics under climate change: persistence criterion and effects of fluctuations*, J. Math. Biol., 84 (2022), p. 42. Paper No. 30.
43. W. Shen and G. T. Vickers, *Spectral theory for general nonautonomous/random dispersal evolution operators*, J. Differential Equations, 235 (2007), pp. 262–297.
44. L. Zhang and X.-Q. Zhao, *Asymptotic behavior of the basic reproduction ratio for periodic reaction-diffusion systems*, SIAM J. Math. Anal., 53 (2021), pp. 6873–6909.
45. L. Zhao, Z.-C. Wang, and L. Zhang, *Propagation dynamics for a time-periodic reaction-diffusion SI epidemic model with periodic recruitment*, Z. Angew. Math. Phys., 72 (2021), p. 20. Paper No. 142.

Chapter 3
The Maximum Principle and the Principal Eigenvalue for Systems

Abstract In this chapter the maximum principle and the comparison principle for cooperative parabolic systems are presented. The existence of the principal eigenvalue for cooperative elliptic/parabolic systems is considered, and the asymptotic behavior of the principal eigenvalues for small diffusion rates is determined. As an application, we apply the theory to a two-species competition system.

3.1 Comparison Principle of Cooperative Parabolic Systems

Let $\Omega \subset \mathbb{R}^N$ be a bounded domain with smooth boundary $\partial\Omega$. Let $\mathbf{n} = (n_i(x))$ be the outer unit normal vector on $\partial\Omega$. Define

$$\Omega_T = \Omega \times (0, T], \quad \overline{\Omega}_T = \overline{\Omega} \times [0, T], \quad S\Omega_T = \partial\Omega \times (0, T].$$

For $k = 1, \ldots, K$, consider the non-divergence form parabolic operator

$$\partial_t + \mathcal{L}_k = \partial_t - a_k^{ij} D_{ij} - b_k^j D_j \quad \text{in } \Omega_T$$

where $a_k^{ij}, b_k^j \in C(\overline{\Omega}_T)$, and the oblique boundary condition

$$\mathcal{B}_k = p_k^i D_i + p_k^0 \quad \text{on } S\Omega_T,$$

where $p_k^i \in C(S\Omega_T)$ satisfies, for each k,

$$p_k^i(x,t) n_i(x) > 0 \quad \text{and} \quad p_k^0 \geq 0 \quad \text{on } S\Omega_T.$$

We recall the notion of a generalized subsolution, as introduced in Chapter 1.

Definition 3.1.1 We say that $u \in C(\overline{\Omega}_T)$ satisfies

$$u_t + \mathcal{L}_k u \leq f(x,t,u,Du) \quad \text{in } \Omega_T \quad \text{and} \quad \mathcal{B}_k u \leq g(x,t) \quad \text{on } S\Omega_T \tag{3.1}$$

K. -Y. Lam, Y. Lou, *Introduction to Reaction-Diffusion Equations*, Lecture Notes on Mathematical Modelling in the Life Sciences, https://doi.org/10.1007/978-3-031-20422-7_3

in the generalized sense if for each $(x_0, t_0) \in \overline{\Omega} \times (0, T]$, there exist a neighborhood U of (x_0, t_0) in $\overline{\Omega}_T$ and a function $\tilde{u} \in C^{2,1}(\overline{\Omega}_T)$ such that $\tilde{u} \geq u$ in U, $\tilde{u}(x_0, t_0) = u(x_0, t_0)$, and $\tilde{u}$ satisfies

$$\tilde{u}_t(x_0, t_0) + \mathcal{L}_k \tilde{u}(x_0, t_0) \leq f(x_0, t_0, \tilde{u}(x_0, t_0), D\tilde{u}(x_0, t_0))$$

if $(x_0, t_0) \in \Omega_T$, or

$$\mathcal{B}\tilde{u}(x_0, t_0) \leq g(x_0, t_0)$$

if $(x_0, t_0) \in S\Omega_T$.

Theorem 3.1.2 *Assume $c_{k\ell} \in C(\overline{\Omega}_T)$ and $c_{k\ell} \geq 0$ for $k \neq \ell$. Suppose that*

1. *for $1 \leq k \leq K$, $u_k(x, t) \leq 0$ in Ω_T,*
2. *for $1 \leq k \leq K$, $(u_k)_t + \mathcal{L}_k u_k \leq c_{k\ell} u_\ell$ in Ω_T in the generalized sense, and*
3. *there exist k_0 and $(x_0, t_0) \in \Omega_T$ such that $u_{k_0}(x_0, t_0) = 0$.*

Then

$$u_{k_0} \equiv 0 \quad \text{in } \Omega \times (0, t_0].$$

Assume, in addition, that $\inf_{\Omega_T} c_{ij} > 0$ whenever $i \neq j$, then for every k,

$$u_k \equiv 0 \quad \text{in } \Omega \times (0, t_0].$$

Proof First, assume $c_{k\ell} \geq 0$ for $k \neq \ell$. Then $v = u_{k_0}$ satisfies $v \leq 0$ in Ω_T, $v(x_0, t_0) = 0$, and

$$v_t + \mathcal{L}_{k_0} v - c_{k_0 k_0} v \leq \sum_{l \neq k_0} c_{k_0 \ell} u_\ell \leq 0 \quad \text{in } \Omega_T \tag{3.2}$$

in the generalized sense. It follows from Lemma 1.1.8 that $v \equiv 0$ in $\Omega \times (0, t_0)$. By continuity, $v \equiv 0$ in $\Omega \times (0, t_0]$.

Next, assume $c_{k\ell} > 0$ in $\overline{\Omega}_T$ for $k \neq \ell$, then setting $u_{k_0} = v \equiv 0$ in (3.2) implies that $u_\ell \equiv 0$ in $\Omega \times (0, t_0]$ for all $\ell \neq k_0$. □

Remark 3.1.3 The condition $\inf_{\Omega_T} c_{k\ell} > 0$ for all $k \neq \ell$ can be relaxed to $c_{k\ell} \geq 0$ in Ω for $k \neq \ell$ and that the matrix $(\sup_{\Omega_T} c_{k\ell})$ is irreducible. See [9, Lemmas 4.1 and 4.3] for the case of autonomous coefficients.

Theorem 3.1.4 (Weak maximum principle) *Let $c_{k\ell} \in C(\overline{\Omega}_T)$ satisfy $c_{k\ell} \geq 0$ whenever $k \neq \ell$ (without assuming irreducible conditions). Assume $(u_k)_{k=1}^K$ satisfies*

$$u_k(x, 0) \leq 0 \quad \text{for } x \in \Omega,\ 1 \leq k \leq K, \tag{3.3}$$

and

$$\begin{cases} (u_k)_t + \mathcal{L}_k u_k \leq c_{k\ell} u_\ell & \text{in } \Omega_T, \\ \mathcal{B}_k u_k \leq 0 & \text{on } S\Omega_T \end{cases} \tag{3.4}$$

hold in the generalized sense. Then $u_k \leq 0$ in Ω_T for all k.

Proof Here we give the proof assuming the differential inequalities in (3.4) are satisfied in the classical sense. The general case follows by obvious modifications (see Chapter 1).

By replacing $u_k(x,t)$ with $e^{Mt}u_k(x,t)$ for some $M > 0$ large, we may assume that for each k,

$$\sum_{l=1}^{K} c_{k\ell} > 0. \tag{3.5}$$

We claim that $\sup_{\Omega_T} u_k \le 0$ in Ω_T for all k. For this purpose, fix $\epsilon > 0$ and define $v_k(x,t) = u_k(x,t) - \epsilon$, then v_k satisfies (3.3) and (3.4), and for each k,

$$\sup_{\Omega} v_k(x,0) < 0. \tag{3.6}$$

We claim that $v_k \le 0$ in Ω_T for all k. Suppose not, then there exist $\epsilon > 0$, k_0, and $(x_0, t_0) \in \Omega_T$ such that $v_{k_0}(x_0, t_0) = 0$ and for all k,

$$v_k \le 0 \quad \text{in } \Omega \times [0, t_0].$$

Since $v_{k_0}(x,0) < 0$ in Ω, it follows from Theorem 3.1.2 that $v_{k_0} < 0$ in $\Omega \times [0, t_0]$ and $v_{k_0}(x_0, t_0) = 0$ for some $x_0 \in \partial\Omega$. Using (3.5), it follows that $(v_{k_0})_t + \mathcal{L}_{k_0} v_{k_0} \le 0$ in $\Omega \times (0, t_0]$, and we may apply the Boundary Point Lemma (Theorem 1.1.7) to deduce that $p^j_{k_0} D_j v_{k_0} > 0$ at (x_0, t_0). Hence,

$$\mathcal{B}_{k_0} u_{k_0} = p^j_{k_0} D_j u_{k_0} + p^0_{k_0} u_{k_0} = p^j_{k_0} D_j v_{k_0} + p^0_{k_0}(v_{k_0} + \epsilon) > 0 \quad \text{at } (x_0, t_0).$$

This is in contradiction with $\mathcal{B}_{k_0} u_{k_0} \le 0$ on $S\Omega_T$. Hence, $u_k - \epsilon \le 0$ in Ω_T for all k. Since $\epsilon > 0$ is arbitrary, we deduce that $u_k \le 0$ in Ω_T for all k. □

Consider the nonlinear cooperative system:

$$\begin{cases} (u_k)_t + \mathcal{L}_k u_k = f_k(x,t,u) & \text{in } \Omega_T, \\ \mathcal{B}_k u_k = g(x,t) & \text{on } S\Omega_T, \\ u_k(x,0) = u_{k,0}(x) & \text{on } B\Omega_T, \end{cases} \tag{3.7}$$

where we assume that for each k,

$$f_k(x,t,u) \quad \text{is nondecreasing in } u_\ell, \text{ for } \ell \in \{1, ..., K\} \setminus \{k\}.$$

Theorem 3.1.5 *Suppose that (u_k) and (v_k) are respectively sub- and supersolutions of* (3.7) *(the PDE and BC in the generalized sense, and IC in the classical sense), then $u_k \le v_k$ hold for all k and $(x,t) \in \Omega_T$.*

Proof Observe that $w_k = u_k - v_k$ satisfies $w_k(x,0) \le 0$ in Ω for all k, and

$$\begin{cases} (w_k)_t + \mathcal{L}_k w_k \le c_{k\ell} w_\ell & \text{in } \Omega_T, \\ \mathcal{B}_k w_k \le 0 & \text{in } S\Omega_T \end{cases} \tag{3.8}$$

in the generalized sense, where

$$c_{k\ell} = \int_0^1 D_\ell f_k(x, t, \tau u + (1 - \tau)v)\, \mathrm{d}\tau.$$

Since $c_{k\ell} \geq 0$ in Ω_T when $k \neq \ell$, the conclusion follows from Theorem 3.1.4. □

3.2 The Principal Eigenvalue of Cooperative Systems

In this section we first establish the existence of the principal eigenvalues for cooperative parabolic systems in Subsection 3.2.1. In Subsection 3.2.2, we prove the asymptotic behavior of the principal eigenvalues of elliptic systems for small diffusion rates.

3.2.1 Existence results

Theorem 3.2.1 *If $a_k^{ij}, b_k^i, c_{k\ell}, p_k^i, p_k^0$ are continuous and T-periodic, and $c_{k\ell} \geq 0$ in Ω_T provided $k \neq \ell$, then the periodic-parabolic eigenvalue problem*

$$\begin{cases} (\varphi_k)_t + \mathcal{L}_k \varphi_k = c_{k\ell}\varphi_\ell + \lambda \varphi_k & \text{in } \Omega_T, \\ \mathcal{B}_k \varphi_k = 0 & \text{on } S\Omega_T, \\ \varphi_k(x, 0) = \varphi_k(x, T) & \text{for } x \in \Omega \end{cases} \tag{3.9}$$

has a principal eigenvalue, denoted as λ_1, in the sense that

1. $\lambda_1 \in \mathbb{R}$ *is an eigenvalue of* (3.9)*;*
2. $\lambda_1 \leq \inf \operatorname{Re} \lambda$, *where the infimum is taken over all eigenvalues of* (3.9)*;*
3. *the eigenfunction corresponding to λ_1 can be chosen componentwise nonnegative in $\overline{\Omega}_T$.*

Proof We will proceed as in the proof of Theorem 2.1.1. Let $\Phi_T : [C(\overline{\Omega})]^K \to [C(\overline{\Omega})]^K$ be given by $\Phi_T(u_{0,1}, \ldots, u_{0,K}) = (u_1(\cdot, T), \ldots, u_K(\cdot, T))$, where (u_k) is the unique solution of

$$\begin{cases} (u_k)_t + \mathcal{L}_k u_k = c_{k\ell} u_\ell & \text{in } \Omega_T, \\ \mathcal{B}_k u_k = 0 & \text{on } S\Omega_T, \\ u_k(x, 0) = u_{0,k}(x) & \text{in } \Omega. \end{cases}$$

We claim that λ is an eigenvalue of (3.9) with eigenfunction $(\varphi_k(x, t))$ if and only if $\mathrm{e}^{-\lambda T}$ is an eigenvalue of Φ_T with eigenfunction $(\varphi_k(x, 0))$. Indeed, suppose that λ is an eigenvalue of (3.9) with eigenfunction $\varphi(x, t) = (\varphi_k(x, t))$, then it follows that $\Phi_T[\varphi(\cdot, 0)] = \mathrm{e}^{-\lambda T}\varphi(\cdot, T) = \mathrm{e}^{-\lambda T}\varphi(\cdot, 0)$. On the other hand, suppose $\Phi_T[\varphi_0] = \Lambda \varphi_0$, then

$$\lambda = -\frac{\log \Lambda}{T} \quad \text{and} \quad \varphi(\cdot,t) = \mathrm{e}^{\lambda t}\Phi_t[\varphi_0] \tag{3.10}$$

defines an eigenpair of (3.9).

Next, observe that the operator Φ_T is compact, and monotone. The compactness follows from the parabolic L^p estimate, whereas the monotonicity follows from Theorem 3.1.4.

Finally, we invoke the Krein–Rutman Theorem (Theorem 1.3.3) to deduce that Φ_T has a principal eigenvalue Λ_1, and the corresponding eigenfunction can be chosen to be componentwise nonnegative in Ω. By virtue of the correspondence between the eigenvalues established in the beginning of the proof, we deduce that the problem (3.9) has the principal eigenvalue $\lambda_1 = -\frac{1}{T}\log \Lambda_1$ with the desired properties. □

Theorem 3.2.2 *If* $a_k^{ij}, b_k^i, c_{k\ell}, p_k^i, p_k^0$ *are continuous and T-periodic, and* $\inf_{\Omega_T} c_{k\ell} > 0$ *provided* $k \neq \ell$*, then the periodic-parabolic eigenvalue problem* (3.9) *has a principal eigenvalue* λ_1*, in the sense that*

1. $\lambda_1 \in \mathbb{R}$ *is a simple eigenvalue of* (3.9)*;*
2. $\lambda_1 < \inf_{\lambda \neq \lambda_1} \operatorname{Re} \lambda$*;*
3. *the eigenfunction corresponding to* λ_1 *can be chosen componentwise strictly positive in* $\overline{\Omega}_T$*;*
4. *if* λ *is an eigenvalue of* (3.9)*, with a componentwise nonnegative eigenfunction, then* $\lambda = \lambda_1$.

Remark 3.2.3 In fact, the same conclusion holds if $\inf_{\Omega_T} c_{k\ell} > 0$ for $k \neq \ell$ is replaced by $c_{k\ell} \geq 0$ for $k \neq \ell$ and the cooperative matrix $(\sup_{\Omega_T} c_{k\ell})$ is irreducible.

Proof Let Φ_T be given as in the proof of Theorem 3.2.1. Then Φ_T is compact and strongly monotone, thanks to Theorems 3.1.2 and 3.1.4. The theorem follows by repeating the proof of Theorem 3.2.1, while using the strong form of the Krein–Rutman Theorem (Theorem 1.3.4). □

Theorem 3.2.4 *If* a^{ij}, b^i, c, p^i, p^0 *are independent of time and* $\inf_\Omega c_{k\ell} \geq 0$ *for* $k \neq l$*, then the elliptic eigenvalue problem*

$$\begin{cases} \mathcal{L}_k \varphi_k = c_{k\ell}\varphi_\ell + \lambda \varphi_k & \text{in } \Omega, \\ \mathcal{B}_k \varphi_k = 0 & \text{on } \partial\Omega \end{cases} \tag{3.11}$$

has a principal eigenvalue λ_1*, in the sense that* λ_1 *has the least real part among all eigenvalues of* (3.11)*, and the corresponding eigenfunction* (φ_k) *can be chosen componentwise nonnegative.*

Proof Let Φ_T be given as in the proof of Theorem 3.2.1. We will take $T_j = 2^{-j}$ for $j \geq 1$. Then for each $j \geq 1$, $\Phi_{2^{-j}}$ is a linear positive compact operator on $[C(\overline{\Omega})]^K$. By the Krein–Rutman Theorem (Theorem 1.3.3), there exist λ_j and $\varphi_j = (\varphi_{j,k})_{k=1}^K$ such that

$$\varphi_j = \exp(\lambda_j 2^{-j})\Phi_{2^{-j}}(\varphi_j),$$

where we used (3.10). Hence, for each $j \geq 1$,

$$\varphi_j = \exp(\lambda_j t)\Phi_t(\varphi_j), \quad t = i2^{-j},\ i \in \mathbb{N}. \tag{3.12}$$

If we normalize φ_j by $\sum_k \|\varphi_{j,k}\|_{C(\overline{\Omega})} = 1$ then we may use the regularity theory to deduce, up to passing to a subsequence, that $\lambda_j \to \tilde{\lambda}$ and $\varphi_j \to \tilde{\varphi} = (\tilde{\varphi}_k)$ uniformly in Ω. By letting $j \to \infty$ in (3.12), we deduce that

$$\tilde{\varphi} = \exp(\tilde{\lambda}t)\Phi_t(\tilde{\varphi}), \quad t = i2^{-j},\ i, j \geq 1.$$

Note that the above equality holds on the set of dyadic numbers of the form $i2^{-j}$, which is a dense subset of $(0, \infty)$. We then conclude, by continuity, that

$$\tilde{\varphi} = \exp(\tilde{\lambda}t)\Phi_t(\tilde{\varphi}), \quad t \geq 0.$$

This implies the existence of the eigenvalue $\tilde{\lambda}$ with the nonnegative eigenfunction $\tilde{\varphi}(x) = (\tilde{\varphi}_k(x))$.

It remains to show that $\tilde{\lambda}$ has the least real part among all eigenvalues of (3.11). Let $\tilde{\lambda}'$ be an eigenvalue of (3.11), then $\exp(-\tilde{\lambda}'t)$ is an eigenvalue of Φ_t. Take $t = 2^{-j}$, then

$$|\exp(-\tilde{\lambda}'2^{-j})| \leq r(\Phi_{2^{-j}}) = \exp(-\lambda_j 2^{-j}), \quad j \geq 1,$$

thanks to the Krein–Rutman Theorem (Theorem 1.3.3). Here $r(T)$ denotes the spectral radius of the operator T. Hence,

$$\lambda_j \leq \mathrm{Re}\,\tilde{\lambda}', \quad j \geq 1.$$

Letting $j \to \infty$, we deduce $\tilde{\lambda} \leq \mathrm{Re}\,\tilde{\lambda}'$. This completes the proof. □

Theorem 3.2.5 *If a^{ij}, b^i, c, p^i, p^0 are independent of time and $\inf_\Omega c_{k\ell} > 0$ for $k \neq l$, then the elliptic eigenvalue problem* (3.11) *has a principal eigenvalue λ_1, in the sense that*

1. *$\lambda_1 \in \mathbb{R}$ is a simple eigenvalue of* (3.11)*;*
2. *$\lambda_1 < \inf_{\lambda \neq \lambda_1} \mathrm{Re}\,\lambda$, where λ denotes an eigenvalue of* (3.11)*;*
3. *the eigenfunction corresponding to λ_1 can be chosen strictly positive in $\overline{\Omega_T}$;*
4. *if λ is an eigenvalue of* (3.11) *with a nonnegative eigenfunction, then $\lambda = \lambda_1$.*

Proof In this case Φ_t is a compact, strongly positive, linear operator, and one can apply the strong form of the Krein–Rutman Theorem. The desired properties follow from the fact that if λ is an eigenvalue of (3.11) then $e^{-\lambda t}$ is an eigenvalue of Φ_t. For instance, λ_1 is a simple eigenvalue of (3.11), since $e^{-\lambda_1 t}$ is a simple eigenvalue of Φ_t. Also, to show $\lambda_1 < \mathrm{Re}\,\lambda$ for any eigenvalue $\lambda \neq \lambda_1$, observe that $e^{-\lambda_1 t}$ (resp. $e^{-\lambda t}$) is the principal eigenvalue (an eigenvalue) of Φ_t. The Krein–Rutman Theorem applied to Φ_t implies that $e^{-\lambda_1 t} > |e^{-\lambda t}|$ holds for $t > 0$, which in turn yields that $\lambda_1 < \mathrm{Re}\,\lambda$. We leave the verification as an exercise. □

3.2.2 Asymptotic behavior of the principal eigenvalue

For the purpose of applications, in this subsection we study the asymptotic behavior of the principal eigenvalue of the linear cooperative elliptic system (3.11) when the diffusion rates are sufficiently small. For the sake of clarity, we set $\mathcal{L}_k = -d_k\Delta$, then (3.11) becomes

$$\begin{cases} -d_k\Delta\varphi_k = c_{k\ell}\varphi_\ell + \lambda\varphi_k & \text{in } \Omega, \\ \mathcal{B}_k\varphi_k = 0 & \text{on } \partial\Omega. \end{cases} \tag{3.13}$$

To formulate the theorem, we recall the Perron–Frobenius Theorem concerning the principal eigenvalue of real $N \times N$ matrices with nonnegative off-diagonal entries.

Theorem 3.2.6 (Perron–Frobenius) *Let $C = (c_{ij}) \in \mathbb{R}^N \times \mathbb{R}^N$ such that $c_{ij} \geq 0$ for all $i \neq j$. Then C has a Perron–Frobenius eigenvalue $\Lambda_1(C) \in \mathbb{R}$ in the sense that $\Lambda_1(C) \geq \operatorname{Re}\lambda$ for any other eigenvalue λ of C.*

If, in addition, C is irreducible (i.e. $(\alpha I + C)^m$ is a componentwise positive matrix for some $m \geq 1$ and $\alpha > 0$), then

1. *$\Lambda_1(C)$ is a simple eigenvalue;*
2. *$\Lambda_1(C) > \operatorname{Re}\lambda$ for any other eigenvalue λ of C;*
3. *the corresponding eigenvector can be chosen componentwise positive;*
4. *if λ is an eigenvalue with a nonnegative eigenvector, then $\lambda = \Lambda_1(C)$.*

Remark 3.2.7 If, in addition, $c_{ii} \geq 0$ for all i, then $\Lambda_1(C)$ is given by the spectral radius of C, i.e. $\Lambda_1(C) = \lim_{m\to\infty} \|C^m\|^{1/m}$, where $\|\cdot\|$ is any matrix norm.

Theorem 3.2.8 *Suppose that $c_{k\ell} \geq 0$ when $k \neq \ell$, then the principal eigenvalue λ_1 of (3.13) satisfies*

$$\lim_{\max\{d_k\}\to 0} \lambda_1 = -\max_{x\in\overline{\Omega}} \Lambda_1(c(x)), \tag{3.14}$$

where for each $x \in \overline{\Omega}$, $\Lambda_1(c(x))$ is the Perron–Frobenius eigenvalue of the matrix $c(x) := (c_{k\ell}(x))$.

This theorem was first established by N. Dancer for the case when $d_k = \epsilon\delta_k$ for some fixed positive vector $(\delta_k)_{k=1}^K$ and $\epsilon \to 0$ [4]. This assumption was later removed in [9]. The generalization to the periodic-parabolic setting is given in [1].

Here we give a proof based on the following two comparison lemmas for principal eigenvalues.

Lemma 3.2.9 *Suppose that the assumptions of Theorem 3.2.5 hold and let λ_1 be the principal eigenvalue of the linear elliptic system (3.11). Suppose, in addition, that there exist a strictly positive function $w = (w_k)_{k=1}^K \in [C(\overline{\Omega})]^K$ and a constant $\underline{\lambda}$ such that $w_k > 0$ in $\overline{\Omega}$ for all k and w is a generalized supersolution, i.e.*

$$\begin{cases} \mathcal{L}_k w_k \geq c_{k\ell} w_\ell + \underline{\lambda} w_k & \text{in } \Omega,\ k \geq 1, \\ \mathcal{B}_k w_k \geq 0 & \text{on } \partial\Omega,\ k \geq 1 \end{cases} \tag{3.15}$$

holds in the generalized sense. Then $\lambda_1 \geq \underline{\lambda}$.

Lemma 3.2.10 *Suppose that the assumptions of Theorem* 3.2.5 *hold and let* λ_1 *be the principal eigenvalue of the linear elliptic system* (3.11)*. Suppose, in addition, that there exist a nonnegative function* $w = (w_k)_{k=1}^K \in [C(\overline{\Omega})]^K$ *and a constant* $\underline{\lambda}$ *such that* w_k *is nonnegative and nontrivial and* w *is a generalized subsolution, i.e.*

$$\begin{cases} \mathcal{L}_k w_k \leq c_{k\ell} w_\ell + \overline{\lambda} w_k & \text{in } \Omega,\ k \geq 1, \\ \mathcal{B}_k w_k \leq 0 & \text{on } \partial\Omega,\ k \geq 1 \end{cases} \tag{3.16}$$

holds in the generalized sense. Then $\lambda_1 \leq \overline{\lambda}$.

The proofs of Lemmas 3.2.9 and 3.2.10 are analogous to that of Lemmas 1.3.12 and 1.3.13, and are omitted.

Proof (of Theorem 3.2.8*)* For each $x \in \overline{\Omega}$, let $\phi(x) = (\phi_k(x))$ be the (componentwise) positive eigenvector of $(c_{k\ell}(x))$ (by Perron–Frobenius Theorem) that is uniquely identified by $\sum_{k=1}^K \phi_k(x) = 1$. Without loss of generality, by replacing c_{kk} by $c_{kk} + B$ for some large positive number B, we may assume that $\Lambda_1(c(x)) > 0$ for all x.

We prove the special case when $(c_{k\ell}(x))$ is C^2 in x, and is irreducible for each x. Thanks to the implicit function theorem, the corresponding (componentwise) positive Perron–Frobenius eigenvector $\phi(x) = (\phi_k(x))$ also depends smoothly on $x \in \overline{\Omega}$, and $\sum_{k=1}^K \|\phi_k\|_{C^2(\overline{\Omega})} \leq K_0$ holds for some positive constant K_0 which is independent of x. We leave it as an exercise for the reader to reduce the general case to this situation.

We first prove the lower bound. Define

$$w_k(x) = \phi_k(x) + A\delta^{-2}(\delta - d(x))_+^3,$$

where $s_+ = \max\{s, 0\}$ and $d(x) = \mathrm{dist}(x, \partial\Omega)$. Since $\partial\Omega$ is smooth, the distance function $d(x)$ is smooth near to the boundary. In particular, $w_k \in C^2(\overline{\Omega})$ if we choose δ small enough.

We claim that there exists an $A > 0$ such that for any $\delta > 0$ small, we have $\mathcal{B}_k w_k \geq 0$ on $\partial\Omega$. Indeed, this follows if we estimate the normal derivative of w_k as follows:

$$\mathbf{n} \cdot \nabla w_k \geq 2A - \|\phi_k\|_{C^1(\overline{\Omega})}.$$

Next, we verify that w_k is a strict positive supersolution:

$$\begin{aligned} d_k \Delta w_k - c_{k\ell} w_\ell &\geq -d_k \|w_k\|_{C^2(\overline{\Omega})} - \Lambda_1(c(x))\phi_k(x) + O(\delta) \\ &\geq -d_k \|w_k\|_{C^2(\overline{\Omega})} - \Lambda_1(c(x)) \frac{\phi_k(x)}{\phi_k(x) + A\delta} w_k + O(\delta) \\ &\geq \left[-\max_{\overline{\Omega}} \Lambda_1(c(x)) \frac{1}{1 + A\delta} - d_k \frac{\|w_k\|_{C^2(\overline{\Omega})}}{w_k} + O(\delta) \right] w_k. \end{aligned}$$

By Lemma 3.2.9, we deduce that

$$\lambda_1 \geq -\max_{\overline{\Omega}} \Lambda_1(c(x))\frac{1}{1+A\delta} - d_k\frac{\|w_k\|_{C^2(\overline{\Omega})}}{\inf_\Omega w_k} + O(\delta).$$

Letting $\max\{d_k\} \searrow 0$ and then $\delta \searrow 0$, we deduce that

$$\liminf_{\max\{d_k\}\to 0} \lambda_1 \geq -\max_{x\in\overline{\Omega}} \Lambda_1(c(x)).$$

This proves the lower bound.

Next, we prove the upper bound. Fix a ball $B_R = B_R(x_0)$ which is contained in Ω, and let μ_R and φ_R be the principal eigenvalue and eigenfunction of

$$-\Delta\varphi_R = \mu_R\varphi_R \quad \text{in } B_R, \quad \varphi_R = 0 \quad \text{on } \partial B_R.$$

Set

$$w(x) = (w_k(x))_{k=1}^K = (a_k\varphi_R(x))_{k=1}^K,$$

where $\mathbf{a} = (a_k)_{k=1}^K$ is the (componentwise) positive eigenvector of the constant matrix $C = (C_{k\ell}) = (\inf_{B_R} c_{k\ell})$, such that for each $1 \leq k \leq K$,

$$\sum_{\ell=1}^{K} C_{k\ell}a_\ell = \Lambda_1(C)a_k.$$

Then it is easy to verify that

$$-d_k\Delta w_k - c_{k\ell}w_\ell \leq d_k\mu_R a_k\varphi_R - C_{k\ell}$$

holds in B_R.

By Lemma 3.2.10, we deduce that

$$\lambda_1 \leq d_k\mu_R - \Lambda_1(C).$$

Taking lim sup as $\max\{d_k\} \to 0$, and then taking $R \to 0$, we deduce that

$$\limsup_{\max\{d_k\}\to 0} \lambda_1 \leq -\Lambda_1(c(x_0)).$$

Since x_0 is arbitrary, we obtain

$$\limsup_{\max\{d_k\}\to 0} \lambda_1 \leq -\max_{x_0\in\overline{\Omega}} \Lambda_1(c(x_0)).$$

This proves the upper bound. □

For the purpose of applications we may further consider the situation where the coefficients $c_{k\ell}$ in (3.13) may depend on the diffusion rates. For each $n \in \mathbb{N}$, consider the set of diffusion rates $\{d_k^n\}_k$ and coefficients $(c_{k\ell}^n)_{k,\ell}$ satisfying $c_{k\ell}^n \geq 0$ when

$k \neq \ell$. Assume

$$\max_k\{d_k^n\} \to 0+, \quad \text{and} \quad c_{k\ell}^n \to c_{k\ell}^* \text{ in } C(\overline{\Omega}) \quad \text{as } n \to \infty.$$

Let $\lambda_1(n)$ be the principal eigenvalue of

$$\begin{cases} d_k^n \Delta\varphi_k + c_{k\ell}^n \varphi_\ell + \lambda_1(n)\varphi_k = 0 & \text{in } \Omega, \\ \mathbf{n} \cdot \nabla\varphi_k = 0 & \text{on } \partial\Omega, \end{cases} \tag{3.17}$$

where $1 \leq k \leq K$.

Corollary 3.2.11 *Under the above setting,*

$$\lim_{n\to\infty} \lambda_1(n) = -\max_{x\in\overline{\Omega}} \Lambda_1(c^*(x)),$$

where for each $x \in \overline{\Omega}$, the real number $\Lambda_1(c^*(x))$ is the Perron–Frobenius eigenvalue of the matrix $c^*(x) := (c_{k\ell}^*(x))$.

Proof For each fixed $N \in \mathbb{N}$, define

$$\overline{c}_{k\ell}^N = \sup_{n\geq N} c_{k\ell}^n \quad \text{and} \quad \underline{c}_{l\ell}^N = \inf_{n\geq N} c_{k\ell}^n \quad \text{for each } k, \ell.$$

Let $\overline{\lambda}_1^N(n)$ be the principal eigenvalue of

$$\begin{cases} d_k^n \Delta\varphi_k + \overline{c}_{k\ell}^N \varphi_\ell + \overline{\lambda}_1^N(n)\varphi_k = 0 & \text{in } \Omega, \\ \mathbf{n} \cdot \nabla\varphi_k = 0 & \text{on } \partial\Omega, \end{cases}$$

where $1 \leq k \leq K$. And let $\underline{\lambda}_1^N(n)$ be the principal eigenvalue of a similar problem with $\overline{c}_{k\ell}^N$ being replaced by $\underline{c}_{k\ell}^N$.

On one hand, by the eigenvalue comparison lemmas (Lemmas 3.2.9 and 3.2.10),

$$\overline{\lambda}_1^N(n) \leq \lambda_1(n) \leq \underline{\lambda}_1^N(n) \quad \text{for } n \geq N. \tag{3.18}$$

On the other hand, Theorem 3.2.8 implies

$$\lim_{n\to\infty} \overline{\lambda}_1^N(n) = -\max_{x\in\overline{\Omega}} \Lambda_1(\overline{c}^N(x)), \quad \text{and} \quad \lim_{n\to\infty} \underline{\lambda}_1^N(n) = -\max_{x\in\overline{\Omega}} \Lambda_1(\underline{c}^N(x)).$$

Hence, we may let $n \to \infty$ in (3.18) to get, for each fixed N,

$$-\max_{x\in\overline{\Omega}} \Lambda_1(\overline{c}^N(x)) \leq \liminf_{n\to\infty} \lambda_1(n) \leq \limsup_{n\to\infty} \lambda_1(n) \leq -\max_{x\in\overline{\Omega}} \Lambda_1(\underline{c}^N(x)). \tag{3.19}$$

Since $\overline{c}_{k\ell}^N \to c_{k\ell}^*$ and $\underline{c}_{k\ell}^N \to c_{k\ell}^*$ uniformly as $N \to \infty$, we may take $N \to \infty$ in (3.19) to conclude. □

3.3 Comparison Principle and Principal Eigenvalue for Competitive Parabolic Systems

In this section we apply the theory developed in previous sections to two-species competitive systems. In this connection we assume $K = 2$ throughout this section.

Theorem 3.3.1 *Let $c_{k\ell} \in C(\overline{\Omega}_T)$ satisfy $c_{12}, c_{21} \leq 0$. Assume (u, v) satisfies, for each k, $u(x,0) \leq 0 \leq v(x,0)$ in Ω, and*

$$\begin{cases} u_t + \mathcal{L}_1 u \leq c_{11}u + c_{12}v & \text{in } \Omega_T, \\ v_t + \mathcal{L}_2 v \geq c_{21}u + c_{22}v & \text{in } \Omega_T, \\ \mathcal{B}_1 u \leq 0 \quad \text{and} \quad \mathcal{B}_2 v \geq 0 & \text{on } S\Omega_T \end{cases} \tag{3.20}$$

in the generalized sense. Then, $u \leq 0$ and $v \geq 0$ in Ω_T.

1. *If $u(x_0, t_0) = 0$ for some $(x_0, t_0) \in \Omega_T$, then $u \equiv 0$ in $\Omega \times [0, t_0]$.*
2. *If $v(x_0, t_0) = 0$ for some $(x_0, t_0) \in \Omega_T$, then $v \equiv 0$ in $\Omega \times [0, t_0]$.*

Assuming in addition that both c_{12}, c_{21} are nonnegative and nontrivial, then $\min\{-u(x_0, t_0), v(x_0, t_0)\} = 0$ for some (x_0, t_0) implies

$$u \equiv v \equiv 0 \quad \text{in } \Omega \times [0, t_0].$$

Proof Observe that $(u_1, u_2) = (u, -v)$ satisfies a cooperative system, so that the desired conclusion follows from Theorems 3.1.2 and 3.1.4. □

By the transformation $(\varphi_1, \varphi_2) = (\varphi, -\psi)$, the following results are special cases of Theorem 3.2.1 and 3.2.2.

Theorem 3.3.2 *Consider the periodic-parabolic eigenvalue problem*

$$\begin{cases} \varphi_t + \mathcal{L}_1 \varphi = c_{11}\varphi + c_{12}\psi + \lambda\varphi & \text{in } \Omega_T, \\ \psi_t + \mathcal{L}_2 \psi = c_{21}\varphi + c_{22}\psi + \lambda\psi & \text{in } \Omega_T, \\ \mathcal{B}_1 \varphi = \mathcal{B}_2 \psi = 0 & \text{on } S\Omega_T, \\ \varphi(x,0) = \varphi(x,T), \quad \psi(x,0) = \psi(x,T) & \text{for } x \in \Omega. \end{cases} \tag{3.21}$$

If $a_k^{ij}, b_k^i, c_{k\ell}, p_k^i, p_k^0$ are continuous and T-periodic, $c_{12} \leq 0$ and $c_{21} \leq 0$ in Ω, then (3.21) *has a principal eigenvalue λ_1 in the sense that*

1. *$\lambda_1 \in \mathbb{R}$ is an eigenvalue of* (3.21)*;*
2. *$\lambda_1 \leq \inf \operatorname{Re} \lambda$, where the infimum is taken over all eigenvalues of* (3.21)*;*
3. *the eigenfunction $(\varphi(x,t), \psi(x,t))$ corresponding to λ_1 can be chosen such that $\varphi \leq 0 \leq \psi$ in $\overline{\Omega}_T$.*

Theorem 3.3.3 *If $a_k^{ij}, b_k^i, c_{k\ell}, p_k^i, p_k^0$ are continuous and T-periodic, and satisfy $\max\{c_{12}, c_{21}\} < 0$ in Ω, then the periodic-parabolic eigenvalue problem* (3.21) *has a principal eigenvalue λ_1, in the sense that*

1. *$\lambda_1 \in \mathbb{R}$ is a simple eigenvalue of* (3.21)*;*
2. *$\lambda_1 < \inf_{\lambda \neq \lambda_1} \operatorname{Re} \lambda$ where the infimum is taken over all eigenvalues of* (3.21)*;*
3. *the eigenfunction $(\varphi(x,t), \psi(x,t))$ corresponding to λ_1 can be chosen such that $\varphi(x,t) < 0 < \psi(x,t)$ in $\overline{\Omega}_T$;*
4. *suppose λ is an eigenvalue of* (3.21) *with eigenfunctions (φ, ψ). If $\varphi \leq 0 \leq \psi$ in Ω_T, then $\lambda = \lambda_1$.*

Consider the following nonlinear competitive system with two species:

$$\begin{cases} u_t + \mathcal{L}_1 u = f_1(x,t,u,v) & \text{in } \Omega_T, \\ v_t + \mathcal{L}_2 v = f_2(x,t,u,v) & \text{in } \Omega_T, \\ \mathcal{B}_1 u = g_1(x,t), \quad \mathcal{B}_2 v = g_2(x,t) & \text{on } S\Omega_T, \end{cases} \tag{3.22}$$

where, $u \mapsto f_2(x,t,u,v)$ and $v \mapsto f_1(x,t,u,v)$ are nonincreasing.

Theorem 3.3.4 *Suppose that (u_1, v_1) and (u_2, v_2) are respectively sub- and super-solutions of* (3.22) *in the generalized sense. If*

$$u_1(x,0) \leq u_2(x,0) \quad \textit{and} \quad v_1(x,0) \geq v_2(x,0) \quad \textit{in } \Omega, \tag{3.23}$$

then

$$u_1(x,t) \leq u_2(x,t) \quad \textit{and} \quad v_1(x,t) \geq v_2(x,t) \quad \textit{in } \Omega_T.$$

Assume, in addition, that $\partial_v f_1 < 0$ and $\partial_u f_2 < 0$ in $\Omega_T \times [0,\infty) \times [0,\infty)$. If (3.23) *holds and $(u_1, v_1) \not\equiv (u_2, v_2)$ in $\Omega \times \{0\}$, then*

$$u_1(x,t) < u_2(x,t) \quad \textit{and} \quad v_1(x,t) > v_2(x,t) \quad \textit{in } \Omega_T.$$

Proof By Theorem 3.1.5, we have

$$u_1 \leq u_2 \quad \text{and} \quad v_1 \geq v_2 \quad \text{in } \Omega_T.$$

For the second part, observe that $w = u_1 - u_2$ satisfies

$$w_t + \mathcal{L}_1 w = cw + f_1(x,t,u_2,v_1) - f_1(x,t,u_2,v_2) \leq cw, \tag{3.24}$$

where

$$c(x,t) = \int_0^1 f_1(x,t,\tau u_1(x,t) + (1-\tau)u_2(x,t), v_1(x,t))\, d\tau.$$

Suppose $w(x_0, t_0) = 0$, then the strong maximum principle for single parabolic equations (Theorem 1.1.9) implies that $w \equiv 0$ in $\Omega \times [0, t_0]$ and equality holds everywhere in (3.24). Since f_1 is strictly increasing in v_1, we further have $v_1 \equiv v_2$ in $\Omega \times [0, t_0]$. In particular $(u_1, v_1) \equiv (u_2, v_2)$ in $\Omega \times \{0\}$, which is a contradiction. Hence $w < 0$ in Ω_T, i.e. $u_1 < u_2$ in Ω_T. □

Theorem 3.3.4 is often referred to as the comparison principle for two species competition models. Biologically it means that for two sets of populations, if a species (with density u_1) is inferior to its counterpart (the species with density u_2) initially, while its competitor (the species with density v_1) is superior to the competitor of its counterpart (the species with density v_2) initially, then the same order continues to hold for all time. This allows us to cast the two-species competition system (3.22) as a monotone dynamical system in proper functional spaces. In contrast, competition systems for three or more species are not necessarily monotone.

3.4 Further Reading

The weak maximum principle and the comparison principle for cooperative systems and applications were discussed in [2, 12]. The existence of the principal eigenvalue of (3.11) is obtained by Sweers [13] via the Krein–Rutman Theorem [8]. Nagel [10] and de Figueiredo–Mitidieri [5] also studied the principal eigenvalue problem, using semigroup theory and the maximum principle, respectively. For a comprehensive treatment of the method of super- and subsolutions for equations and systems, we refer to the monograph by Pao [11].

The asymptotic behavior of the principal eigenvalues of the cooperative system (3.11) can be motivated by the study on the global dynamics of the two-species competition model with diffusion and spatial heterogeneity in [7]. A more general question of Hutson is to determine the global dynamics of reaction diffusion systems with small diffusion, provided that the corresponding kinetic ODE system has a unique, globally attracting equilibrium. We refer to [7, 9] for further discussions.

The strong maximum principle for weakly coupled parabolic and elliptic systems was considered in [15]. It was later extended to strongly coupled quasilinear parabolic systems in [14]; see [3] for the weak maximum principle of quasilinear parabolic systems and [6] for more recent developments.

Problems

3.4.1 Justify Remark 3.1.3 by showing that the conclusion of Theorem 3.1.2 holds if the assumption $c_{k\ell} > 0$ in $\overline{\Omega}_T$ for all $k \neq \ell$ is replaced by the weaker assumption that $c_{k\ell} \geq 0$ in $\overline{\Omega}_T$ for all $k \neq \ell$, and that the matrix $(c_{k\ell}(x))$ is continuous and irreducible for each $(x, t) \in \overline{\Omega}_T$.

3.4.2 Let $\lambda_{1,R} \in \mathbb{R}$ be the principal eigenvalue of the elliptic system with Robin boundary condition

$$\begin{cases} d_k \Delta \varphi + c_{k\ell}\varphi_\ell + \lambda \varphi_k = 0 & \text{in } \Omega, \text{ for } 1 \leq k \leq K, \\ \mathbf{n} \cdot \nabla \varphi_k + p^0 \varphi_k = 0 & \text{on } \partial\Omega, \text{ for } 1 \leq k \leq K, \end{cases} \tag{3.25}$$

where $d_k > 0$ and $c_{k\ell} \geq 0$ for $k \neq \ell$, and $p^0 \geq 0$ on $\partial\Omega$ is independent of k. Suppose d_k and $c_{k\ell}$ are constants, and that $c_{k\ell} \geq 0$ if $k \neq \ell$. Show that $\lambda_{1,R} = -\Lambda_1(-\mu_1\mathbf{D}+\mathbf{c})$, where μ_1 is the principal eigenvalue of the Laplacian in Ω with Robin boundary condition,

$$\Delta\phi + \mu_1\phi = 0 \quad \text{in } \Omega, \quad \mathbf{n}\cdot\nabla\phi + p^0\phi = 0 \quad \text{on } \partial\Omega,$$

and $\mathbf{D}$ and $\mathbf{c}$ are the constant matrices

$$\mathbf{D}_{k\ell} = \begin{cases} d_k & \text{if } k = \ell, \\ 0 & \text{otherwise,} \end{cases} \quad \text{and} \quad \mathbf{c}_{k\ell} = c_{k\ell}.$$

[Recall that $\Lambda_1(M)$ denotes the Perron–Frobenius eigenvalue for a given cooperative matrix $M \in \mathbb{R}^{n\times n}$.]

3.4.3 Let $\lambda_{1,R}$ be the principal eigenvalue of (3.25), and let $\lambda_{1,D}$ be the principal eigenvalue of the elliptic system

$$\begin{cases} d_k\Delta\varphi + c_{k\ell}\varphi_\ell + \lambda\varphi_k = 0 & \text{in } \Omega, \ \text{for } 1 \leq k \leq n, \\ \varphi_k = 0 & \text{on } \partial\Omega, \ \text{for } 1 \leq k \leq n, \end{cases}$$

where $d_k > 0$ and $c_{k\ell} \geq 0$ for $k \neq \ell$. Show that $\lambda_{1,D} > \lambda_{1,R}$. [In the special case when d_k, $c_{k\ell}$ are constants, this follows from the fact that the principal eigenvalue of the Dirichlet Laplacian is greater than the Neumann/Robin Laplacian. See Problem 1.4.6.]

References

1. X. Bai and X. He, *Asymptotic behavior of the principal eigenvalue for cooperative periodic-parabolic systems and applications*, J. Differential Equations, 269 (2020), pp. 9868–9903.
2. R. S. Cantrell and C. Cosner, *Spatial ecology via reaction-diffusion equations*, Wiley Series in Mathematical and Computational Biology, John Wiley & Sons, Ltd., Chichester, 2003.
3. K. N. Chueh, C. C. Conley, and J. A. Smoller, *Positively invariant regions for systems of nonlinear diffusion equations*, Indiana Univ. Math. J., 26 (1977), pp. 373–392.
4. E. N. Dancer, *On the principal eigenvalue of linear cooperating elliptic systems with small diffusion*, J. Evol. Equ., 9 (2009), pp. 419–428.
5. D. G. de Figueiredo and E. Mitidieri, *Maximum principles for linear elliptic systems*, Rend. Istit. Mat. Univ. Trieste, 22 (1990), pp. 36–66 (1992).
6. L. C. Evans, *A strong maximum principle for parabolic systems in a convex set with arbitrary boundary*, Proc. Amer. Math. Soc., 138 (2010), pp. 3179–3185.
7. V. Hutson, Y. Lou, and K. Mischaikow, *Convergence in competition models with small diffusion coefficients*, J. Differential Equations, 211 (2005), pp. 135–261.
8. M. G. Kreĭn and M. A. Rutman, *Linear operators leaving invariant a cone in a Banach space*, Uspehi Matem. Nauk (N. S.), 3 (1948), pp. 3–95.
9. K.-Y. Lam and Y. Lou, *Asymptotic behavior of the principal eigenvalue for cooperative elliptic systems and applications*, J. Dynam. Differential Equations, 28 (2016), pp. 29–48.
10. R. Nagel, *Operator matrices and reaction-diffusion systems*, Rend. Sem. Mat. Fis. Milano, 59 (1989), pp. 185–196 (1992).
11. C. V. Pao, *Nonlinear parabolic and elliptic equations*, Plenum Press, New York, 1992.

12. M. H. PROTTER AND H. F. WEINBERGER, *Maximum principles in differential equations*, Springer-Verlag, New York, 1984. Corrected reprint of the 1967 original.
13. G. SWEERS, *Strong positivity in $C(\overline{\Omega})$ for elliptic systems*, Math. Z., 209 (1992), pp. 251–271.
14. X. WANG, *A remark on strong maximum principle for parabolic and elliptic systems*, Proc. Amer. Math. Soc., 109 (1990), pp. 343–348.
15. H. F. WEINBERGER, *Invariant sets for weakly coupled parabolic and elliptic systems*, Rend. Mat. (6), 8 (1975), pp. 295–310.

Chapter 4
The Principal Floquet Bundle for Parabolic Equations

Abstract In this chapter, we establish the existence and uniqueness of the principal Floquet bundle for scalar parabolic equations with bounded coefficients, as a natural generalization of the principal eigenvalue when the coefficients for the equations are time-independent or time-periodic. Next, we give two proofs of the exponential separation property. One is based on the Harnack principle, and the other one relies on the generalized relative entropy. Finally, we introduce the notion of the normalized principal Floquet bundle, and derive the smooth dependence of the bundle on the coefficients in appropriate settings.

4.1 Existence Results for Non-Divergence Form Parabolic Equations

Let $\Omega \subset \mathbb{R}^N$ be a smooth bounded domain, and consider the linear parabolic operator of non-divergence form:

$$\partial_t \psi + \mathcal{L}_t \psi = \partial_t \psi - a^{ij}(x,t) D_{ij}\psi - b^j(x,t) D_j \psi - c(x,t)\psi, \tag{4.1}$$

with the oblique boundary condition

$$\mathcal{B}_t \psi = p^i(x,t) D_i \psi + p^0(x,t)\psi, \tag{4.2}$$

where repeated indices are summed from 1 to N; the coefficients a^{ij}, b^j, c, p^i are continuous in x, t and satisfy, for some $\Lambda > 1$,

$$\begin{cases} \frac{1}{\Lambda}|\xi|^2 \le a^{ij}(x,t)\xi_i \xi_j \le \Lambda |\xi|^2 & \text{for } x \in \Omega,\ t \in \mathbb{R},\ \xi \in \mathbb{R}^N, \\ \inf\limits_{\partial\Omega \times \mathbb{R}} p_0(x,t) \ge 0 \quad \text{and} & \inf\limits_{\partial\Omega\times\mathbb{R}} n_i(x) p^i(x,t) > 0 \end{cases} \tag{4.3}$$

where $\mathbf{n}(x) = (n_i(x))_{i=1}^N$ is the outward unit normal vector on $\partial\Omega$.

K. -Y. Lam, Y. Lou, *Introduction to Reaction-Diffusion Equations*, Lecture Notes on Mathematical Modelling in the Life Sciences, https://doi.org/10.1007/978-3-031-20422-7_4

Definition 4.1.1 We say that $\phi_1(x,t) \in C^{2,1}_{loc}(\overline{\Omega} \times \mathbb{R})$ is a principal Floquet bundle with respect to $(\mathcal{L}_t, \mathcal{B}_t)$ if it is a positive solution of

$$\partial_t \phi + \mathcal{L}_t \phi = 0 \text{ in } \Omega \times \mathbb{R}, \quad \text{and} \quad \mathcal{B}_t \phi = 0 \text{ on } \partial\Omega \times \mathbb{R}.$$

We use the term bundle in the sense that for each t, $\text{span}\{\phi_1(\cdot,t)\}$ defines a one-dimensional subspace that plays the role of the principal eigenfunction in the case when the coefficients are independent of t. In the latter case, the principal Floquet bundle is represented by $\phi_1(x,t) = e^{-\mu_1 t}\tilde{\phi}_1(x)$, where μ_1 and $\tilde{\phi}_1(x)$ are the principal eigenvalue and positive eigenfunction of the elliptic problem

$$-a^{ij}D_{ij}\tilde{\phi} - b^j D_j\tilde{\phi} - c\tilde{\phi} = \mu\tilde{\phi} \quad \text{in } \Omega, \quad p^i D_i\tilde{\phi} + p^0\tilde{\phi} = 0 \quad \text{on } \partial\Omega.$$

A similar remark holds when the coefficients are T-periodic in time for some $T > 0$. See Remark 4.2.3 for further discussions.

The principal Floquet bundle is a generalization of the notion of principal eigenvalue for elliptic or periodic parabolic problems (see Remark 4.2.3), and it can be useful to study semilinear equations which are nonautonomous and nonperiodic in time; see Problem 4.4.2. Furthermore, it is also useful to study models with multiple species that may not admit a comparison principle [2, 7]. In Chapter 8, we present a result in [1] concerning the competition of N phytoplankton species, for $N \geq 3$.

To begin, we recall the existence and uniqueness of the principal Floquet bundle.

Theorem 4.1.2 *Suppose that for some* $\beta \in (0,1)$,

$$a^{ij}, b^j, c \in C^{\beta,\beta/2}(\overline{\Omega} \times \mathbb{R}) \quad \textit{and} \quad p^i \in C^{1+\beta,(1+\beta)/2}(\partial\Omega \times \mathbb{R}).$$

Then there exists a unique positive solution $\phi_1 \in C^{2+\beta,1+\beta/2}_{loc}(\overline{\Omega} \times \mathbb{R})$ *satisfying, in the classical sense,*

$$\begin{cases} \partial_t \phi + \mathcal{L}_t \phi = 0 & \textit{for } x \in \Omega,\ t \in \mathbb{R}, \\ \mathcal{B}_t \phi = 0 & \textit{for } x \in \partial\Omega,\ t \in \mathbb{R}, \\ \phi(x,t) > 0 & \textit{for } x \in \Omega,\ t \in \mathbb{R}, \\ \int_\Omega \phi(x,0)\,dx = 1. \end{cases} \tag{4.4}$$

Proof First, we prove the existence of at least one positive solution to (4.4). To this end, we claim that there exist $C_1, \gamma > 0$ such that for any $k \in \mathbb{N}$ and positive solution w of

$$\begin{cases} \partial_t w + \mathcal{L}_t w = 0 & \text{in } \Omega \times [-k-1, k+1], \\ \mathcal{B}_t w = 0 & \text{on } \partial\Omega \times [-k-1, k+1], \end{cases} \tag{4.5}$$

we have

$$\frac{1}{C_1}e^{-\gamma|t|} \leq \frac{w(x,t)}{\sup_\Omega w(\cdot,0)} \leq C_1 e^{\gamma|t|} \quad \text{for } t \in [-k,k]. \tag{4.6}$$

Indeed, by the Harnack inequality (see Problem 1.4.7), there exists a positive constant C_2 such that for any $t_0 \in [-k, k-1]$,

$$w(x,t) \le C_2 w(y,s) \quad \text{for any } x, y \in \Omega, \text{ and } t, s \in [t_0, t_0 + 2].$$

This implies (4.6) by taking $C_1 = C_2$ and $\gamma = \log C_2$. (See Problem 4.4.3.)

For each k, let λ_k and Φ_k be the principal eigenvalue and the corresponding positive eigenfunction of

$$\begin{cases} \partial_t \Phi + \mathcal{L}_t \Phi = \lambda \Phi & \text{in } \Omega \times (-k-1, k+1), \\ \mathcal{B}_t \Phi = 0 & \text{on } \partial\Omega \times [-k-1, k+1], \\ \Phi(x, -k-1) = \Phi(x, k+1) & \text{in } \Omega. \end{cases}$$

Then

$$\phi_k(x,t) := \mathrm{e}^{-\lambda_k t} \frac{\Phi_k(x,t)}{\sup_\Omega \Phi_k(\cdot, 0)}$$

is a positive solution of (4.5). Therefore, there exist $C_1, \gamma > 0$ independent of k such that

$$\frac{1}{C_1} \mathrm{e}^{-\gamma|t|} \le \phi_k(x,t) \le C_1 \mathrm{e}^{\gamma|t|} \quad \text{in } \Omega \times [-k, k].$$

By standard parabolic L^p estimates [9, Chapter VII], $\{\phi_k\}$ is uniformly bounded in $W^{2,1,p}_{loc}(\Omega \times \mathbb{R})$, for any $p > 1$. We may pass to a subsequence to obtain $\phi_k \to \phi$ weakly in $W^{2,1,p}_{loc}(\Omega \times \mathbb{R})$. Thanks to Sobolev estimates [9, Chapter VI], $\{\phi_k\}$ also converges strongly in $C_{loc}(\overline{\Omega} \times \mathbb{R})$. Since $\sup_\Omega \phi(\cdot, 0) = \lim_{k\to\infty} \sup_\Omega \phi_k(\cdot, 0) = 1$, the limit function ϕ satisfies (4.4), up to multiplication by some constant. Finally, $\phi \in C^{2+\beta, 1+\beta/2}_{loc}(\overline{\Omega} \times \mathbb{R})$ thanks to the Schauder estimates [9, Chapter IV]. This proves the existence.

For the uniqueness, let ϕ and $\tilde{\phi}$ be solutions of (4.4), we will show that $\phi = \tilde{\phi}$. Let

$$\rho_{\min}(t) := \inf_\Omega \frac{\phi(\cdot, t)}{\tilde{\phi}(\cdot, t)} \qquad \text{for } t \in \mathbb{R}.$$

We claim that for each $t < 0$, there exists an $x(t) \in \Omega$ such that $\phi(x(t), t) = \tilde{\phi}(x(t), t)$. Suppose not, then the strong maximum principle implies that either $\phi(x, 0) > \tilde{\phi}(x, 0)$ in Ω or $\phi(x, 0) < \tilde{\phi}(x, 0)$ in Ω. This is impossible since $\int_\Omega \phi(x, 0)\,\mathrm{d}x = \int_\Omega \tilde{\phi}(x, 0)\,\mathrm{d}x = 1$.

Recall that, by the Harnack principle (Theorem 1.2.6), there exists a constant $C > 1$ such that

$$\sup_\Omega \phi(\cdot, t) \le C \inf_\Omega \phi(\cdot, t) \quad \text{and} \quad \sup_\Omega \tilde{\phi}(\cdot, t) \le C \inf_\Omega \tilde{\phi}(\cdot, t) \quad \text{for } t \in \mathbb{R}.$$

Now, for $t < 0$,

$$\begin{aligned} 1 = \frac{\phi(x(t), t)}{\tilde{\phi}(x(t), t)} &\ge \rho_{\min}(t) = \inf_\Omega \frac{\phi(\cdot, t)}{\tilde{\phi}(\cdot, t)} \ge \frac{\inf_\Omega \phi(\cdot, t)}{\sup_\Omega \tilde{\phi}(\cdot, t)} \ge \frac{1}{C^2} \frac{\sup_\Omega \phi(\cdot, t)}{\inf_\Omega \tilde{\phi}(\cdot, t)} \\ &\ge \frac{1}{C^2} \frac{\phi(x(t), t)}{\tilde{\phi}(x(t), t)} = \frac{1}{C^2}. \end{aligned}$$

It follows that $1/C^2 \le \rho_{\min}(t) \le 1$ for $t < 0$. Hence, the constant $\rho_0 := \inf_{t<0} \rho_{\min}$ is well-defined and satisfies $1/C^2 \le \rho_0 \le 1$. By the definition of ρ_0, $v(x,t) = \phi(x,t) - \rho_0\tilde{\phi}(x,t)$ is a nonnegative solution of $v_t + \mathcal{L}_t v = 0$ in $\Omega \times \mathbb{R}$ and $\mathcal{B}_t v = 0$ on $\partial\Omega \times \mathbb{R}$. By the Harnack principle (Theorem 1.2.6) again, we have for $t < 0$,

$$\begin{aligned}\inf_{\Omega} v(\cdot,t) &\ge \frac{1}{C}\sup_{\Omega} v(\cdot,t) \ge \frac{1}{C} v(x(t),t) = \frac{1}{C}(\phi(x(t),t) - \rho_0\tilde{\phi}(x(t),t))\\ &= \frac{1-\rho_0}{C}\tilde{\phi}(x(t),t) \ge \frac{1-\rho_0}{C^2}\sup_{\Omega}\tilde{\phi}(\cdot,t).\end{aligned} \tag{4.7}$$

Since $v(x,t) = \phi(x,t) - \rho_0\tilde{\phi}(x,t)$, this implies that

$$\phi(x,t) \ge \rho_0\tilde{\phi}(x,t) + \frac{1-\rho_0}{C^2}\tilde{\phi}(x,t) \quad \text{in } \Omega \times (-\infty, 0).$$

Dividing by $\tilde{\phi}(x,t)$, and taking the infimum over $\Omega \times (-\infty, 0)$, we obtain

$$\rho_0 \ge \rho_0 + \frac{1-\rho_0}{C^2}.$$

This implies $\rho_0 \ge 1$. Since $\rho_0 \le 1$, we obtain $\rho_0 = 1$. This means that $\phi \ge \tilde{\phi}$. The opposite inequality can be proved by interchanging $\phi, \tilde{\phi}$. This proves the uniqueness. □

To prove the smooth dependence of the principal Floquet bundle, we introduce the notion of a normalized principal Floquet bundle.

Definition 4.1.3 We say that the pair $(\psi_1(x,t), H_1(t)) \in C^{2,1}(\overline{\Omega} \times \mathbb{R}) \times C(\mathbb{R})$ is the normalized principal Floquet bundle corresponding to $(\mathcal{L}_t, \mathcal{B}_t)$ if it satisfies, in the classical sense,

$$\begin{cases} \partial_t\psi + \mathcal{L}_t\psi = H(t)\psi & \text{for } x \in \Omega,\ t \in \mathbb{R},\\ \mathcal{B}_t\psi(x,t) = 0 & \text{for } x \in \partial\Omega,\ t \in \mathbb{R},\\ \int_\Omega \psi(x,t)\,\mathrm{d}x = 1 & \text{for } t \in \mathbb{R},\\ \psi(x,t) > 0 & \text{for } x \in \Omega,\ t \in \mathbb{R}.\end{cases} \tag{4.8}$$

We derive the existence and uniqueness of the normalized principal Floquet bundle.

Theorem 4.1.4 *Suppose that for some* $\beta \in (0,1)$,

$$a^{ij}, b^j, c \in C^{\beta,\beta/2}(\overline{\Omega} \times \mathbb{R}) \quad \textit{and} \quad p^i \in C^{1+\beta,(1+\beta)/2}(\partial\Omega \times \mathbb{R}).$$

Then there exists a unique normalized principal Floquet bundle in

$$C^{2+\beta,1+\beta/2}(\overline{\Omega} \times \mathbb{R}) \times C^{\beta/2}(\mathbb{R}),$$

i.e., a unique pair

$$(\psi_1(x,t), H_1(t)) \in C^{2+\beta,1+\beta/2}(\overline{\Omega} \times \mathbb{R}) \times C^{\beta/2}(\mathbb{R})$$

satisfying (4.8) *in the classical sense.*

Furthermore, there exists a $C_0 > 1$ *independent of t such that*

$$\frac{1}{C_0} \le \psi_1(x,t) \le C_0 \quad \textit{for all } x \in \Omega,\ t \in \mathbb{R}. \tag{4.9}$$

Proof By Theorem 4.1.2, the problem (4.4) has a unique positive solution $\phi \in C^{2+\beta,1+\beta/2}_{loc}(\overline{\Omega} \times \mathbb{R})$. Furthermore, the uniform Harnack principle (1.2.6) holds, i.e. there exists a C such that

$$\sup_{x\in\Omega} \phi(x,t) \le C \inf_{x\in\Omega} \phi(x,t) \quad \text{for all } t \in \mathbb{R}. \tag{4.10}$$

We proceed to define the normalize principal Floquet bundle (ψ_1, H_1) by

$$H_1(t) := -\frac{\mathrm{d}}{\mathrm{d}t}\left[\log \left\|\phi(\cdot,t)\right\|_{L^1(\Omega)}\right] = -\frac{\int_\Omega \partial_t \phi \,\mathrm{d}x}{\int_\Omega \phi \,\mathrm{d}x}$$

and

$$\psi_1(x,t) := \exp\left(\int_0^t H_1(s)\,\mathrm{d}s\right)\phi(x,t). \tag{4.11}$$

Then it is immediate that $H_1 \in C^{\beta/2}_{loc}(\mathbb{R})$ and $\psi_1 \in C^{2+\beta,1+\beta/2}_{loc}(\overline{\Omega} \times \mathbb{R})$ and that (ψ_1, H_1) satisfies (4.8). To conclude the proof, it remains to show that

$$\|H_1\|_{C^{\beta/2}(\mathbb{R})} \le C \quad \text{and} \quad \|\psi_1\|_{C^{2+\beta,1+\beta/2}(\overline{\Omega}\times\mathbb{R})} \le C. \tag{4.12}$$

By the Harnack principle (see Problem 1.4.7), there exists a C independent of $t_0 \in \mathbb{R}$ such that

$$\sup_{\Omega\times[t_0,t_0+1]} \phi(x,t) \le C \inf_{x\in\Omega} \phi(x, t_0+1). \tag{4.13}$$

By parabolic estimates, there exists a C independent of $t \in \mathbb{R}$ such that

$$\|\phi\|_{C^{\beta,\beta/2}(\overline{\Omega}\times[t-1/2,t])} + \|\partial_t\phi\|_{C^{\beta,\beta/2}(\overline{\Omega}\times[t-1/2,t])} \le C\|\phi\|_{L^\infty(\Omega\times(t-1,t))}. \tag{4.14}$$

Combining with (4.13), we have

$$\|\phi\|_{C^{\beta,\beta/2}(\overline{\Omega}\times[t-1/2,t])} + \|\partial_t\phi\|_{C^{\beta,\beta/2}(\overline{\Omega}\times[t-1/2,t])} \le C\int_\Omega \phi(x,t)\,\mathrm{d}x$$

for some constant C that is independent of $t \in \mathbb{R}$. In particular, if we define

$$F(t) := -\int_\Omega \partial_t\phi(x,t)\,\mathrm{d}x \quad \text{and} \quad G(t) := \int_\Omega \phi(x,t)\,\mathrm{d}x,$$

then there is a C independent of t such that

$$|F(t)| + \frac{|F(t) - F(s)|}{|t - s|^{\beta/2}} + \frac{|G(t) - G(s)|}{|t - s|^{\beta/2}} \le CG(t) \quad \text{for } s \in [t - 1/2, t).$$

Since $H_1(t) = F(t)/G(t)$, we obtain $\|H_1\|_{C(\mathbb{R})} \le C$ and

$$\frac{|H_1(t) - H_1(s)|}{|t - s|^{\beta/2}} \le \frac{1}{G(t)} \frac{|F(t) - F(s)|}{|t - s|^{\beta/2}} + \frac{|F(s)|}{G(s)G(t)} \frac{|G(t) - G(s)|}{|t - s|^{\beta/2}} \le C'$$

for $s \in [t - 1/2, t)$. This proves the first half of (4.12).

Next, it follows from (4.10) and (4.11) that

$$\frac{1}{C}\psi_1(y, t) \le \psi_1(x, t) \le C\psi_1(y, t) \quad \text{for all } x, y \in \Omega,\ t \in \mathbb{R},$$

where C is independent of x, y, t. By fixing x, t and integrating over $y \in \Omega$, we obtain (4.9).

Finally, the second half of (4.12) follows by applying the Schauder estimates on (4.8), using $H_1 \in C^{\beta/2}(\mathbb{R})$ (the first half of (4.12)) and $\|\psi_1\|_{L^\infty(\Omega\times\mathbb{R})} \le C$ (from (4.9)). □

4.2 Existence Results for Divergence Form Parabolic Equations

Let Ω be a bounded domain in $\mathbb{R}^N$ with smooth boundary $\partial\Omega$ and outward unit normal vector $\mathbf{n}(x)$. Consider the linear elliptic operator of divergence form[1] (repeated indices are added from 1 to N):

$$\mathcal{L}\varphi := -\partial_{x_i}(a^{ij}(x, t)\partial_{x_j}\varphi) + \partial_{x_i}(b^i(x, t)\varphi) \quad \text{for } x \in \Omega, \tag{4.15}$$

endowed with the conormal boundary operator:

$$\mathcal{B}\varphi := n_i(x)\left[a^{ij}(x, t)\partial_{x_j}\varphi - b^i(x, t)\varphi\right] + p^0(x, t)\varphi \quad \text{for } x \in \partial\Omega, \tag{4.16}$$

where $p^0 \in C^{1+\beta,(1+\beta)/2}(\partial\Omega \times \mathbb{R})$ is nonnegative, $a^{ij}(x), b^i \in C^{1+\beta,(1+\beta)/2}(\overline{\Omega} \times \mathbb{R})$ for some $0 < \beta < 1$, and (a^{ij}) is symmetric and satisfies, for some $\Lambda > 1$

$$\frac{1}{\Lambda}|\xi|^2 \le a^{ij}(x, t)\xi_i\xi_j \le \Lambda|\xi|^2 \quad \text{for } x \in \Omega,\ t \in \mathbb{R},\ \xi \in \mathbb{R}^N. \tag{4.17}$$

[1] The smooth dependence on parameters relies on the decomposition $L^2(\Omega) = X^1(t) \oplus X^2(t)$ given in (4.24). This decomposition, and also the smooth dependence on parameters, is available even for non-divergence form operators (see [2] for details). Here we prove the simpler case of divergence form operators, for which the complementary bundle $X^2(t)$ admits a simple characterization based on the unique positive solution of the adjoint problem. For the divergence form problem with Dirichlet boundary conditions, the decomposition was treated in [5]. Here we provide a complete proof for the conormal boundary condition.

Definition 4.2.1 We say that $(\varphi(x,t), H(t)) \in C^{2,1}(\overline{\Omega}\times\mathbb{R})\times C(\mathbb{R})$ is the *normalized principal Floquet bundle* corresponding to $\mathcal{A} = (a^{ij}, b^i, c, p^0)$ if they satisfy

$$\begin{cases} \partial_t \varphi + \mathcal{L}\varphi = c(x,t)\varphi + H(t)\varphi & \text{in } \Omega\times\mathbb{R}, \\ \mathcal{B}\varphi = 0 & \text{on } \partial\Omega\times\mathbb{R}, \\ \varphi > 0 \quad \text{in } \overline{\Omega}\times\mathbb{R} \quad \text{and} \quad \int_\Omega \varphi(x,t)\,\mathrm{d}x = 1 & \text{for all } t\in\mathbb{R}. \end{cases} \tag{4.18}$$

We say that $\psi(x,t) \in C^{2,1}(\overline{\Omega}\times\mathbb{R})$ is the *adjoint bundle* if it satisfies

$$\begin{cases} -\partial_t \psi + \mathcal{L}^*\psi = c(x,t)\psi + H(t)\psi & \text{in } \Omega\times\mathbb{R}, \\ \mathcal{B}^*\psi = 0 & \text{on } \partial\Omega\times\mathbb{R}, \\ \int_\Omega \varphi(x,t)\psi(x,t)\,\mathrm{d}x = 1 \quad \text{and} \quad \psi > 0 & \text{in } \overline{\Omega}\times\mathbb{R}, \end{cases} \tag{4.19}$$

where

$$\mathcal{L}^*\psi = -\partial_{x_j}(a^{ij}\partial_{x_i}\psi) - b^i\partial_{x_i}\psi, \qquad \text{and} \qquad \mathcal{B}^*\psi = n_j a^{ij}\partial_{x_i}\psi + p^0\psi.$$

By a rescaling, we will assume without loss of generality that $|\Omega| = 1$ throughout the rest of this chapter. Since the choices of Ω, a^{ij} are fixed throughout this chapter, we sometimes suppress the dependence of various constants on Ω and a^{ij}. Denote, for any function $g(x)$, the spatial average by

$$\fint_\Omega g(x)\,\mathrm{d}x = \frac{1}{|\Omega|}\int_\Omega g(x)\,\mathrm{d}x.$$

Similarly, for any function $c(x,t)$, we denote its spatial average by

$$\bar{c}(t) := \fint_\Omega c(x,t)\,\mathrm{d}x.$$

Define the space

$$X_{\text{coeff}} = \{(a^{ij}, b^i, c, p^0) \in \hat{X} : \, p^0 \geq 0 \text{ and (4.17) holds}\}, \tag{4.20}$$

where

$$\hat{X} = [C^{1+\beta,(1+\beta)/2}(\overline{\Omega}\times\mathbb{R})]^{N^2+N} \times C^{\beta,\beta/2}(\overline{\Omega}\times\mathbb{R}) \times C^{1+\beta,(1+\beta)/2}(\partial\Omega\times\mathbb{R}).$$

Theorem 4.2.2 *Given* $(a^{ij}, b^i, c, p^0) \in X_{\text{coeff}}$, *there is a unique ordered triple*

$$(\varphi(x,t), \psi(x,t), H(t)) \in [C^{2+\beta,1+\beta/2}(\overline{\Omega}\times\mathbb{R})]^2 \times C^{\beta/2}(\mathbb{R})$$

satisfying (4.18)–(4.19) *with the following properties.*

1. *(Harnack principle) There exists a* $C_1 > 1$ *independent of* t *such that*

$$\frac{1}{C_1} \leq \varphi(x,t) \leq C_1 \quad \text{and} \quad \frac{1}{C_1} \leq \psi(x,t) \leq C_1 \quad \text{in } \Omega\times\mathbb{R}. \tag{4.21}$$

2. *(Decomposition) For $t \in \mathbb{R}$, set*

$$\begin{aligned} X^1(t) &:= \operatorname{span}\{\varphi(\cdot,t)\}, \\ X^2(t) &:= \{v_0 \in L^2(\Omega) : \textstyle\int_\Omega \psi(x,t) v_0(x)\,\mathrm{d}x = 0\}. \end{aligned} \tag{4.22}$$

These spaces are forward-invariant in the sense that for $i \in \{1,2\}$,

$$u_0 \in X^i(t_0) \quad \Longrightarrow \quad u(\cdot,t;u_0) \in X^i(t) \quad \textit{for all } t \in [t_0,\infty),$$

where $u = u(x,t;u_0)$ is the classical solution of

$$u_t + \mathcal{L}_t u = cu + H(t)u \ \text{ in } \Omega \times [t_0,\infty), \ \text{ and } \ \mathcal{B}_t u = 0 \ \text{ on } \partial\Omega \times [t_0,\infty), \tag{4.23}$$

with initial value $u(\cdot,t_0) = u_0$. Moreover, $X^1(t), X^2(t)$ are complementary subspaces of $L^2(\Omega)$:

$$L^2(\Omega) = X^1(t) \oplus X^2(t) \quad \textit{for each } t \in \mathbb{R}. \tag{4.24}$$

3. *(Exponential separation) There are constants $C_2, \gamma > 0$ such that for any $t, t_0 \in \mathbb{R}$ such that $t > t_0$, and any $u_0 \in X^2(t_0) \cap C(\overline{\Omega})$ one has*

$$\|u(\cdot,t;u_0)\|_{C(\overline{\Omega})} \le C_2 \mathrm{e}^{-\gamma(t-t_0)} \|u_0\|_{C(\overline{\Omega})}. \tag{4.25}$$

Proof In view of the regularity of a^{ij}, we can write (4.18) and (4.19) in non-divergence form. The existence and uniqueness of $(\varphi(x,t), H(t))$ satisfying (4.18) follows from Theorem 4.1.4. Next, by reversing time if necessary, Theorem 4.1.2 gives the existence and uniqueness of $\psi(x,t) > 0$ satisfying (4.19) except for the constraint

$$\int_\Omega \varphi(x,t)\psi(x,t)\,\mathrm{d}x = 1 \quad \text{for } t \in \mathbb{R}. \tag{4.26}$$

We claim that $\rho(t) := \int_\Omega \varphi(x,t)\psi(x,t)\,\mathrm{d}x$ is independent of t. Indeed,

$$\frac{\mathrm{d}}{\mathrm{d}t}\rho(t) = \int_\Omega (\varphi_t\psi + \varphi\psi_t)\,\mathrm{d}x = \int_\Omega (\mathcal{L}_t\varphi)\psi\,\mathrm{d}x - \int_\Omega \varphi(\mathcal{L}_t\psi)\,\mathrm{d}x = 0,$$

where we used the boundary conditions to integrate by parts for the last equality. Hence, we may multiply ψ by a positive constant to ensure (4.26) holds. This proves the existence and uniqueness of the ordered triple (φ, ψ, H) satisfying (4.18) and (4.19).

Next, we observe that assertion 1 is a consequence of the normalizations

$$\int_\Omega \varphi(x,t)\,\mathrm{d}x = 1, \quad \text{and} \quad \int_\Omega \varphi(x,t)\psi(x,t)\,\mathrm{d}x = 1 \quad \text{for } t \in \mathbb{R},$$

and that, by the Harnack principle (Theorem 1.2.6),

$$\sup_\Omega \varphi(\cdot,t) \le C \inf_\Omega \varphi(\cdot,t) \quad \text{and} \quad \sup_\Omega \psi(\cdot,t) \le C \inf_\Omega \psi(\cdot,t) \quad \text{for all } t \in \mathbb{R}.$$

Since this is similar to the proof of $1/C_1 \le \phi \le C_1$ in Theorem 4.1.4, we omit the details.

Let $X^i(t)$ be given in (4.22), then $X^1(t) \oplus X^2(t) = L^2(\Omega)$ by construction. In fact, each $f \in L^2(\Omega)$ can be written as

$$f(x) = \left(\int_\Omega f(y)\psi(y,t)\,\mathrm{d}y\right)\phi(x,t) + \left[f(x) - \left(\int_\Omega f(y)\psi(y,t)\,\mathrm{d}y\right)\phi(x,t)\right]. \tag{4.27}$$

See Problem 4.4.1. It is easy to see that $X^1(t)$ is forward-invariant, since if $u_0 \in X^1(t_0)$, then $u_0(x) = k\varphi(x,t_0)$ for some k, and hence $u(x,t;u_0) = k\varphi(x,t)$ for $(x,t) \in \Omega \times [t_0,\infty)$. On the other hand, if $u_0 \in X^2(t_0)$, then $\int_\Omega u_0(x)\psi(x,t_0)\,\mathrm{d}x = 0$. It follows from integrating the equations that $\int_\Omega u(x,t;u_0)\psi(x,t)\,\mathrm{d}x$ is constant in t, so that $u(\cdot,t;u_0) \in X^2(t)$ for all $t \in [t_0,\infty)$. This proves assertion 2.

Next, we give a proof of assertion 3 based on Harnack's principle. Another proof, based on generalized relative entropy, will be given in the next section. Observe that for any $u_0 \in X^2(t) \cap C(\overline{\Omega})$, the solution $u(x,t) = u(x,t;u_0)$ satisfies

$$\|u(\cdot,t)\|_\infty \le \mathrm{e}^{\|c\|_\infty(t-t_0)}\|u_0\|_\infty \quad \text{for } t \ge t_0. \tag{4.28}$$

Indeed, this follows by applying the maximum principle to $\mathrm{e}^{\|c\|_\infty(t-t_0)}\|u_0\|_\infty \pm u(x,t;u_0)$.

Since $\int_\Omega u(x,t)\psi(x,t)\,\mathrm{d}x = 0$ for all $t \ge t_0$ and $\psi(x,t) > 0$, it follows that there is an $x(t) \in \Omega$ such that $u(x(t),t) = 0$.

We assume, for simplicity, that $t_0 = 0$, and remark that the general case follows in a similar manner. For $t \in [0,\infty)$, we define

$$\rho_+(t) := \inf\{a > 0 : a\varphi(x,t) - u(x,t) \ge 0 \text{ in } \Omega\}.$$

Note that $\rho_+(t)$ is well-defined since by assertion 1, the function φ is bounded above and below by a constant. Also, there exists an $\tilde{x}(t) \in \overline{\Omega}$ such that $\rho_+(t)\varphi(\tilde{x}(t),t) = u(\tilde{x}(t),t)$, and hence

$$\rho_+(t) \le \frac{\sup_\Omega u(\cdot,t)}{\inf_\Omega \varphi(\cdot,t)}. \tag{4.29}$$

Fix $t_1 \in [0,\infty)$ and define $w(x,t) = \rho_+(t_1)\varphi(x,t) - u(x,t)$. It follows that w is a nonnegative solution of (4.23) and

$$\begin{aligned} w(\tilde{x}(t_1+1),t_1+1) &= \rho_+(t_1)\varphi(\tilde{x}(t_1+1),t_1+1) - u(\tilde{x}(t_1+1),t_1+1) \\ &= (\rho_+(t_1) - \rho_+(t_1+1))\varphi(\tilde{x}(t_1+1),t_1+1). \end{aligned} \tag{4.30}$$

Since $u(\cdot,t) \in X^2(t)$, there exists a $y(t) \in \Omega$ such that $u(y(t),t) = 0$, i.e. $w(y(t),t) = \rho_+(t_1)\varphi(y(t),t)$. This allows us to apply the Harnack principle (Theorem 1.2.6) repeatedly to get

$$\begin{aligned}\inf_{\Omega} w(\cdot, t_1+1) &\geq C^{-1} \sup_{\Omega} w(\cdot, t_1+1) \geq C^{-1} w(x(t_1+1), t_1+1) \\ &= C^{-1}\rho_+(t_1)\varphi(y(t_1+1), t_1+1) \geq C^{-2}\rho_+(t_1) \sup_{\Omega} \varphi(\cdot, t_1+1).\end{aligned}$$

Combining with (4.30), we get

$$(\rho_+(t_1) - \rho_+(t_1+1))\varphi(\tilde{x}(t_1+1), t_1+1) \geq C^{-2}\rho_+(t_1) \sup_{\Omega} \varphi(\cdot, t_1+1).$$

This implies that

$$\rho_+(t_1+1) \leq \kappa\rho_+(t_1) \quad \text{for all } t_1 \in \mathbb{R}, \tag{4.31}$$

where $\kappa = 1 - C^{-2} \in (0, 1)$. Hence, for $t > 0$, we have

$$\begin{aligned}\sup_{\Omega} u(\cdot, t) &\leq \rho_+(t) \sup_{\Omega} \varphi(\cdot, t) \\ &\leq C_1\rho_+(t) \\ &\leq C_1\kappa^{[t]}\rho_+(t - [t]) \\ &\leq C_1\kappa^{t-1} \frac{\sup_{\Omega} u(\cdot, t - [t])}{\inf_{\Omega} \varphi(\cdot, t - [t])} \\ &\leq C_1\kappa^{t-1} \frac{\mathrm{e}^{\|c\|} \sup_{\Omega} u(\cdot, 0)}{1/C_1} = \frac{C_1^2\mathrm{e}^{\|c\|}}{\kappa} \mathrm{e}^{t \log \kappa} \sup_{\Omega} u(\cdot, 0),\end{aligned}$$

where $[t]$ is the greatest integer less than t. Note that we used (4.21) for the second inequality, (4.31) for the third inequality, (4.29) for the fourth inequality, (4.21) and (4.28) for the last inequality. By estimating $-u$ in the same way, we have

$$\sup_{\Omega} |u(\cdot, t)| \leq \frac{C_1^2\mathrm{e}^{\|c\|}}{\kappa} \mathrm{e}^{t \log \kappa} \sup_{\Omega} |u(\cdot, 0)|.$$

This proves assertion 3 with $C_2 = C_1^2\mathrm{e}^{\|c\|}/\kappa$, and $\gamma = -\log \kappa > 0$. □

Remark 4.2.3 If the coefficients of $\mathcal{L}$, $\mathcal{B}$, and c are independent of t, then X_1 and X_2 are independent of t. In fact,

$$X_1 = \text{span}\{\phi_1\}, \quad X_2 = \left\{v_0 \in L^2(\Omega) : \int_{\Omega} \phi_1 v_0 \, \mathrm{d}x = 0\right\}, \quad H(t) = \mathrm{e}^{-\lambda_1 t}$$

where λ_1 and ϕ_1 are the principal eigenvalue and positive eigenfunction of

$$\mathcal{L}\phi_1 = c(x)\phi_1 + \lambda_1\phi_1 \quad \text{in } \Omega, \quad \mathcal{B}\phi_1 = 0 \quad \text{on } \partial\Omega,$$

and ψ_1 is the principal eigenfunction of the adjoint problem

$$\mathcal{L}^*\psi_1 = c(x)\psi_1 + \lambda_1\psi_1 \quad \text{in } \Omega, \quad \mathcal{B}^*\psi_1 = 0 \quad \text{on } \partial\Omega.$$

In particular, there are constants $C_2, \gamma > 0$ such that for any $t > 0$ and any $u_0 \in X^2 \cap C(\overline{\Omega})$, the solution $u(\cdot, t) = \mathrm{e}^{-(\mathcal{L}-c)t} u_0$ satisfies

$$\|u(\cdot, t)\|_{C(\overline{\Omega})} \leq C_2 \mathrm{e}^{-(\lambda_1+\gamma)t} \|u_0\|_{C(\overline{\Omega})} \qquad \text{for } t \geq 0.$$

A similar conclusion holds when $\mathcal{L}$, $\mathcal{B}$ and c are all periodic in time.

4.3 The Generalized Relative Entropy

Let $(\varphi, \psi, H(t))$ be the principal Floquet bundle and the corresponding adjoint bundle satisfying (4.18)–(4.19). Following [12, Chapter 6], we introduce the generalized relative entropy

$$\int \psi(x,t)\varphi(x,t) G\left(\frac{u(x,t)}{\varphi(x,t)}\right) \mathrm{d}x.$$

As we shall see, for arbitrary convex functions $G(s)$, all of these entropies decay in time, and not only for the special case $G(s) = s \log s$.

Lemma 4.3.1 *Given an arbitrary convex C^2 function $G : \mathbb{R} \to \mathbb{R}$, and a positive solution $u(x,t) = u(x,t;u_0)$ of* (4.23), *then*

$$\frac{\mathrm{d}}{\mathrm{d}t} \int_\Omega \psi\varphi G\left(\frac{u}{\varphi}\right) \mathrm{d}x = -\int_\Omega \psi\varphi G''\left(\frac{u}{\varphi}\right) a^{ij} \partial_{x_i}\left(\frac{u}{\varphi}\right) \partial_{x_j}\left(\frac{u}{\varphi}\right) \mathrm{d}x \tag{4.32}$$

for $t > t_0$. In particular, the integral $\int_\Omega \psi(x,t)\varphi(x,t) G\left(\frac{u(x,t)}{\varphi(x,t)}\right) \mathrm{d}x$ is nonincreasing in time.

Proof By replacing $c(x,t)$ with $c(x,t) - H(t)$, we may assume $H(t) \equiv 0$ without loss of generality. We calculate

$$\partial_t\left(\frac{u}{\varphi}\right) - \partial_{x_i}\left[a^{ij}\partial_{x_j}\left(\frac{u}{\varphi}\right)\right] + 2\varphi a^{ij}\partial_{x_i}\left(\frac{u}{\varphi}\right)\partial_{x_j}\left(\frac{1}{\varphi}\right) + b^i\partial_{x_i}\left(\frac{u}{\varphi}\right) = 0.$$

Therefore, for any C^2 function G, we obtain

$$\begin{aligned}\partial_t G\left(\frac{u}{\varphi}\right) - \partial_{x_i}\left[a^{ij}\partial_{x_j} G\left(\frac{u}{\varphi}\right)\right] + 2\varphi a^{ij}\partial_{x_i} G\left(\frac{u}{\varphi}\right)\partial_{x_j}\left(\frac{1}{\varphi}\right)\\ + b^i\partial_{x_i} G\left(\frac{u}{\varphi}\right) + G''\left(\frac{u}{\varphi}\right) a^{ij}\partial_{x_i}\left(\frac{u}{\varphi}\right)\partial_{x_j}\left(\frac{u}{\varphi}\right) = 0.\end{aligned}$$

Next, we also consider the equation for $\varphi G\left(\frac{u}{\varphi}\right)$:

$$\partial_t\left[\varphi G\left(\frac{u}{\varphi}\right)\right]-\partial_{x_i}\left[a^{ij}\partial_{x_j}\left(\varphi G\left(\frac{u}{\varphi}\right)\right)\right]+\varphi G''\left(\frac{u}{\varphi}\right)a^{ij}\partial_{x_i}\left(\frac{u}{\varphi}\right)\partial_{x_j}\left(\frac{u}{\varphi}\right)$$
$$+\partial_{x_i}\left[b^i\varphi G\left(\frac{u}{\varphi}\right)\right]-c\varphi G\left(\frac{u}{\varphi}\right)=0.$$

Combining with (4.19) (the equation of ψ), we have

$$\partial_t\left[\psi\varphi G\left(\frac{u}{\varphi}\right)\right]-\partial_{x_i}\left[\psi a^{ij}\partial_{x_j}\left(\varphi G\left(\frac{u}{\varphi}\right)\right)-\psi b^i\varphi G\left(\frac{u}{\varphi}\right)\right]$$
$$+\partial_{x_i}\left[\varphi G\left(\frac{u}{\varphi}\right)a^{ij}\partial_{x_j}\psi\right]+\psi\varphi G''\left(\frac{u}{\varphi}\right)a^{ij}\partial_{x_i}\left(\frac{u}{\varphi}\right)\partial_{x_j}\left(\frac{u}{\varphi}\right)=0.$$

To get the desired result, we note that the above can be integrated, using the following boundary conditions

$$n_i\left\{a^{ij}\partial_{x_j}\left[\varphi G\left(\frac{u}{\varphi}\right)\right]-b^i\varphi G\left(\frac{u}{\varphi}\right)\right\}=0=n_i\left(a^{ij}\partial_{x_j}\psi\right)\quad\text{on }\partial\Omega\times(t_0,\infty).$$

This completes the proof. □

Remark 4.3.2 For a general convex function $G:\mathbb{R}\to\mathbb{R}$, it follows that G' is nondecreasing, and G'' is defined a.e. In that case, (4.32) holds with the equality sign being replaced by a less than or equal to sign; see Problem 4.4.4.

One can derive the exponential separation (assertion 3 of Theorem 4.2.2) from generalized relative entropy and Poincaré's inequality (see Problem 2.4.5).

Corollary 4.3.3 *There exist constants $C_2,\gamma>0$ such that for any $t>t_0$, and any $u_0\in X^2(t_0)\cap C(\overline{\Omega})$ one has*

$$\|u(\cdot,t;u_0)\|_{C(\overline{\Omega})}\le C_2\mathrm{e}^{-\gamma(t-t_0)}\|u_0\|_{C(\overline{\Omega})}.$$

Proof Take $G(s)=s^2$ in Lemma 4.3.1, then we have

$$\frac{\mathrm{d}}{\mathrm{d}t}\int_\Omega\psi\varphi\left|\frac{u}{\varphi}\right|^2\,\mathrm{d}x\le-C\int_\Omega\psi\varphi\left|\nabla\frac{u}{\varphi}\right|^2\,\mathrm{d}x.$$

Since $u=u(x,t;u_0)$ is solution to (4.23) with initial data $u_0\in X^2$, we have

$$\int_\Omega\psi(x,t)\varphi(x,t)\frac{u(x,t)}{\varphi(x,t)}\,\mathrm{d}x=\int_\Omega\psi(x,t)u(x,t)\,\mathrm{d}x=0\quad\text{for each }t>0.$$

Hence, by Poincare's inequality (see Problem 2.4.5), there exists a $C_t>0$ such that

$$\int_\Omega\psi\varphi\left|\frac{u}{\varphi}\right|^2\,\mathrm{d}x\le C_t\int_\Omega\psi\varphi\left|\nabla\left(\frac{u}{\varphi}\right)\right|^2\,\mathrm{d}x.$$

By a compactness argument, it is not difficult to show that $\sup_{t\in\mathbb{R}} C_t < +\infty$. Hence, we deduce that

$$\frac{\mathrm{d}}{\mathrm{d}t}\int_\Omega \psi\varphi\left|\frac{u}{\varphi}\right|^2 \mathrm{d}x \le -C\int_\Omega \psi\varphi\left|\frac{u}{\varphi}\right|^2 \mathrm{d}x.$$

This shows exponential decay in the L^2 norm. It is standard to deduce the exponential decay in the $C(\overline{\Omega})$ norm via parabolic regularity. □

Theorem 4.3.4 *The following mapping is smooth:*

$$\begin{array}{ccc} X_{\mathrm{coeff}} & \to & [C^{2+\beta,1+\beta/2}(\overline{\Omega}\times\mathbb{R})]^2 \times C^{\beta/2}(\mathbb{R}), \\ \mathcal{A} = (a^{ij}, b^i, c, p^0) & \mapsto & (\varphi, \psi, H), \end{array}$$

where X_{coeff} is defined in (4.20).

Proof We denote $\varphi = \varphi(x,t;\mathcal{A})$, $\psi = \psi(x,t;\mathcal{A})$ and $H = H(t;\mathcal{A})$ to stress the dependence of the normalized principal Floquet bundle on the coefficients $\mathcal{A} = (a^{ij}, b^i, c, p^i)$ of $\mathcal{L}_t$ and $\mathcal{B}_t$. The continuous dependence of (φ, ψ, H) on $\mathcal{A}$ follows from the uniqueness of the pair and parabolic regularity theory (see [4] for details). In the following we will directly prove the smooth dependence on $\mathcal{A}$. We only prove the smooth dependence of (φ, H) as the smooth dependence on ψ is analogous.

Consider the mapping

$$\mathcal{F} : C^{2+\beta,1+\beta/2}_{\mathcal{B}}(\overline{\Omega}\times\mathbb{R}) \times C^{\beta/2}(\mathbb{R}) \times X_{\mathrm{coeff}} \to C^{\beta,\beta/2}(\overline{\Omega}\times\mathbb{R}) \times C^{1+\beta/2}(\mathbb{R})$$

that is defined by

$$\mathcal{F}(\varphi, H, \mathcal{A}) := \begin{pmatrix} \partial_t\varphi - \mathcal{L}_t\varphi - H\varphi \\ \int_\Omega \varphi\,\mathrm{d}x - 1 \end{pmatrix},$$

where

$$C^{2+\beta,1+\beta/2}_{\mathcal{B}}(\overline{\Omega}\times\mathbb{R}) = \{u \in C^{2+\beta,1+\beta/2}(\overline{\Omega}\times\mathbb{R}) : \mathcal{B}_t u = 0 \text{ on } \partial\Omega\times\mathbb{R}\}.$$

Then, for each fixed $\mathcal{A} = (a^{ij}, b^i, c, p^0)$,

$$\mathcal{F}(\varphi(\cdot,\cdot;\mathcal{A}), H(\cdot;\mathcal{A}), \mathcal{A}) = 0.$$

To prove the smooth dependence on $\mathcal{A}$, it suffices to show that

$$D_{(\varphi,H)}\mathcal{F} = D_{(\varphi,H)}\mathcal{F}(\varphi(\cdot,\cdot;\mathcal{A}), H(\cdot;\mathcal{A}), \mathcal{A}) \tag{4.33}$$

is invertible as a mapping from

$$C^{2+\beta,1+\beta/2}_{\mathcal{B}}(\overline{\Omega}\times\mathbb{R}) \times C^{\beta/2}(\mathbb{R}) \to C^{\beta,\beta/2}(\overline{\Omega}\times\mathbb{R}) \times C^{1+\beta/2}(\mathbb{R}).$$

To this end, given $(f(x,t), G(t)) \in C^{\beta,\beta/2}(\overline{\Omega}\times\mathbb{R})\times C^{1+\beta/2}(\mathbb{R})$, we need to prove the existence and uniqueness of $(w(x,t), Y(t))$ in the class $C^{2+\beta,1+\beta/2}_{\mathcal{B}}(\overline{\Omega}\times\mathbb{R})\times C^{\beta/2}(\mathbb{R})$ such that

$$\begin{cases} \partial_t w - \mathcal{L}_t w - H w - Y(t)\varphi = f(x,t) & \text{for } t \in \mathbb{R},\ x \in \Omega, \\ \int_\Omega w(x,t)\,\mathrm{d}x = G(t) & \text{for } t \in \mathbb{R}, \end{cases} \tag{4.34}$$

where $H = H(t;\mathcal{A})$ and $\varphi = \varphi(x,t;\mathcal{A})$. First, we show the existence. We begin by defining the projections $P^i : L^2(\Omega) \to L^2(\Omega)$ onto $X^i(t)$ by

$$P^1(t)[u_0] = \left(\int_\Omega u_0(z)\psi(z,t)\,\mathrm{d}z\right)\varphi(x,t), \quad \text{and} \quad P^2(t)[u_0] = u_0 - P^1(t)[u_0].$$

Indeed, $P^i(t)[u_0] \in X^i(t)$ for $i = 1,2$, since $P^1(t)[u_0] = \operatorname{span}\{\varphi(\cdot,t)\}$ and $\int_\Omega \psi(x,t)P^2(t)[u_0]\,\mathrm{d}x = 0$ for any $u_0 \in L^2(\Omega)$ and $t \in \mathbb{R}$. Observe that there is a constant C such that

$$\|P^i(t)[u_0]\|_{L^\infty(\Omega)} \le C\|u_0\|_{L^\infty(\Omega)} \quad \text{for } t \in \mathbb{R},\ u_0 \in L^\infty(\Omega), \tag{4.35}$$

and that

$$f \in C^{\beta,\beta/2} \implies P^i(t)[f(\cdot,t)] \in C^{\beta,\beta/2}. \tag{4.36}$$

Next, we choose $w^\perp$ as

$$w^\perp(x,t) = \int_{-\infty}^t U(t,\tau)[P^2(\tau)f(\cdot,\tau)]\,\mathrm{d}\tau,$$

where, for $t > t_0$, $U(t,t_0)$ is the evolution operator of the initial value problem (4.23), i.e.

$$U(t,t_0)[u_0] = u(x,t;u_0,t_0),$$

where u is the solution of (4.23) with initial data u_0. We claim that $w^\perp$ is well-defined. Indeed,

$$\begin{aligned} \sup_{t\in\mathbb{R}} \|w^\perp(\cdot,t)\|_{L^\infty(\Omega)} &\le \sup_{t\in\mathbb{R}} \left\| \int_{-\infty}^t U(t,\tau)\left[P^2(\tau)[f(\cdot,\tau]\right]\,\mathrm{d}\tau \right\|_{L^\infty(\Omega)} \\ &\le C \sup_{t\in\mathbb{R}} \int_{-\infty}^t \mathrm{e}^{-\gamma(t-s)} \|f(\cdot,\tau)\|_{L^\infty(\Omega)}\,\mathrm{d}\tau \\ &\le C \sup_{t\in\mathbb{R}} \|f(\cdot,t)\|_{L^\infty(\Omega)} < \infty, \end{aligned} \tag{4.37}$$

where we used assertion 3 of Theorem 4.2.2 and (4.35) for the second inequality. Moreover, since $w^\perp$ defines a solution of $\partial_t w^\perp + \mathcal{L}_t w^\perp - (c+H)w^\perp = P^2(t)[f(\cdot,t)]$ with homogeneous oblique boundary condition, and the right-hand side $P^2(t)[f(\cdot,t)]$ belongs to $C^{\beta,\beta/2}(\overline{\Omega}\times\mathbb{R})$ (by (4.36)), it follows by parabolic regularity estimates [9, Chapter IV, Theorem 4.30] that $w^\perp \in C^{2+\beta,1+\beta/2}(\overline{\Omega}\times\mathbb{R})$. Next, define $w \in C^{2+\beta,1+\beta/2}(\overline{\Omega}\times\mathbb{R})$ by

$$w(x,t) = w^\perp(x,t) + \left[-\int_\Omega w^\perp(y,t)\,\mathrm{d}y + G(t)\right]\varphi(x,t).$$

Then w satisfies the second part of (4.34). Moreover, we have

$$\begin{aligned}&\partial_t w + \mathcal{L}_t w - cw - H(t)w\\&= P^2(t)[f(\cdot,t)] + \left\{-\frac{\mathrm{d}}{\mathrm{d}t}\left[\int_\Omega w^\perp(y,t)\,\mathrm{d}y\right] + G'(t)\right\}\varphi(x,t).\end{aligned}$$

It therefore suffices to choose $Y(t)$ such that

$$Y(t)\varphi(x,t) + P^1(t)[f(\cdot,t)] = \left\{-\frac{\mathrm{d}}{\mathrm{d}t}\left[\int_\Omega w^\perp(y,t)\,\mathrm{d}y\right] + G'(t)\right\}\varphi(x,t).$$

Note that $Y \in C^{\beta/2}(\mathbb{R})$ by using the $C^{2+\beta,1+\beta/2}$ regularity of $w^\perp$ and (4.36). This proves the existence.

For the uniqueness, set $f = 0$ and $G = 0$, then using the variation of constants formula on $\partial_t w + \mathcal{L}_t w - (c+H)w = Y(t)\varphi(x,t)$, we get

$$\begin{aligned}w(\cdot,t) &= U(t,s)w(\cdot,s) + \int_s^t U(t,\tau)[Y(\tau)\varphi(\cdot,\tau)]\,\mathrm{d}\tau\\&= U(t,s)w(\cdot,s) + \int_s^t Y(\tau)\{U(t,\tau)[\varphi(\cdot,\tau)]\}\,\mathrm{d}\tau\\&= U(t,s)w(\cdot,s) + \int_s^t Y(\tau)\varphi(\cdot,t)\,\mathrm{d}\tau\\&= U(t,s)w(\cdot,s) + \left[\int_s^t Y(\tau)\,\mathrm{d}\tau\right]\varphi(\cdot,t) \quad \text{for } t > s,\end{aligned}$$

where we used the fact that $U(t,s)\varphi(\cdot,s) = \varphi(\cdot,t)$ for $t > s$ for the third equality. Hence, we deduce

$$w(\cdot,t) = U(t,s)w(\cdot,s) + \left[\int_s^t Y(\tau)\,\mathrm{d}\tau\right]\varphi(\cdot,t) \quad \text{for any } t > s. \tag{4.38}$$

Next, apply the projection $P^2(t)$ on both sides of (4.38),

$$P^2(t)[w(\cdot,t)] = P^2(t)[U(t,s)w(\cdot,s)] = U(t,s)\left[P^2(s)[w(\cdot,s)]\right],$$

provided $t, s \in \mathbb{R}$ and $t > s$. This implies

$$\begin{aligned}\left\|P^2(t)[w(\cdot,t)]\right\|_{L^\infty(\Omega)} &\le Ce^{-\gamma(t-s)}\left\|P^2(t)[w(\cdot,s)]\right\|_{L^\infty(\Omega)}\\&\le Ce^{-\gamma(t-s)}\|w(\cdot,s)\|_{L^\infty(\Omega)} \le Ce^{-\gamma(t-s)},\end{aligned} \tag{4.39}$$

where we used assertion 3 of Theorem 4.2.2 for the first inequality, (4.35) for the second one, and the fact that w is bounded uniformly in time (in fact, $w \in C^{2+\beta,1+\beta/2}(\overline{\Omega}\times\mathbb{R})$) for the third one. Letting $s \to -\infty$ in (4.39), we deduce that $P^2(t)[w(\cdot,t)] = 0$ for each $t \in \mathbb{R}$. Hence, $w(\cdot,t) \in X^1(t)$ and thus $w(\cdot,t) = \sigma(t)\varphi(\cdot,t)$ for some function $\sigma(t)$. Now, using $G(t) \equiv 0$, the second equation in

(4.34) gives

$$0 = \int_{\Omega} w(x,t)\,dx = \sigma(t) \int_{\Omega} \varphi(x,t)\,dx = \sigma(t) \qquad \text{for } t \in \mathbb{R}.$$

This implies $w(x,t) = 0$. Substituting into (4.38), we have

$$\int_s^t Y(\tau)\,d\tau = 0 \quad \text{for any } t, s \in \mathbb{R},\ t > s.$$

This means $Y(t) \equiv 0$ as well. This proves the uniqueness.

Having shown that $D_{(\varphi,H)}\mathcal{F}$ given in (4.33) is an isomorphism, we may apply the implicit function theorem to conclude the smooth dependence of the normalized principal Floquet bundle $(\varphi(x,t), H(t))$ on the coefficients $\mathcal{A} = (a^{ij}, b^i, c, p^0)$. This concludes the proof. □

4.4 Further Reading

For linear parabolic equations in one-dimensional space, the existence and uniqueness of Floquet bundles, characterized by nodal properties of solutions as in the classical Sturm-–Liouville theory, was obtained by Chow et al. [3]. Subsequently Mierczyński [10] generalized the existence and uniqueness of the principal Floquet bundle when the spatial dimension is greater than one, by invoking the general exponential separation results of Poláčik and Tereščák [13]. Later on, Huska and collaborators [4, 5, 6] significantly weakened the smoothness assumptions on the coefficients, and proved the continuous dependence of the principal Floquet bundle on the coefficients of the linear problem. The smooth dependence on coefficients was more recently obtained in [2]. These results generalize the smooth dependence on coefficients of the principal eigenvalue and eigenfunctions of uniformly elliptic, or periodic-parabolic operators.

The notion of generalized relative entropy is a versatile tool to derive exponential separation in a wide range of linear models, including linear ODEs with a cooperative matrix that is possibly time-dependent; see [12, Chapter 6] for a more complete account.

In Problem 4.4.2, we give an elementary application of the principal Floquet bundle in determining the persistence of a population modeled by a logistic model. This is a generalization of the result for the autonomous problem (see Theorem 5.2.2). The principal Floquet bundle can also be applied to determine the dynamics of multi-species models; see [2, 7] where the theory is applied to analyze the competition dynamics of a large number of species, and [8] for an application to a selection-mutation model with spatial structure that exhibits moving Dirac concentrations. For further developments and applications, we refer to the recent monograph by Mierczyński and Shen [11].

Problems

4.4.1 (Decomposition of $L^2(\Omega)$ by the principal bundle and adjoint bundle) Let φ and ψ satisfy (4.18)–(4.19). Show that

$$f - \left(\int_\Omega f(y)\psi(y,t)\, dy \right) \varphi(\cdot, t) \in X^2(t) \quad \text{for every } t \in \mathbb{R}, \text{ and } f \in L^2(\Omega),$$

where $X^2(t)$ is given by (4.22). In particular, (4.27) gives the decomposition of $L^2(\Omega) = X^1(t) \oplus X^2(t)$.

4.4.2 (Persistence criterion for nonautonomous diffusive logistic model) Let Ω be a smooth bounded domain in $\mathbb{R}^N$ and let (ψ_1, H_1) be the normalized principal Floquet bundle corresponding to $(\mathcal{L}_t, \mathcal{B}_t)$, where $(\mathcal{L}_t, \mathcal{B}_t)$ are given as in (4.1)–(4.3). Let $u(x,t)$ be a nonnegative, classical solution of the logistic problem

$$\begin{cases} \partial_t u + \mathcal{L}_t u = -u^2 & \text{in } \Omega \times (0,\infty), \\ \mathcal{B}_t u = 0 & \text{on } \partial\Omega \times (0,\infty), \\ u(x,0) = u_0(x) & \text{in } \Omega. \end{cases} \tag{4.40}$$

Show that the following statements hold.

1. If $\limsup_{t\to\infty} \int_0^t H(s)\, ds < 0$, then $\limsup_{t\to\infty} \left[\sup_{x\in\Omega} u(x,t)\right] = 0$.
2. If $\liminf_{t\to\infty} \int_0^t H(s)\, ds > 0$, then $\liminf_{t\to\infty} \left[\inf_{x\in\Omega} u(x,t)\right] > 0$.

4.4.3 Let $w(x,t) \in C(\overline{\Omega} \times \mathbb{R})$ be given. Suppose that there is a constant C_0 such that for any $t_0 \in \mathbb{R}$, we have

$$w(x,t) \le C_0 w(y,s) \quad \text{for any } x, y \in \Omega, \text{ and } t, s \in [t_0, t_0 + 2].$$

Show that

$$\frac{1}{C_0} e^{-\gamma|t|} \le \frac{w(x,t)}{\sup_\Omega w(\cdot, 0)} \le C_0 e^{\gamma|t|} \quad \text{for } t \in \mathbb{R},$$

with $\gamma = \log C_0$.

4.4.4 Let $(\varphi, \psi, H(t))$ be the principal Floquet bundle and the corresponding adjoint bundle satisfying (4.18)–(4.19), u be a positive solution of (4.23) defined on $\overline{\Omega} \times [t_0, \infty)$, and let $G : \mathbb{R} \to \mathbb{R}$ be convex and continuous. Show that

$$\frac{d}{dt} \int_\Omega \psi\varphi G\left(\frac{u}{\varphi}\right) dx \le -\int_\Omega \psi\varphi G''\left(\frac{u}{\varphi}\right) a^{ij} \partial_{x_i}\left(\frac{u}{\varphi}\right) \partial_{x_j}\left(\frac{u}{\varphi}\right) dx \quad \text{for } t > t_0,$$

where G'' is the second derivative of G. [For a general convex function G, its derivative G' is monotone, so G'' exists a.e. in $\mathbb{R}$.]

References

1. R. S. Cantrell and K.-Y. Lam, *Competitive exclusion in phytoplankton communities in a eutrophic water column*, Discrete Contin. Dyn. Syst. Ser. B, 26 (2021), pp. 1783–1795.
2. ———, *On the evolution of slow dispersal in multispecies communities*, SIAM J. Math. Anal., 53 (2021), pp. 4933–4964.
3. S.-N. Chow, K. Lu, and J. Mallet-Paret, *Floquet bundles for scalar parabolic equations*, Arch. Rational Mech. Anal., 129 (1995), pp. 245–304.
4. J. Húska, *Harnack inequality and exponential separation for oblique derivative problems on Lipschitz domains*, J. Differential Equations, 226 (2006), pp. 541–557.
5. J. Húska and P. Poláčik, *The principal Floquet bundle and exponential separation for linear parabolic equations*, J. Dynam. Differential Equations, 16 (2004), pp. 347–375.
6. J. Húska, P. Poláčik, and M. V. Safonov, *Harnack inequalities, exponential separation, and perturbations of principal Floquet bundles for linear parabolic equations*, Ann. Inst. H. Poincaré C Anal. Non Linéaire, 24 (2007), pp. 711–739.
7. K.-Y. Lam and Y. Lou, The principal Floquet bundle and the dynamics of fast diffusing communities. Manuscript submitted for publication, 2022.
8. K.-Y. Lam, Y. Lou, and B. Perthame, *A Hamilton–Jacobi approach to evolution of dispersal*, 2022. arXiv:2205.05534 [math.AP].
9. G. M. Lieberman, *Second order parabolic differential equations*, World Scientific Publishing Co., Inc., River Edge, NJ, 1996.
10. J. Mierczyński, *Globally positive solutions of linear parabolic PDEs of second order with Robin boundary conditions*, J. Math. Anal. Appl., 209 (1997), pp. 47–59.
11. J. Mierczyński and W. Shen, *Spectral theory for random and nonautonomous parabolic equations and applications*, vol. 139 of Chapman & Hall/CRC Monographs and Surveys in Pure and Applied Mathematics, CRC Press, Boca Raton, FL, 2008.
12. B. Perthame, *Transport Equations in Biology*, Frontiers in Mathematics, Birkhäuser Verlag, Basel, 2007.
13. P. Poláčik and I. Tereşčák, *Exponential separation and invariant bundles for maps in ordered Banach spaces with applications to parabolic equations*, J. Dynam. Differential Equations, 5 (1993), pp. 279–303.

Part II
Ecological Dynamics

Chapter 5
The Logistic Equation With Diffusion

Abstract In this chapter, we consider the dynamics of a single species in bounded regions, as modeled by a nonlinear scalar parabolic equation, with proper boundary conditions. We introduce the notion of super- and subequilibrium, which leads to concrete a priori L^∞ bounds of the solution. Next, we define the notion of linear stability (resp. instability) of equilibrium as characterized by the positivity (resp. negativity) of the principal eigenvalue for the corresponding linearized elliptic operator. We show that linear stability (resp. instability) implies nonlinear stability (resp. instability) for equilibrium solutions. For the case of a single logistic equation with diffusion, we show the convergence of the time-dependent solution to the unique positive equilibrium as time tends to infinity, and further discuss various qualitative estimates of the positive equilibrium as the diffusion coefficient tends to zero or infinity. Finally, we prove the existence of the critical domain size for the persistence of a single species in rivers, as modeled by a scalar reaction-diffusion-advection equation in an interval.

5.1 A Reaction-Diffusion Model for a Single Species

Let Ω be a bounded domain in $\mathbb{R}^N$ with smooth boundary $\partial\Omega$ and outer unit normal vector $\mathbf{n}$. Consider the semilinear parabolic equation

$$\begin{cases} \partial_t u + \mathcal{L}u = F(x,u) & \text{in } \Omega \times (0,T), \\ \mathcal{B}u = 0 & \text{on } \partial\Omega \times (0,T), \\ u(x,0) = u_0(x) & \text{in } \Omega, \end{cases} \tag{5.1}$$

where $\mathcal{L} = -a^{ij}(x)D_{ij} - b^j(x)D_j - c(x)$ is a time-independent elliptic operator of non-divergence form satisfying the uniform ellipticity condition. Namely, $a^{ij}, b^j, c \in C^\alpha(\overline{\Omega})$ and

$$a^{ij}(x)\xi_i\xi_j \geq \nu|\xi|^2, \quad \text{for } x \in \overline{\Omega},\ \xi \in \mathbb{R}^n, \tag{5.2}$$

K. -Y. Lam, Y. Lou, *Introduction to Reaction-Diffusion Equations*, Lecture Notes on Mathematical Modelling in the Life Sciences, https://doi.org/10.1007/978-3-031-20422-7_5

for some positive constant ν. Moreover, $\mathcal{B}$ is the first order oblique derivative operator $\mathcal{B} = p^i D_i + p^0$ such that $p^i \in C^{1+\beta}(\overline{\Omega})$ satisfies

$$p^i(x)n_i(x) > 0 \quad \text{and} \quad p^0 \geq 0 \quad \text{on } \partial\Omega. \tag{5.3}$$

Remark 5.1.1 Some particular examples include the logistic equation

$$F(x, u) = r(x)\left(1 - \frac{u}{K(x)}\right)u,$$

and the harvesting model

$$F(x, u) = r(x)\left(1 - \frac{u}{K(x)}\right)u - h(x)u.$$

Theorem 5.1.2 *Suppose $F(x, 0) \equiv 0$ in Ω and there exist $\beta \in (0, 1)$ and $C_1 = C_1(R)$ such that*

$$|F(x_1, s_1) - F(x_2, s_2)| \leq C_1(|x_1 - x_2|^\beta + |s_1 - s_2|) \quad \text{for } x_i \in \Omega,\ 0 \leq s_i \leq R. \tag{5.4}$$

Then for each $0 \leq u_0 \in C(\overline{\Omega})$, there exists a $T_{\max} \in (0, \infty]$ such that model (5.1) *has a unique classical solution $u \in C(\overline{\Omega} \times [0, T_{\max})) \cap C^{2,1}(\overline{\Omega} \times (0, T_{\max}))$.*

Moreover, $T_{\max} = \infty$ if, for each $M, T > 0$ there exists a constant $C(M, T)$ such that for any solution u of model (5.1) *such that $0 \leq u_0(x) \leq M$ in Ω, we have*

$$0 \leq u(x, t) \leq C(M, T) \quad \text{for } (x, t) \in \Omega_T. \tag{5.5}$$

Proof Set $X = C(\overline{\Omega})$, and

$$\begin{cases} D(\mathcal{L}) = \{\phi \in \cap_{p>1} W^{2,p}(\Omega) : \mathcal{L}\phi \in C(\overline{\Omega}),\ \mathcal{B}\phi|_{\partial\Omega} = 0\}, \\ X_{1/2} = \{\phi \in C^1(\overline{\Omega}) : \mathcal{B}\phi|_{\partial\Omega} = 0\} \end{cases} \tag{5.6}$$

and set

$$f : X_{1/2} \to X, \quad f(\phi) = F(\cdot, \phi(\cdot)).$$

For every $v_i \in X_{1/2}$ $(i = 1, 2)$ such that $\|v_i\|_{C(\overline{\Omega})} \leq R$, it holds that

$$\|f(v_1) - f(v_2)\|_{C(\overline{\Omega})} \leq C_1 \|v_1 - v_2\|_{C(\overline{\Omega})}.$$

Note also that $\overline{D(\mathcal{L})} = X$. By [19, Theorem 7.1.5 and Proposition 7.1.10], it follows that for every $u_0 \in C(\overline{\Omega})$, there exists a $T > 0$ such that the problem has a classical solution $u \in C([0, T]; X) \cap C^1((0, T]; X) \cap C((0, T]; D(\mathcal{L}))$. Moreover, the solution u satisfies

$$\sup_{0<t<\min\{1,T/2\}} t^{1/2} \|u(\cdot, t)\|_{C^1(\overline{\Omega})} < \infty.$$

By observing that $u, F(x,u) \in C^{\beta,\beta/2}(\Omega \times [\delta, T))$ for every $\delta \in (0,T)$, it follows from the Schauder estimates that $u \in C^{2,1}(\Omega \times [\delta, T))$ for each $0 < \delta < T$. The uniqueness now follows by the maximum principle.

Finally, assume (5.5). Then for any nonnegative initial data u_0 the corresponding solution remains bounded in X. Having ruled out blow up in finite time, the global existence follows from [19, Proposition 7.1.8]. □

Remark 5.1.3 A sufficient condition for (5.5) is

$$F(x,0) = 0, \quad \text{and} \quad F(x,u) \le C|u|$$

for some constant $C > 0$ independent of x and u (See Problem 5.4.1).

Remark 5.1.4 The above proof also yields

$$\sup_{0<t<\min\{1,T_{\max}\}} t^{1/2} \|\mathcal{L}^{1/2} u(\cdot,t)\|_{C(\overline{\Omega})} < \infty,$$

where $\mathcal{L}^{1/2}$ is the fractional power of $\mathcal{L}$ regarded as a sectorial operator with domain $\mathrm{D}(\mathcal{L})$ given in (5.6).

Definition 5.1.5 We say that $\theta \in C^2(\Omega) \cap C^1(\overline{\Omega})$ is an equilibrium solution of (5.1) if θ is a classical solution to

$$\begin{cases} \mathcal{L}\theta = F(x,\theta) & \text{in } \Omega, \\ \mathcal{B}\theta = 0 & \text{on } \partial\Omega. \end{cases} \tag{5.7}$$

If, in addition, $\inf_\Omega \theta > 0$, then we say that θ is a positive equilibrium solution.

We say that $\theta \in W^{2,p}(\Omega)$ is a strong solution to (5.7) if $p > N$ and $\mathcal{L}\theta = F(x,\theta)$ holds almost everywhere in Ω and $\mathcal{B}u = 0$ holds everywhere on $\partial\Omega$. Note that $W^{2,p}(\Omega) \subset C^{1+\gamma}(\overline{\Omega})$ with $\gamma = 1 - N/p$.

We also recall the notion of super- and subequilibria from Definition 1.2.2.

Proposition 5.1.6 *Suppose* $u \in C^{2,1}(\Omega_T) \cap C^{1,0}(\overline{\Omega} \times (0,T]) \cap C(\overline{\Omega}_T)$ *is a classical solution of* (5.1) *such that the initial condition* u_0 *is a superequilibrium (resp. subequilibrium) in the generalized sense* [1], *then* u *is nonincreasing (resp. nondecreasing) in* t. *Moreover, if* $\limsup_{t\to\infty} \|u(\cdot,t)\| < \infty$, *then* $u_\infty(x) = \liminf_{t\to\infty} u(x,t)$ *is an equilibrium of* (5.7).

Proof See Corollary 1.2.4. □

Next, we give a proposition from the perspective of dynamical systems.

Definition 5.1.7 Given $u_1, u_2 \in C(\overline{\Omega})$ such that $u_1 \le u_2$ in Ω. We define the order interval

$$[u_1, u_2] := \{\tilde{u} \in C(\overline{\Omega}) : u_1(x) \le \tilde{u}(x) \le u_2(x) \ \text{ in } \Omega\}.$$

[1] See Definition 1.1.1.

Proposition 5.1.8 *Suppose u_1, u_2 are respectively the subequilibrium and superequilibrium of* (5.7) *in the generalized sense, then the following statements hold.*

1. *Let u be a solution of* (5.1). *If $u(x,0) = u_1$ (resp. u_2), then u is increasing (resp. decreasing) in $t \geq 0$.*
2. *The set $[u_1, u_2]$ is forward-invariant; i.e. for any solution u of* (5.1) *such that $u_1 \leq u(\cdot,0) \leq u_2$ in Ω, we have $u_1 \leq u(\cdot,t) \leq u_2$ in Ω for all $t > 0$.*
3. *There exist a maximal equilibrium u_M and a minimal equilibrium u_m in $[u_1, u_2]$, in the sense that if $\theta \in [u_1, u_2]$ is an equilibrium, then $u_m \leq \theta \leq u_M$ in Ω.*
4. *The set $[u_m, u_M]$ attracts $[u_1, u_2]$, in the sense that*

$$u_m(x) \leq \liminf_{t\to\infty} u(x,t) \leq \limsup_{t\to\infty} u(x,t) \leq u_M(x),$$

   for any solution u of (5.1) *with initial condition in $[u_1, u_2]$.*
5. *Suppose, in addition, that $u_m = u_M$, and denote their common value to be θ. Then the equilibrium θ is the unique equilibrium in $[u_1, u_2]$ and it attracts every trajectory initiating in $[u_1, u_2]$.*

Proof This follows from Proposition 5.1.6 and the comparison principle. □

When (5.1) has a unique equilibrium, then one can obtain explicit bounds of $\theta(x)$ by constructing super- or subequilibria.

Corollary 5.1.9 *Suppose* (5.7) *has a unique positive equilibrium θ which is globally asymptotically stable among all nonnegative, nontrivial solutions of* (5.1), *i.e.*

$$\lim_{t\to\infty} \sup_{x\in\Omega} |u(x,t) - \theta(x)| = 0$$

for every nonnegative, nontrivial solution u of (5.1). *Then the following statements hold:*

1. *if $\underline{\theta}$ is a nonnegative subequilibrium in the classical or generalized sense, then*

$$\theta(x) \geq \underline{\theta}(x) \quad \text{in } \Omega;$$

2. *if $\overline{\theta}$ is a positive superequilibrium in the classical or generalized sense, then*

$$\theta(x) \leq \overline{\theta}(x) \quad \text{in } \Omega.$$

Remark 5.1.10 It is well known that reaction-diffusion-advection models with logistic nonlinearity have at most one positive equilibria; see e.g. Theorem 5.2.2. Hence, the asymptotic profile of the equilibrium solution, as the parameters become large or small, can be determined by way of constructing explicit super- and subequilibria and invoking Corollary 5.1.9. See [4, 6] for concentration phenomena, and [15] for the existence of sharp transition layers.

Proof Suppose $\underline{\theta}$ is a nonnegative subequilibrium, and let u be the solution of (5.1) with initial data $\underline{\theta}$. Then by Corollary 1.2.4, u is nondecreasing in t. By the hypothesis, we also have $u(x,t) \to \theta(x)$. Hence,

$$\underline{\theta}(x) = u(x,0) \le \lim_{t\to\infty} u(x,t) \le \theta(x).$$

This proves assertion 1. Assertion 2 is similar. □

Next, we consider the linear stability of equilibrium solutions. For this purpose, consider the linearized system

$$\begin{cases} \mathcal{L}\phi = F_u(x,\theta(x))\phi + \mu\phi & \text{in } \Omega, \\ \mathcal{B}\phi = 0 & \text{on } \partial\Omega. \end{cases} \tag{5.8}$$

By Theorem 1.3.6, (5.8) has a principal eigenvalue μ_1 such that it is real, simple, and is associated with a positive eigenfunction $\phi_1(x)$.

Definition 5.1.11 Let θ be an equilibrium of (5.1). We say that θ is linearly stable (resp. linearly unstable) if $\mu_1 > 0$ (resp. $\mu_1 < 0$). If $\mu_1 = 0$, we say that θ is linearly neutrally stable.

First, we show that linear stability implies nonlinear asymptotic stability (i.e. local asymptotic stability).

Proposition 5.1.12 *Let θ be an equilibrium of* (5.1). *If θ is linearly stable, then there exists a $\delta > 0$ such that the solution $u(x,t)$ of* (5.1) *satisfies $u(x,t) \longrightarrow \theta(x)$ provided*

$$\|u(x,0) - \theta(x)\| < \delta.$$

In particular, θ is an isolated equilibrium.

Proof Let $\mu_1 > 0$ denote the principal eigenvalue of (5.8), and we choose an eigenfunction ϕ_1 such that $\phi_1 > 0$ and $\sup_\Omega \phi_1 = 1$. For $\epsilon > 0$, define

$$u_\epsilon^\pm = \theta \pm \epsilon\phi_1.$$

Step 1. We claim that there exists an $\epsilon_0 > 0$ such that for $\epsilon \in (0,\epsilon_0]$, u_ϵ^+ (resp. u_ϵ^-) is a superequilibrium (resp. subequilibrium) of (5.7) in the classical sense.

Indeed, it is immediate that $\mathcal{B}u_\epsilon^\pm = 0$ on $\partial\Omega$. Moreover,

$$\begin{aligned} \mathcal{L}u_\epsilon^+ - F(x,u_\epsilon^+) &= F(x,\theta) + \epsilon\phi_1[F_u(x,\theta) + \mu_1] - F(x,\theta+\epsilon\phi_1) \\ &= \epsilon[\mu_1\phi_1 + O(\epsilon)] > 0 \quad \text{provided } 0 < \epsilon \ll 1. \end{aligned}$$

Note that we used $F(x,\theta+\epsilon\phi_1) = F(x,\theta) + F_u(x,\theta)\epsilon\phi_1 + O(\epsilon^2)$ for the last equality, and $\inf_\Omega \phi_1 > 0$ in the inequality. This proves that u^+ is a superequilibrium of (5.7) in the classical sense. It can be similarly established that u_ϵ^- is a subequilibrium of (5.7) in the classical sense.

Step 2. Fix $\epsilon \in (0,\epsilon_0]$ and define

$$I_0 = [u_{\epsilon_0}^-, u_{\epsilon_0}^+] = \{u_0 \in C(\overline{\Omega}) : u_{\epsilon_0}^- \le u_0 \le u_{\epsilon_0}^+ \text{ in } \Omega\}.$$

We claim that θ is the only equilibrium of (5.7) in I_0. Suppose $\tilde{\theta}$ is another equilibrium solution in I_0, then there exists an $\epsilon_1 \in (0,\epsilon_0)$ such that either

$$\tilde{\theta} \le u^+_{\epsilon_1} \quad \text{in } \overline{\Omega}, \quad \text{and equality holds at some } x_1 \in \overline{\Omega} \tag{5.9}$$

or

$$\tilde{\theta} \ge u^-_{\epsilon_1} \quad \text{in } \overline{\Omega}, \quad \text{and equality holds at some } x_1 \in \overline{\Omega}. \tag{5.10}$$

For definiteness, suppose (5.9) holds. Then $w = \tilde{\theta} - \theta - \epsilon_1\phi_1$ satisfies

$$\begin{cases} \mathcal{L}w < \tilde{F}(x)w & \text{in } \Omega, \\ \mathcal{B}w = 0 & \text{on } \partial\Omega, \\ w \le 0 \quad \text{in } \Omega, \quad w(x_1) = 0, \end{cases}$$

where $\tilde{F}(x) = \int_0^1 F_u(x, \tau\tilde{\theta}(x) + (1-\tau)u^+_{\epsilon_1})\,d\tau$. Now, $w \not\equiv 0$, since $u^+_{\epsilon_1}$ is not an equilibrium. By regarding w as a strict subsolution of

$$w_t + \mathcal{L}w = \tilde{F}(x)w \quad \text{in } \Omega_T, \quad \mathcal{B}w = 0 \quad \text{on } S\Omega_T,$$

the strong maximum principle says that either $w \equiv 0$ or $\sup_\Omega w < 0$. The former is incompatible with $\mathcal{L}w < \tilde{F}(x)w$ and the latter is incompatible with $w(x_1) = 0$. This contradiction completes Step 2.

Finally, the desired conclusion follows from Proposition 5.1.8. □

Similarly, one can establish that linear instability implies nonlinear instability.

Proposition 5.1.13 *Let θ be an equilibrium of* (5.1). *If θ is linearly unstable, then there exists an $\epsilon_0 > 0$ such that for any $\delta > 0$, there exists a solution u of* (5.1) *such that*

$$\|u(x,0) - \theta(x)\| < \delta \quad \textit{and} \quad \liminf_{t\to\infty} \|u(\cdot,t) - \theta\| \ge \epsilon_0.$$

Proof See Problem 5.4.2. □

5.2 The Logistic Equation

In this section, we specialize to the logistic equation.

$$\begin{cases} \partial_t u - d\Delta u = u(m(x) - u) & \text{in } \Omega \times (0,\infty), \\ \mathbf{n}\cdot\nabla u = 0 & \text{on } \partial\Omega \times (0,\infty), \\ u(x,0) = u_0(x) & \text{in } \Omega, \end{cases} \tag{5.11}$$

where $d \in (0,\infty)$ and $m \in C^\alpha(\overline{\Omega})$.

The stability of the trivial solution $u \equiv 0$ is determined by the following eigenvalue problem:

$$\begin{cases} -d\Delta\phi = m(x)\phi + \mu\phi & \text{in } \Omega, \\ \mathbf{n}\cdot\nabla\phi = 0 & \text{on } \partial\Omega. \end{cases} \tag{5.12}$$

See Theorem 1.3.6 for the existence and properties of the principal eigenvalue μ_1.

It is well understood that the large time asymptotics of this model can be characterized by the stability of the trivial solution [2, 7]; see Theorem 5.2.2 below. First, we start with the existence of a positive equilibrium.

Proposition 5.2.1 *Model* (5.11) *has a nonnegative, nontrivial equilibrium θ if and only if $\mu_1 < 0$.*

Proof Let μ_1 be the principal eigenvalue of (5.12) with an associated positive eigenfunction $\phi_1 > 0$ such that $\|\phi_1\|_\infty = 1$.

First, suppose (5.11) has an equilibrium $\theta(x) \geq, \not\equiv 0$, i.e.

$$-d\Delta\theta = \theta(m-\theta) \quad \text{in } \Omega, \quad \text{and} \quad \mathbf{n}\cdot\nabla\theta = 0 \quad \text{on } \partial\Omega. \tag{5.13}$$

Then $\theta > 0$ in $\overline{\Omega}$, by the strong maximum principle. Multiplying the above equation by ϕ_1 and integrating by parts, we have

$$\int_\Omega \theta^2\phi_1 = \int_\Omega \phi_1(d\Delta\theta + m\theta) = \int_\Omega \theta(d\Delta\phi_1 + m\phi_1) = -\mu_1\int_\Omega \theta\phi_1.$$

Since $\theta > 0$ and $\phi_1 > 0$ in Ω, we deduce that $\mu_1 < 0$. This proves sufficiency.

Now, suppose $\mu_1 < 0$. We will demonstrate that (5.11) has at least one positive equilibrium θ. To this end, define

$$\underline{\theta} = \epsilon\phi_1 \quad \text{and} \quad \overline{\theta} = M.$$

We claim that, for $\epsilon > 0$ sufficiently small and $M > 0$ sufficiently large, $\underline{\theta}$ and $\overline{\theta}$ form a pair of sub- and superequilibria in the classical sense. Since both functions satisfy the Neumann boundary condition, it remains to check the differential inequalities:

$$\begin{cases} -\Delta(\epsilon\phi_1) - \epsilon\phi_1(m-\epsilon\phi_1) = \epsilon\phi_1(\mu_1 + \epsilon\phi_1) < 0 & \text{in } \Omega, \\ -\Delta M - M(m-M) \geq 0 \quad \text{in } \Omega, \end{cases}$$

provided $0 < \epsilon < -\mu_1$ and $M \geq \sup_\Omega m$. The existence of a positive equilibrium θ follows from Proposition 5.1.8. □

The main theorem of this section is as follows.

Theorem 5.2.2 *Let u be a solution of* (5.11).

1. *If $\mu_1 \geq 0$, then $u(x,t) \to 0$ regardless of the initial conditions.*
2. *If $\mu_1 < 0$, then* (5.11) *has a unique positive equilibrium θ_d. Moreover, θ_d satisfies*

$$0 < \theta_d(x) \leq \sup_\Omega m \quad \textit{for } x \in \overline{\Omega} \tag{5.14}$$

and is globally asymptotically stable among all nonnegative, nontrivial solutions of (5.11).

Remark 5.2.3 The dichotomy of Theorem 5.2.2 holds for general elliptic operators $\mathcal{L}$ and general oblique boundary conditions $\mathcal{B}$. See Problem 5.4.3.

Proof Suppose $\mu_1 < 0$, then Proposition 5.1.8 implies that there is a maximal and minimal positive equilibrium θ_M and θ_m, such that $0 < \theta_m \leq \theta_M$ and

$$\theta_m \leq \theta \leq \theta_M \quad \text{in } \Omega, \text{ for any positive equilibrium } \theta.$$

(Observe that the equilibrium problem associated to (5.11) is

$$d\Delta\theta + \theta(m - \theta) = 0 \quad \text{in } \Omega, \quad \text{and} \quad \mathbf{n} \cdot \theta = 0 \quad \text{on } \partial\Omega,$$

which is a special case of (5.7).) Multiplying the equation of θ_m by θ_M and then integrating by parts, we have

$$d \int_\Omega \nabla\theta_M \cdot \nabla\theta_m = \theta_M\theta_m(m - \theta_m). \tag{5.15}$$

Reversing the roles of θ_m, θ_M, we obtain

$$d \int_\Omega \nabla\theta_M \cdot \nabla\theta_m = \int_\Omega \theta_M\theta_m(m - \theta_M). \tag{5.16}$$

Subtracting (5.16) from (5.15), we have

$$0 = \int_\Omega \theta_M\theta_m(\theta_M - \theta_m).$$

Since $\theta_M \geq \theta_m$, we must have $\theta_M = \theta_m$. Hence, (5.11) has a unique positive equilibrium θ in $[\epsilon\phi_1, M]$. Note that θ is necessarily independent of ϵ and M. By Proposition 5.1.8, the positive equilibrium attracts all solutions with initial data in $[\epsilon\phi_1, M]$. Since this is true for all $0 < \epsilon \ll 1$ and $M \gg 1$, we deduce that (5.11) has a unique positive equilibrium, which we denote hereafter by θ_d. Moreover, θ_d attracts all solutions with strictly positive initial data. Finally, if $u(x, 0)$ is nonnegative and nontrivial, then the strong maximum principle asserts that $u(x, t) > 0$ in $\overline{\Omega} \times (0, \infty)$. From the above argument, we have again $u(x, t) \to \theta_d$. This proves the case $\mu_1 < 0$.

Next, suppose $\mu_1 \geq 0$. We will show that every nonnegative solution u of (5.11) converges to 0 as $t \to \infty$. Indeed, fix $M > 0$ such that $M \geq \max\{m(x), u(x, 0)\}$ for $x \in \Omega$, and observe that $\overline{u} := M$ and $\underline{u} = 0$ form a pair of super- and subsolutions for (5.11). By Theorem 5.2.2, $\theta = 0$ is the only equilibrium of (5.11) satisfying $0 \leq \theta \leq M$. Hence, we can conclude by Proposition 5.1.8, that u converges to 0 as $t \to \infty$. □

Consider the diffusive logistic equation (5.11) with diffusion rate $d > 0$ and intrinsic growth rate $m \in C(\overline{\Omega})$. If $m(x) \equiv m_0$ is a constant, then the principal eigenvalue μ_1 of (5.12) satisfies $\mu_1 = -m_0$. By Theorem 5.2.2, the population persists if $m_0 > 0$, and goes to extinction if $m_0 \leq 0$, i.e. diffusion does not play a significant role in the long-time dynamics here. However, when $m(x)$ is nonconstant, there is generally a critical diffusion rate for the persistence of the population, as the following result illustrates.

Theorem 5.2.4 *Consider the time-dependent problem* (5.11) *with a given nonconstant intrinsic growth rate* $m \in C(\overline{\Omega})$.

1. *If* $\int_\Omega m\,\mathrm{d}x \ge 0$, *then* (5.11) *has a unique positive equilibrium* θ_d *for any* $d > 0$. *Moreover,* θ_d *is globally asymptotically stable among all nonnegative, nontrivial solutions.*
2. *If* $\max_{\overline{\Omega}} m \le 0$, *then every nonnegative solution* u *satisfies* $u(x,t) \to 0$ *as* $t \to \infty$.
3. *If* $\int_\Omega m\,\mathrm{d}x < 0$ *and* $\max_{\overline{\Omega}} m > 0$, *then there exists a critical diffusion rate* $\hat{d} \in (0,\infty)$ *such that the conclusion of part 1 holds if* $d \in (0,\hat{d})$ *and the conclusion of part 2 holds if* $d \in [\hat{d},\infty)$.

Proof Let μ_1 be the principal eigenvalue of (5.12) with a positive eigenfunction ϕ. Recall the dichotomy result from Theorem 5.2.2, which states that the trivial solution is globally asymptotically stable if $\mu_1 \ge 0$, and that the unique positive equilibrium exists and is globally asymptotically stable if $\mu_1 < 0$.

To prove assertion 1, suppose $\int_\Omega m \ge 0$. We need to show that $\mu_1 < 0$ for all $d > 0$. Indeed, $\phi > 0$ in $\overline{\Omega}$, thanks to the strong maximum principle. We may divide (5.12) by ϕ and integrate by parts to get

$$-d\int_\Omega \frac{|\nabla\phi|^2}{\phi^2} = -d\int_\Omega \frac{\Delta\phi}{\phi} = \int_\Omega (m+\mu_1).$$

Using the facts $\int_\Omega m \ge 0$ and that m, and hence ϕ, are nonconstant functions, we deduce from the above that $\mu_1 < 0$. This proves assertion 1.

To prove assertion 2, suppose that $m \le 0$ in $\overline{\Omega}$, we need to show that $\mu_1 \ge 0$ for all $d > 0$. Indeed, integrating (5.12) over Ω, and using the Neumann boundary condition, we obtain

$$0 = -d\int_\Omega \Delta\phi = \int_\Omega (m+\mu_1)\phi.$$

Since $m \le 0$ in $\overline{\Omega}$, we deduce that $\mu_1 \ge 0$. This proves assertion 2.

For assertion 3, we recall from Propositions 1.3.16 and 1.3.19 that

$$\lim_{d\to 0+} \mu_1 = -\max_{\overline{\Omega}} m < 0 \quad \text{and} \quad \lim_{d\to\infty} \mu_1 = -\frac{1}{|\Omega|}\int_\Omega m\,\mathrm{d}x > 0.$$

Moreover, μ_1 is differentiable and strictly increasing in $d \in (0,\infty)$ thanks to Proposition 1.3.15. Hence, there exists a $\hat{d} \in (0,\infty)$ such that $\mu_1 < 0$ when $0 < d < \hat{d}$, and $\mu_1 \ge 0$ when $d \ge \hat{d}$. This proves assertion 3. □

Theorem 5.2.5 *Denote the unique positive equilibrium of* (5.11) *by* θ_d, *if it exists.*

1. *If* $m > 0$ *somewhere, then* θ_d *exists for all* $0 < d \ll 1$ *and* $\lim_{d\to 0} \theta_d = m_+$.
2. *If* $\int_\Omega m\,dx > 0$, *then* θ_d *exists for all* $d > 0$ *and* $\lim_{d\to\infty} \theta_d = \fint_\Omega m\,dx$.

Proof Without loss of generality, we may perform a rescaling and assume that $|\Omega| = 1$. For simplicity, we prove the case when $\inf_\Omega m > 0$ and $m \in C^2(\overline{\Omega})$. We leave the general case as an exercise.
Step 1. For $n \in \mathbb{N}$, define

$$\theta_n^\pm = m(x) \pm \left[\frac{2}{n} + Cn^2 \max\left\{\frac{1}{n} - \mathrm{dist}(x, \partial\Omega), 0\right\}^3\right],$$

where we take $C \gg 1$ such that for each $n \ge 1$, we have

$$\mathbf{n}\cdot\nabla\theta^+ \ge 0 \ge \mathbf{n}\cdot\nabla\theta^- \quad \text{on } \partial\Omega.$$

Step 2. For each fixed n, there exists a $\hat{d}_n$ such that for $d \in (0, \hat{d}_n)$, the functions θ_n^+ and θ_n^- form a pair of super- and subsolutions. By Corollary 5.1.9, the unique equilibrium θ_d must stay between $\theta_n^\pm$. Hence

$$m - \frac{2+C}{n} \le \theta_n^-(x) \le \theta_d(x) \le \theta_n^+(x) \le m + \frac{2+C}{n} \quad \text{in } \Omega, \text{ for } 0 < d < \hat{d}_n.$$

Letting $d \to 0+$ and then $n \to \infty$, we complete the proof of part 1.

To prove part 2, we first divide (5.13) by θ_d, and integrate by parts to get

$$0 \ge -\int_\Omega \frac{|\nabla\theta_d|^2}{|\theta_d|^2} = \int_\Omega (m - \theta_d). \tag{5.17}$$

This implies

$$\int_\Omega \theta_d \ge \int_\Omega m > 0. \tag{5.18}$$

Rewriting the equation of θ_d as

$$-\Delta\theta_d = \frac{1}{d}\theta_d(m - \theta_d), \tag{5.19}$$

upon multiplying by θ_d and integrating by parts, we have

$$\int_\Omega |\nabla\theta_d|^2 \le \frac{C}{d},$$

where C is independent of $d \ge 1$. It then follows by Poincaré's inequality that

$$\int_\Omega \left|\theta_d - \int_\Omega \theta_d\right|^2 \le \frac{C'}{d}, \tag{5.20}$$

for some constant C' independent of $d \geq 1$. Now, integrating (5.13), we have

$$0 = \int_\Omega \theta_d(m - \theta_d). \tag{5.21}$$

We can rewrite the above as

$$-\int_\Omega \left(\theta_d - \int_\Omega \theta_d\right)(m - \theta_d) = \left[\int_\Omega \theta_d\right]\left[\int_\Omega (m - \theta_d)\right] = \left[\int_\Omega \theta_d\right]\left[\int_\Omega m - \int_\Omega \theta_d\right]. \tag{5.22}$$

By (5.18) and (5.20), we may pass to the limit in (5.22) and deduce that $\int_\Omega \theta_d \to \int_\Omega m > 0$ as $d \to \infty$. In view of (5.20), we deduce that $\theta_d \to \int_\Omega m$ in $L^2(\Omega)$ as $d \to \infty$.

It remains to improve the convergence of $\theta_d \to \int_\Omega m$. To this end, observe by the maximum principle that

$$\sup_\Omega \theta_d \leq \sup_\Omega m, \tag{5.23}$$

so that the right-hand side of (5.19) is bounded in L^∞. It then follows by elliptic L^p estimates that $\sup_{d\geq 1} \|\theta_d\|_{W^{2,p}(\Omega)}$ is finite for each $p > 1$. Therefore, as $d \to \infty$, we have $\theta_d \to \int_\Omega m$ weakly in $W^{2,p}(\Omega)$ and hence strongly in $C^{1+\alpha}(\overline{\Omega})$. □

It is of biological interest to study the total biomass of a single species at equilibrium, which is given by $\int_\Omega \theta_d \,\mathrm{d}x$, for the model (5.11). It follows from Theorem 5.2.5 and (5.18) that

Corollary 5.2.6 *Suppose that m is nonnegative and not identically zero. Then for all $d > 0$ it holds that*

$$\int_\Omega \theta_d \,\mathrm{d}x > \int_\Omega m \,\mathrm{d}x = \lim_{d\to 0+} \int_\Omega \theta_d \,\mathrm{d}x = \lim_{d\to\infty} \int_\Omega \theta_d \,\mathrm{d}x. \tag{5.24}$$

That is, as shown in [17], the total biomass is minimized at $d = 0$ and $d = \infty$, and it is maximized at some intermediate values of diffusion rates. Later it was found that the total biomass, as a function of the diffusion rate, can have multiple local maxima [16]. In [8] and [10] the authors further studied the effect of the diffusion rate on the total biomass, for more general reaction-diffusion models of single species.

5.3 Critical Domain Size

Reaction-diffusion models can predict the critical domain size needed to sustain a population in an environment that is surrounded by a hostile exterior. In the following, we give a proof of existence of critical domain size for a reaction-diffusion-advection model. Consider

$$\begin{cases} u_t - au_{xx} - bu_x = u(c - \beta(x)u) & \text{in } [0, L] \times [0, \infty), \\ -u_x(0, t) + p_-u(0, t) = u_x(L, t) + p_+u(0, t) = 0 & \text{on } \{0, L\} \times [0, \infty), \\ u(x, 0) = u_0(x) & \text{in } [0, L], \end{cases} \tag{5.25}$$

where $\beta \in C(\overline{\Omega})$ is strictly positive in $[0, L]$, and $a, b, c, p_\pm$ are real constants such that

$$a > 0, \quad c > 0, \quad p_\pm \geq 0, \quad p_+ + p_- > 0. \tag{5.26}$$

By Theorem 5.1.2, for any nonnegative initial data $u_0 \in C([0, L])$, the problem (5.25) has a unique classical solution $u \in C^{2,1}([0, L] \times (0, \infty)) \cap C([0, L] \times [0, \infty))$. Furthermore, it can be proved (along the lines of Proposition 5.2.1) that (5.25) has a positive equilibrium if and only if the trivial solution is unstable, i.e. the principal eigenvalue μ_1 of

$$\begin{cases} a\phi'' + b\phi' + c\phi + \mu_1\phi = 0 & \text{in } [0, L], \\ -\phi'(0) + p_-\phi(0) = \phi'(L) + p_+\phi(L) = 0 \end{cases} \tag{5.27}$$

is negative.

On the one hand, if (5.25) has no positive equilibria, then the trivial solution attracts all nonnegative solutions of (5.25). On the other hand, if a positive solution θ exists, it is unique and attracts all nonnegative, nontrivial solutions of (5.25). See Problem 5.4.3.

The following result shows the existence of critical domain size $L^* \in (0, \infty]$.

Proposition 5.3.1 *Suppose* (5.26) *holds, then there exists* $L^* = L^*(a, b, c, p_-, p_+)$ *such that the trivial solution is linearly unstable if and only if* $L > L^*$.

Proof For simplicity, we prove the case $b = 0$ here and leave the general case as an exercise. Let μ_1 be the principal eigenvalue of

$$\begin{cases} a\phi'' + c\phi + \mu_1\phi = 0 & \text{in } [0, L], \\ -\phi'(0) + p_-\phi(0) = \phi'(L) + p_+\phi(L) = 0. \end{cases} \tag{5.28}$$

Step 1. We claim that $\mu_1 \neq -c$. Suppose the contrary, then $\phi = Ax + B$ for some A, B such that $B \geq 0$ and $AL + B \geq 0$. Now, the boundary condition implies

$$-A + p_-B = 0 \quad \text{and} \quad A + p_+(AL + B) = 0.$$

Adding the equalities, we get $(p_- + p_+)B + p_+AL = 0$, which implies $A = B = 0$. This is a contradiction.

Step 2. We claim that $\mu_1 > -c$. Indeed, choose x_0 such that $\phi(x_0) = \sup_{[0,L]} \phi$. If $x_0 \in (0, L)$, then

$$\phi''(x_0) \leq 0, \quad \phi'(x_0) = 0, \quad \phi(x_0) > 0. \tag{5.29}$$

Substituting the above into $a\phi'' + (c + \mu_1)\phi = 0$, we obtain $c + \mu_1 \geq 0$. On the other hand, if $x_0 = 0$ or $x_0 = L$, then we can use the boundary condition, and $p_\pm \geq 0$ to

conclude that $\phi'(x_0) = 0$, and hence (5.29) again holds. Hence we once again obtain $c + \mu_1 \geq 0$. Since $\mu_1 \neq -c$ by Step 1, we have $\mu_1 > -c$.
Step 3. Suppose μ_1 is the principal eigenvalue of (5.28), then

$$L = \sqrt{\frac{a}{c+\mu_1}} \left[\arctan\left(\frac{p_+}{\sqrt{(c+\mu_1)/a}} \right) - \arctan\left(\frac{-p_-}{\sqrt{(c+\mu_1)/a}} \right) \right] := G(\mu_1). \quad (5.30)$$

Indeed, since $\mu_1 + c > 0$ (Step 2), we have

$$\phi(x) = A \cos\left(\sqrt{\frac{c+\mu_1}{a}} (x - \eta) \right), \qquad \text{for some } A > 0,\ |\eta| < \sqrt{\frac{a}{c+\mu_1}} \frac{\pi}{2}.$$

Then the boundary conditions yield

$$\begin{cases} -p_- = -\frac{\phi'(0)}{\phi(0)} = \sqrt{\frac{c+\mu_1}{a}} \tan\left(\sqrt{\frac{c+\mu_1}{a}} (-\eta) \right), \\ p_+ = -\frac{\phi'(L)}{\phi(L)} = \sqrt{\frac{c+\mu_1}{a}} \tan\left(\sqrt{\frac{c+\mu_1}{a}} (L-\eta) \right). \end{cases}$$

Since ϕ is positive on $[0, L]$, and $p_- + p_+ > 0$, we observe that η, L can be uniquely determined, and we may solve for L in these expressions to obtain (5.30). This completes Step 3.
Step 4. We claim that the mapping $L \mapsto \mu_1(L)$ is a homeomorphism from $(0, L) \to (-c, \infty)$.

By Step 2, the range of the mapping is contained in $(-c, \infty)$. Moreover, the mapping is injective since if $\mu_1(L) = \mu_1(\tilde{L})$, we can denote by $\hat{\mu}$ their common value, and apply Step 3 to conclude that $L = \tilde{L} = G(\hat{\mu})$. It is surjective since, for any $\hat{\mu} \in (-c, \infty)$, we have $\mu_1(\hat{L}) = \hat{\mu}$, where $\hat{L} = G(\hat{\mu}) > 0$. (Note that the positivity of $\hat{L}$ follows from (5.30) and the fact that $p_- + p_+ > 0$.) Thus $L \mapsto \mu_1(L)$ is bijective from $(0, \infty)$ to $(-c, \infty)$, and the inverse is given by G. Finally, $L \mapsto \mu_1(L)$ is continuous, since G is.
Step 5. The mapping $L \mapsto \mu_1(L)$ is strictly decreasing.

It follows from $L = G(\mu_1(L))$ and (5.30) that $\mu_1(L) \searrow -c$ as $L \to \infty$ and $\mu_1(L) \nearrow \infty$ as $L \searrow 0$. The mapping $L \mapsto \mu_1(L)$, being a homeomorphism of $(0, \infty)$ to $(-c, \infty)$, must then be strictly decreasing.

It then follows that there exists an L^* such that

$$\mu_1 < 0 \quad \text{if and only if} \quad L > L^*.$$

This completes the proof. □

The concept of critical domain size gives a mathematical description of the so-called drift paradox. The latter refers to the persistence of populations in streams even when they are constantly subject to washing out. This problem is intensified when the habitat quality at the downstream end is poor. Once the species are washed downstream, the chance of survival is greatly reduced. This problem, termed as the "drift paradox", has received considerable attention in ecology. The work of Speirs and Gurney [29] offered an explanation based upon a reaction-diffusion model. The following closely related model of a single species population inhabiting a river habitat is considered in [11, 18].

$$\begin{cases} \partial_t \tilde{u} - \tilde{d}\partial_{xx}^2 \tilde{u} + \tilde{q}\partial_x \tilde{u} = \tilde{u}(\tilde{r} - \tilde{u}) & \text{for } 0 < x < \tilde{L},\ t > 0, \\ \tilde{d}\partial_x \tilde{u}(0,t) - \tilde{q}\tilde{u}(0,t) = 0 & \text{for } t > 0, \\ \tilde{d}\partial_x \tilde{u}(\tilde{L},t) - \tilde{q}\tilde{u}(\tilde{L},t) = -\tilde{b}\tilde{q}\tilde{u}(\tilde{L},t) & \text{for } t > 0, \\ \tilde{u}(x,0) = \tilde{u}_0(x) & \text{for } 0 < x < L, \end{cases} \tag{5.31}$$

where $\tilde{d}, \tilde{q}, \tilde{r}, \tilde{L}$ are the diffusion rate, river flow rate, the local intrinsic growth rate and the domain size, respectively. The nonnegative parameter $\tilde{b}$ represents the rate of population loss at the downstream boundary, while a no-flux condition is imposed at the upstream boundary. They are assumed to be given positive constants. By a nondimensionalization,

$$u(x,t) = \frac{1}{\tilde{r}}\tilde{u}\left(\frac{\tilde{q}x}{\tilde{r}}, \frac{t}{\tilde{r}}\right), \quad d = \frac{\tilde{d}\tilde{r}}{\tilde{q}^2}, \quad L = \frac{\tilde{r}\tilde{L}}{\tilde{q}},$$

and $v(x,t) = \mathrm{e}^{-x/d}u(x,t)$ we obtain the simplified problem

$$\begin{cases} \partial_t v - d\partial_{xx}^2 v - \partial_x v = v(1 - \mathrm{e}^{x/d}v) & \text{for } 0 < x < L,\ t > 0, \\ \partial_x v(0,t) = 0 & \text{for } t > 0, \\ d\partial_x v(L,t) + \tilde{b}v(L,t) = 0 & \text{for } t > 0, \\ v(x,0) = v_0(x) & \text{for } 0 < x < L. \end{cases}$$

In such a case, we obtain (5.25) with

$$\beta(x) = \mathrm{e}^{x/d}, \quad a = d, \quad b = 1, \quad c = 1, \quad p_- = 0, \quad p_+ = \tilde{b} \geq 0,$$

and denote the critical domain size by $L^*(d, \tilde{b})$.

Considering the diffusion rate as a phenotype of the species, we are interested in the range of phenotypes that are able to persist in a given environment. Precisely, given the domain size L and downstream loss rate $\tilde{b} \geq 0$ what is the set of diffusion rates d that leads to persistence? Obviously, the species persists if and only if $L > L^*(d, \tilde{b})$. The answer thus depends on the (monotonic) dependence of L^* on the parameter d.

Proposition 5.3.2 *Let $L^*(d, \tilde{b})$ be given as above. Define*

$$d_{\min}(\tilde{b}) = \tilde{b}(1 - \tilde{b}) \quad \text{if } 0 \leq \tilde{b} \leq \tfrac{1}{2}, \quad \text{and} \quad d_{\min}(\tilde{b}) = \tfrac{1}{4} \quad \text{otherwise}.$$

Then the following statements hold.

1. $L^*(d, \tilde{b}) = +\infty$ *for* $d \in (0, d_{\min}(\tilde{b})]$.
2. $\lim_{d \to d_{\min}(\tilde{b})} L^*(d, \tilde{b}) = +\infty$ *and* $\lim_{d \to +\infty} L^*(d, \tilde{b}) = \tilde{b}$.

3. *Let $\tilde{b} \in (0, \frac{3}{2}]$. Then $d \mapsto L^*(d, \tilde{b})$ is strictly decreasing on $(d_{\min}(b), \infty)$. In particular, $L^*(d, \tilde{b}) > \tilde{b}$ for all $d \in (d_{\min}, \infty)$, such that the following statements hold:*

   a. *If $L \leq \tilde{b}$, then the single species goes extinct for all $d > 0$.*
   b. *If $L > \tilde{b}$, then there exists $\underline{d}(L, \tilde{b}) > d_{\min}(\tilde{b})$ such that the single species persists if and only if $d \in (\underline{d}(L, \tilde{b}), +\infty)$.*

4. *Let $b > \frac{3}{2}$. Then there exists a $\hat{d} \in (\frac{1}{4}, +\infty)$ such that $d \mapsto L^*(d, \tilde{b})$ is strictly decreasing in $(\frac{1}{4}, \hat{d}]$ and strictly increasing in $[\hat{d}, +\infty)$. Also, the quantity*

$$L_{\min}(\tilde{b}) = \min_{d > 1/4} L^*(d, \tilde{b}) = L^*(\hat{d}, \tilde{b})$$

   satisfies $L_{\min}(\tilde{b}) < \tilde{b}$ so that

   a. *If $L \leq L_{\min}(\tilde{b})$, then the single species goes extinct for all $d > 0$.*
   b. *If $L \in (L_{\min}(\tilde{b}), \tilde{b})$, then there exist finite positive numbers $\frac{1}{4} < \underline{d}(L, \tilde{b}) < \overline{d}(L, \tilde{b})$ such that the single species persists if and only if $d \in (\underline{d}, \overline{d})$.*
   c. *If $L \geq \tilde{b}$, then there exists $\underline{d}(L, \tilde{b}) \in (\frac{1}{4}, +\infty)$ such that the single species persists if and only if $d \in (\underline{d}, +\infty)$.*

Proof Parts 1 and 2 are proved in [30] and [18, Theorem 2.1]. Parts 3 and 4 are proved in [11, Proposition 1.3]. □

Note also that for $\tilde{b} > \frac{3}{2}$, there exist choices of L for which only the phenotypes with intermediate diffusion rate persist, whereas those which are too slow or too fast go to extinction. In other words, only an intermediate diffusion rate is viable, and neither a slow or fast diffusion rate is selected for. See Figure 5.1 for an illustration of Proposition 5.3.2.

5.4 Further Reading

The results in this chapter indicate that for the logistic equation, or more general single species equations, the long time dynamics is largely determined by the existence and stability of equilibrium solutions. For a more detailed account of the theory, we refer to the classical text [3], which also contains a bifurcation analysis of the equilibrium problem, among many other things.

The optimal arrangement of resource so as to maximize the total biomass for the steady state of (5.11), subject to L^1 and L^∞ constraints, was investigated in Ding et al. [9]. It was conjectured there that any global maximizer of the total biomass must be of the "bang-bang" type. Namely, that the optimal distribution of resource should be a positive scalar multiple of an indicator function supported on a suitable subset of the spatial domain. This has been proved independently by Nagahara and Yanagida [27] and Mazari et al. [23]. However, the characterization of the optimal set is a more delicate issue: When the diffusion rate is large, the support of the optimizer is

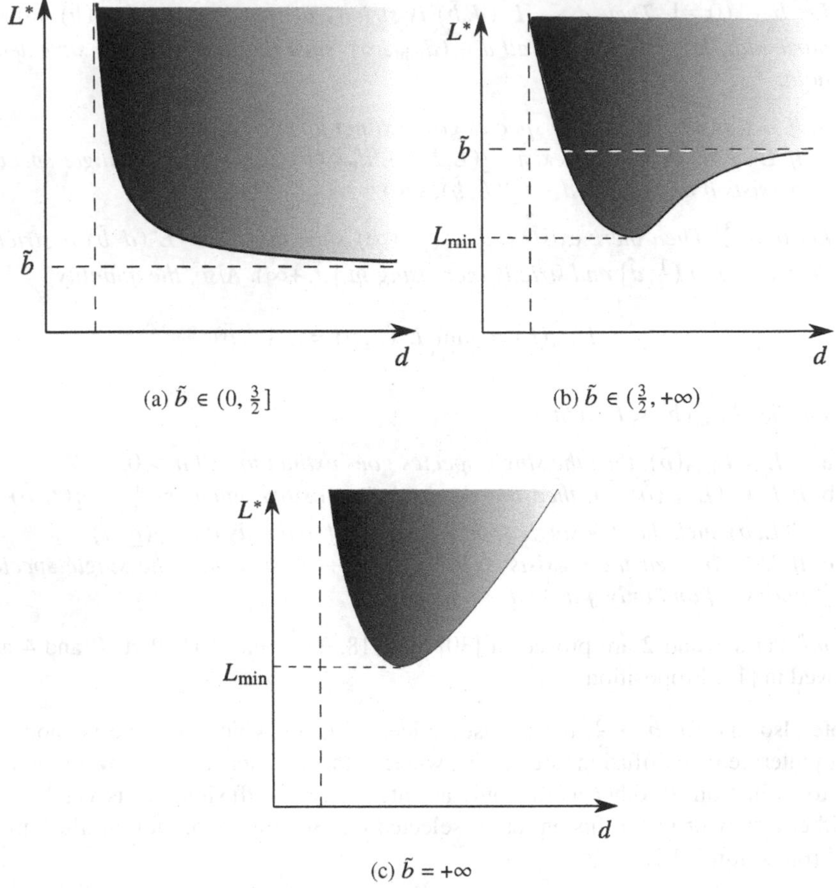

Fig. 5.1: Three graphs illustrating Proposition 5.3.2 concerning the dependence of the critical domain size L^* on the diffusion rate d. The shaded region corresponds to the parameter value where the species persists. Observe that the shaded region is decreasing in $\tilde{b}$. When $\tilde{b} \in (0, \frac{3}{2}]$, L^* is decreasing in d; when $\tilde{b} \in (\frac{3}{2}, +\infty)$, L^* is first decreasing in d and then increasing in d, with $L^* \to \tilde{b}$ as $d \to +\infty$; when $\tilde{b} = +\infty$ (which corresponds to a Dirichlet boundary condition at $x = L$), we have $L^* \to +\infty$ as $d \to \infty$. Note also that for $\tilde{b} > \frac{3}{2}$, there exist choices of L for which only the phenotypes with intermediate diffusion rate persist, whereas those which are too slow or too fast go to extinction.

connected in space. In contrast, if the diffusion rate is small, then the optimal resource distributions tend to alternate spatially, i.e. the fragmentation phenomenon occurs. We refer the interested readers to [22, 23, 25] for further details. See also [26] for the study of a discrete patch model with similar fragmentation phenomenon. We also refer to [1, 12, 13] for some results concerning the upper bounds of the total biomass in terms of the total resource distribution. For some biased movement models with advective terms, the optimal distribution of the resources for maximizing the survival ability of a single species was considered in [24]. They showed that the problem has "regular" solutions only when the domain is a ball and the optimal distribution can be characterized in this case; see [5] for the one-dimensional case.

The idea that reaction-diffusion models predict the minimal patch size needed to sustain a population was introduced by Skellam [28] and Kierstead and Slobodkin [14]; See also [3]. The incorporation of advection originates from Speirs and Gurney [29], who considered the reaction-diffusion-advection model (5.25) when the upstream end is a no-flux or closed boundary, while at the downstream end the zero Dirichlet boundary condition or lethal boundary condition is imposed. They established necessary and sufficient conditions for the persistence of the species. Vasilyeva and Lutscher considered the case when the free-flow condition is imposed at the downstream end [30], and the general boundary conditions at the downstream end are studied in [18]. A complete persistence criteria is found for general boundary conditions in the work [11], where the critical river length can be either monotone decreasing in the diffusion rate or it can first decrease and then increase as the diffusion rate increases, depending on the loss rate at the downstream end. We refer to [20, 21] for related works on the persistence problem in rivers. Some qualitative properties of the population density at equilibrium are investigated in [15].

Problems

5.4.1 Suppose $F \in W^{1,\infty}(\overline{\Omega} \times [0, \infty))$ and

$$F(x,0) = 0, \quad \text{and} \quad F(x,u) \le C_0 u$$

for some constant C_0. For any $M, T > 0$, show that there exists a constant $C(M,T) > 0$ such that any solution u of (5.1) such that $0 \le u_0 \le M$ satisfies

$$0 \le u(x,t) \le C(M,T) \quad \text{in } \Omega_T.$$

5.4.2 Prove Proposition 5.1.13.

5.4.3 Consider

$$\begin{cases} u_t + \mathcal{L}u = uf(x,u) & \text{in } \Omega \times (0,\infty), \\ \mathcal{B}u = 0 & \text{on } \partial\Omega \times (0,\infty), \\ u(x,0) = u_0(x) \ge 0 & \text{in } \Omega, \end{cases} \tag{5.32}$$

and let μ_1 be the principal eigenvalue of

$$\begin{cases} \mathcal{L}\phi = f(x,0)\phi + \mu\phi & \text{in } \Omega, \\ \mathcal{B}\phi = 0 & \text{on } \partial\Omega, \end{cases} \tag{5.33}$$

where $\mathcal{B}$ is the first-order oblique derivative operator on $\partial\Omega$. Suppose $u \mapsto f(x,u)$ is strictly decreasing and there exists an $M_0 > 0$ such that $f(x, M_0) < 0$ for all $x \in \Omega$, then the following threshold-type result holds:

1. If $\mu_1 \geq 0$, then $u(x,t) \to 0$ for any nonnegative solution u of (5.32).
2. If $\mu_1 < 0$, then (5.11) has a unique positive equilibrium θ. Moreover, θ is globally asymptotically stable among all nonnegative, nontrivial solutions of (5.32).

[Hint: Besides following the arguments of this chapter, one can alternatively use the theory of subhomogeneous operators; see Appendix D.]

5.4.4 If the oblique boundary condition in (5.32) and (5.33) is replaced by the Dirichlet boundary condition, show the corresponding threshold-type result.

5.4.5 (Prove Theorem 5.2.5 without the additional assumption $\inf_\Omega m > 0$.) Let $m \in C^1(\overline{\Omega})$ and suppose it is sign-changing in Ω. Show that $\theta_d \to m_+$ in $C(\overline{\Omega})$ as $d \to 0$. Can this be improved to $C^\alpha(\overline{\Omega})$ for some $0 < \alpha < 1$?

5.4.6 Prove Proposition 5.3.1 without the assumption that $b = 0$.

References

1. X. Bai, X. He, and F. Li, *An optimization problem and its application in population dynamics*, Proc. Amer. Math. Soc., 144 (2016), pp. 2161–2170.
2. R. S. Cantrell and C. Cosner, *Diffusive logistic equations with indefinite weights: population models in disrupted environments*, Proc. Roy. Soc. Edinburgh Sect. A, 112 (1989), pp. 293–318.
3. ———, *Spatial ecology via reaction-diffusion equations*, Wiley Series in Mathematical and Computational Biology, John Wiley & Sons, Ltd., Chichester, 2003.
4. R. S. Cantrell, C. Cosner, and Y. Lou, *Advection-mediated coexistence of competing species*, Proc. Roy. Soc. Edinburgh Sect. A, 137 (2007), pp. 497–518.
5. F. Caubet, T. Deheuvels, and Y. Privat, *Optimal location of resources for biased movement of species: the 1D case*, SIAM J. Appl. Math., 77 (2017), pp. 1876–1903.
6. X. Chen, K.-Y. Lam, and Y. Lou, *Dynamics of a reaction-diffusion-advection model for two competing species*, Discrete Contin. Dyn. Syst., 32 (2012), pp. 3841–3859.
7. C. Cosner and A. C. Lazer, *Stable coexistence states in the Volterra-Lotka competition model with diffusion*, SIAM J. Appl. Math., 44 (1984), pp. 1112–1132.
8. D. L. DeAngelis, W.-M. Ni, and B. Zhang, *Dispersal and spatial heterogeneity: single species*, J. Math. Biol., 72 (2016), pp. 239–254.
9. W. Ding, H. Finotti, S. Lenhart, Y. Lou, and Q. Ye, *Optimal control of growth coefficient on a steady-state population model*, Nonlinear Anal. Real World Appl., 11 (2010), pp. 688–704.
10. Q. Guo, X. He, and W.-M. Ni, *On the effects of carrying capacity and intrinsic growth rate on single and multiple species in spatially heterogeneous environments*, J. Math. Biol., 81 (2020), pp. 403–433.

11. W. Hao, K.-Y. Lam, and Y. Lou, *Ecological and evolutionary dynamics in advective environments: critical domain size and boundary conditions*, Discrete Contin. Dyn. Syst. Ser. B, 26 (2021), pp. 367–400.
12. J. Inoue, *Limiting profile for stationary solutions maximizing the total population of a diffusive logistic equation*, Proc. Amer. Math. Soc., 149 (2021), pp. 5153–5168.
13. J. Inoue and K. Kuto, *On the unboundedness of the ratio of species and resources for the diffusive logistic equation*, Discrete Contin. Dyn. Syst. Ser. B, 26 (2021), pp. 2441–2450.
14. H. Kierstead and L. B. Slobodkin, *The size of water masses containing plankton bloom*, J. Mar. Res., 12 (1953), pp. 141–147.
15. K.-Y. Lam, Y. Lou, and F. Lutscher, *The emergence of range limits in advective environments*, SIAM J. Appl. Math., 76 (2016), pp. 641–662.
16. S. Liang and Y. Lou, *On the dependence of population size upon random dispersal rate*, Discrete Contin. Dyn. Syst. Ser. B, 17 (2012), pp. 2771–2788.
17. Y. Lou, *On the effects of migration and spatial heterogeneity on single and multiple species*, J. Differential Equations, 223 (2006), pp. 400–426.
18. Y. Lou and P. Zhou, *Evolution of dispersal in advective homogeneous environment: the effect of boundary conditions*, J. Differential Equations, 259 (2015), pp. 141–171.
19. A. Lunardi, *Analytic semigroups and optimal regularity in parabolic problems*, Modern Birkhäuser Classics, Birkhäuser/Springer Basel AG, Basel, 1995. [2013 reprint of the 1995 original] [MR1329547].
20. F. Lutscher, M. A. Lewis, and E. McCauley, *Effects of heterogeneity on spread and persistence in rivers*, Bull. Math. Biol., 68 (2006), pp. 2129–2160.
21. F. Lutscher, E. Pachepsky, and M. A. Lewis, *The effect of dispersal patterns on stream populations*, SIAM Rev., 47 (2005), pp. 749–772.
22. I. Mazari, G. Nadin, and Y. Privat, *Optimal location of resources maximizing the total population size in logistic models*, J. Math. Pures Appl. (9), 134 (2020), pp. 1–35.
23. ———, *Optimisation of the total population size for logistic diffusive equations: bang-bang property and fragmentation rate*, Communications in Partial Differential Equations, 47 (2022), pp. 797–828.
24. ———, *Shape optimization of a weighted two-phase Dirichlet eigenvalue*, Arch. Ration. Mech. Anal., 243 (2022), pp. 95–137.
25. I. Mazari and D. Ruiz-Balet, *A fragmentation phenomenon for a nonenergetic optimal control problem: Optimization of the total population size in logistic diffusive models*, SIAM Journal on Applied Mathematics, 81 (2021), pp. 153–172.
26. K. Nagahara, Y. Lou, and E. Yanagida, *Maximizing the total population with logistic growth in a patchy environment*, J. Math. Biol., 82 (2021), p. 50. Paper No. 2.
27. K. Nagahara and E. Yanagida, *Maximization of the total population in a reaction-diffusion model with logistic growth*, Calc. Var. Partial Differential Equations, 57 (2018), p. 14. Paper No. 80.
28. J. G. Skellam, *Random dispersal in theoretical populations*, Biometrika, 38 (1951), pp. 196–218.
29. D. C. Speirs and W. S. Gurney, *Population persistence in rivers and estuaries*, Ecology, 82 (2001), pp. 1219–1237.
30. O. Vasilyeva and F. Lutscher, *Population dynamics in rivers: analysis of steady states*, Can Appl Math Q, 18 (2011), pp. 439–469.

Chapter 6
Spreading in Homogeneous and Shifting Environments

Abstract In this chapter, we discuss the Fisher–KPP model in $\mathbb{R}^1$. First, we introduce the notion of spreading speed, and establish the classical result for homogeneous environments where the spreading speed is 2 times the square root of the product of the diffusion rate and local intrinsic growth rate. Next, we discuss the case of a shifting habitat which is motivated by the persistence of species in climate change. In the latter case, we determine the formula for the spreading speed, and show that the spreading speed is nonlocally determined. The main tool is the construction of several generalized super- and subsolutions by gluing elementary functions suitably.

6.1 The Fisher–KPP Equation and the Definition of Spreading Speed

Consider the Fisher–KPP equation in $\mathbb{R}$.

$$\begin{cases} \partial_t u - \partial_{xx} u = u(r(x,t) - u) & \text{for } x \in \mathbb{R},\ t > 0, \\ u(x,0) = u_0(x) & \text{for } x \in \mathbb{R}, \end{cases} \tag{6.1}$$

where $u_0 \in C(\mathbb{R})$ is bounded, nonnegative, nontrivial, and is compactly supported.

The above model was introduced by R. A. Fisher in a slightly different form in the study of the spreading of an advantageous gene in population genetics. It is also suitable for describing the invasion of a single population species, and in that case $r(x,t)$ denotes the intrinsic growth rate of the population. For simplicity, we have scaled the spatial variable so that the diffusion rate is taken to be 1.

In this chapter, we apply the method of generalized super/subsolutions to discuss the asymptotic spreading of the population. We introduce the notion of spreading speed to be the speed $c_* > 0$ such that an observer moving at a speed higher than c_* will outrun the population, whereas an observer moving at a slower speed will be outrun by the population.

K. -Y. Lam, Y. Lou, *Introduction to Reaction-Diffusion Equations*, Lecture Notes on Mathematical Modelling in the Life Sciences, https://doi.org/10.1007/978-3-031-20422-7_6

Definition 6.1.1 We say that a given population with density $u(x,t)$ has spreading speed $c_* > 0$ if the following statements hold:

$$\lim_{t\to\infty} \sup_{x>ct} |u(x,t)| = 0 \quad \text{for each } c \in (c_*, \infty),$$

and

$$\liminf_{t\to\infty} \inf_{0<x<ct} u(x,t) > 0 \quad \text{for each } c \in (0, c_*).$$

We will discuss the spreading speed of (6.1) in the following two cases:

- (Homogeneous environment) $r(x,t) \equiv r_0$ for some constant $r_0 > 0$;
- (Shifting environment) $r(x,t) = g(x - c_1 t)$ for some monotone function g and constant $c_1 > 0$.

6.2 A Maximum Principle for Unbounded Domains

We will estimate the spreading speed from above and below by constructing appropriate generalized super- and subsolutions. To this end, we need a weak maximum principle for unbounded domain. The proof is adapted from [52, Chapter 3, Section 6, Theorem 10].

Theorem 6.2.1 (Weak maximum principle for unbounded domains) *Let Ω be an unbounded domain in $\mathbb{R}^N$ with (possibly empty) smooth boundary $\partial\Omega$. Consider the operator $\mathcal{L} = -a^{ij}D_{ij} - b^i D_i - c$, where $a^{ij}, b^i \in C(\overline{\Omega} \times [0,T])$ are uniformly bounded, $\max_{\overline{\Omega}\times[0,T]} c < +\infty$, and a^{ij} are uniformly elliptic. If u is a generalized subsolution of*

$$\begin{cases} \partial_t u + \mathcal{L}u \le 0 & \text{in } \Omega \times (0,T), \\ u(x,t) \le 0 & \text{on } \partial\Omega \times (0,T), \\ u(x,0) \le 0 & \text{in } \Omega, \end{cases}$$

satisfying the decay condition

$$\liminf_{R\to\infty} \mathrm{e}^{-kR^2} \left[\max_{\substack{x\in\Omega,\ |x|=R \\ 0\le t\le T}} u(x,t) \right] \le 0 \tag{6.2}$$

for some positive constant k, then $u \le 0$ in $\Omega \times [0,T]$.

Remark 6.2.2 For the definition of generalized super- and subsolutions, see Definition 1.1.1 from Chapter 1.

Remark 6.2.3 Suppose $\Omega = \mathbb{R}^N$ (i.e. $\partial\Omega$ is empty) and u is uniformly bounded in x and t. Then it suffices to check the differential inequality $\partial_t u + \mathcal{L}u \le 0$ in $\Omega \times (0,T)$ and that the initial data satisfies $u(x,0) \le 0$ for all x.

Proof Write $x = (x_1, ..., x_N)$, $r^2 = \sum_{i=1}^N x_i^2$, and define the function

$$v(x,t) = u(x,t)e^{-\frac{k\gamma r^2}{\gamma - kt} - \beta t},$$

where k is the constant in (6.2), and β, γ are constants to be specified later. By a direct computation,

$$\begin{aligned} &\partial_t v - a^{ij} D_{ij} v - \left(b^i + \frac{4k\gamma}{\gamma - kt} a^{ij} x_j \right) D_i v - (c(x,t) + H(x,t))v \\ &= e^{-\frac{k\gamma r^2}{\gamma - kt} - \beta t} \left(\partial_t u - a^{ij} D_{ij} u - b^i D_i u - cu \right) \leq 0 \end{aligned} \tag{6.3}$$

in the generalized sense in $\Omega \times [0, \gamma/(2k)]$, where

$$H(x,t) = \frac{4k^2\gamma^2}{(\gamma - kt)^2} a^{ij} x_i x_j + \frac{2k\gamma}{\gamma - kt}(a^{ii} + b^i x_i) - \frac{k^2 \gamma r^2}{(\gamma - kt)^2} - \beta.$$

Next, we claim that

$$c(x,t) + H(x,t) < 0 \quad \text{in } \Omega \times \left[0, \tfrac{\gamma}{2k}\right], \tag{6.4}$$

provided $0 \leq \gamma \ll 1$ and $\beta \gg 1$. Indeed, fix $M > 0$ such that $a^{ij} x_i x_j \leq M r^2$ for all x, then

$$H(x,t) \leq -\frac{k^2 \gamma r^2}{(\gamma - kt)^2}(1 - 4\gamma M) + \frac{2k\gamma}{\gamma - kt}(a^{ii} + b_i x_i).$$

Then, for $t \in [0, \gamma/(2k)]$, we have $\gamma - kt > 0$ and

$$\begin{aligned} (c+H)(x,t) \leq &-\frac{k^2 \gamma r^2}{(\gamma - kt)^2}\left[1 - 4\gamma M - \frac{2(\gamma - kt)}{k} B\right] \\ &- \left[\beta - c - \frac{2k\gamma}{\gamma - kt}(A + a^{ii})\right], \end{aligned}$$

where

$$A = \sup_{0 \leq r \leq 1} |b^i x_i|, \quad \text{and} \quad B = \sup_{r \geq 1} \frac{b^i x_i}{r^2}.$$

Hence, we can choose $\gamma > 0$ small such that the expression of the first square bracket is positive, and then choose β large so that the expression in the second square bracket is positive, uniformly in $t \in [0, \gamma/(2k)]$. This proves (6.4).

Next, define for $R > 0$, the subdomain $\tilde{\Omega}_R = \{x \in \Omega : |x| < R\}$. Having proved that $c + H < 0$ everywhere, we can conclude that the generalized subsolution v of

$$v_t - a^{ij} D_{ij} v - \tilde{b}^i D_i v - (c + H)v \leq 0 \quad \text{in } \Omega \times (0, \gamma/(2k))$$

cannot attain a positive local maximum in the interior (see Lemma 1.1.6). Hence,

$$\sup_{\tilde{\Omega}_R \times (0,\gamma/(2k))} v \le \sup_{\substack{(x,t)\in\Omega \\ |x|=R}} \max\{v, 0\}, \tag{6.5}$$

where we used the fact that u and v are nonpositive on $\partial\Omega \times [0, T]$. Using (6.2), we pass to the limit in (6.5) along a sequence $R = R_k \to \infty$, and conclude that $v \le 0$ in $\Omega \times [0, \gamma/(2k)]$. We can then repeat the proof to prove that $v \le 0$ in $[0, m\gamma/(2k)] \cap [0, T]$ for $m = 2, 3....$ This completes the proof. □

Corollary 6.2.4 *Suppose that $\overline{u}$, $\underline{u}$ are continuous in $\overline{\Omega} \times [0, T]$ and are respectively generalized super- and subsolutions of*

$$u_t + \mathcal{L}u = F(x, t, u) \quad \text{in } \Omega \times (0, T), \tag{6.6}$$

where $\mathcal{L} = -a^{ij}D_{ij} - b^i D_i - c$ is as in Theorem 6.2.1, F is Hölder continuous in (x, t) and locally Lipschitz continuous in the second variable. If

$$\underline{u} \le \overline{u} \quad \text{on} \quad (\Omega \times \{0\}) \cup (\partial\Omega \times [0, T])$$

and both $\underline{u}$, $\overline{u}$ satisfy the decay condition

$$\liminf_{R\to\infty} \mathrm{e}^{-kR^2} \left[\max_{\substack{x\in\Omega,\ |x|=R \\ 0\le t\le T}} (\underline{u}(x, t) - \overline{u}(x, t)) \right] \le 0,$$

then

$$\underline{u} \le \overline{u} \quad \text{in } \Omega \times [0, T].$$

Proof It is straightforward to verify that $u = \underline{u} - \overline{u}$ is a generalized subsolution of

$$u_t - a^{ij}D_{ij}u - b^i D_i u - \tilde{c}u = 0 \quad \text{in } \Omega \times (0, T), \tag{6.7}$$

where $\tilde{c}(x, t) = c(x, t) + \int_0^1 \partial_s F(x, t, \tau\underline{u}(x, t) + (1-\tau)\overline{u}(x, t))\, \mathrm{d}\tau$ is uniformly bounded in $\Omega \times (0, T)$. It follows from Theorem 6.2.1 that $\underline{u} \le \overline{u}$ in $\Omega \times [0, T]$. □

Remark 6.2.5 In particular, Corollary 6.2.4 holds if the super- and subsolutions $\overline{u}$, $\underline{u}$ are uniformly bounded.

6.3 Homogeneous Environments

Throughout this section, we assume $r(x, t) \equiv r_0$ for some constant $r_0 > 0$.

Traveling wave solutions

It is well known that (6.1) admits a family of traveling wave solutions $\{U_c(x,t) : c \geq 2\sqrt{r_0}\}$ such that

$$U_c(x,t) = p(x - ct),$$

where $p(\xi)$ satisfies

$$\begin{cases} -cp' - p'' = p(r_0 - p) & \text{for } \xi \in \mathbb{R}, \\ p(-\infty) = r_0, \quad \text{and} \quad p(+\infty) = 0. \end{cases} \tag{6.8}$$

Moreover, for $c \in [0, 2\sqrt{r_0})$, there are no traveling wave solutions. In particular, $2\sqrt{r_0}$ is the minimal speed of traveling wave solution.

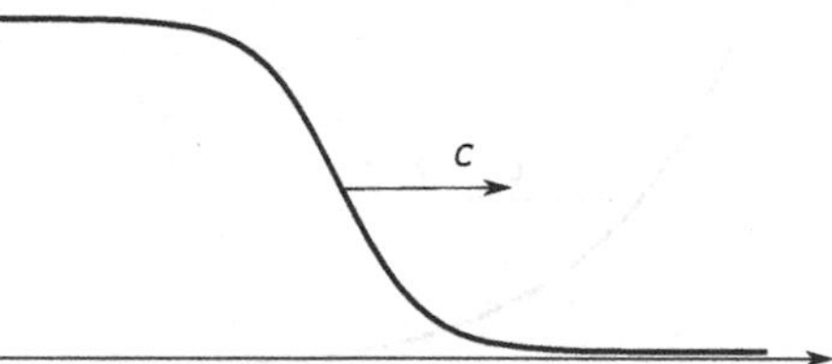

Fig. 6.1: A diagram illustrating the profile of a traveling solution $U_c(x,t)$.

Proof One may prove the existence/nonexistence of a traveling wave with the phase-plane method, by considering

$$p' = q, \quad q' = -cq - q(r_0 - q),$$

which is a system of two first-order differential equations. See, e.g. [66]. □

It is remarkable that, in homogeneous environments, the spreading speed c_* of (6.1) coincides with the minimal wave speed $2\sqrt{r_0}$; See [21, 37].

Theorem 6.3.1 *Suppose $r(x,t) \equiv r_0$ for some positive constant r_0, then $c_* = 2\sqrt{r_0}$.*

Remark 6.3.2 The coincidence of c_* with the minimal speed of traveling wave solutions holds in much greater generality; see [44, 63].

Remark 6.3.3 In fact, the solution initiating from nontrivial, compactly supported initial data converges to a traveling wave profile, after a logarithmic correction due to Bramson [12], i.e., there exists a C_0 depending on initial data, such that

$$\sup_{x>0} \left| u(x,t) - \phi\left(x - 2\sqrt{r_0}t + \frac{3}{2\sqrt{r_0}} \log t + C_0\right) \right| \to 0 \quad \text{as } t \to \infty.$$

Bramson's proof is based on probabilistic arguments. We refer to [28, 50] for a proof based on PDE arguments.

We will give a proof of Theorem 6.3.1 by constructing generalized super/sub-solutions.

Definition 6.3.4 For $r_0 > 0$ and $\lambda > 0$, we define

$$\phi_{r_0,\lambda}(x,t) = \mathrm{e}^{-\lambda(x-ct)}, \quad \text{where} \quad c = \lambda + \frac{r_0}{\lambda}.$$

Lemma 6.3.5 *If $r(x,t) \le r_0$, then $\phi_{r_0,\lambda}$ is a (classical) supersolution to* (6.1).

Proof Indeed, let $\phi(x,t) = \phi_{r_0,\lambda}(x,t)$, then

$$\phi_t - \phi_{xx} - \phi(r(x,t) - \phi) \ge \phi_t - \phi_{xx} - r_0\phi = \phi(\lambda c - \lambda^2 - r_0) = 0.$$

This completes the proof. □

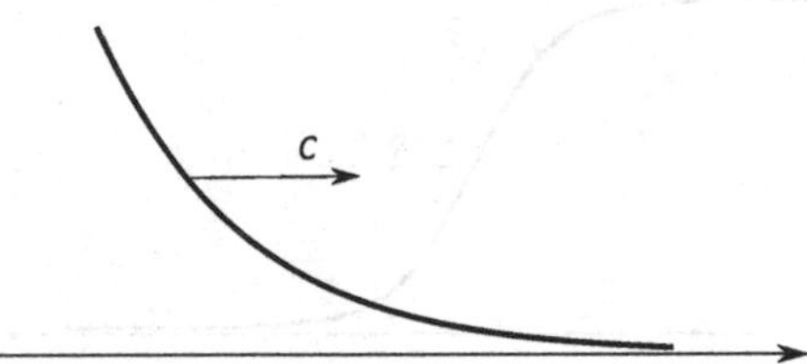

Fig. 6.2: A diagram illustrating $\phi_{r_0,\lambda}(x,t)$.

Definition 6.3.6 For $\alpha, c \ge 0$, and $R > 0$, define

$$\psi_{\alpha,c,R}(x,t) = \begin{cases} \exp\left(-\alpha t - \frac{c}{2}(x-ct)\right)\sin\left(\frac{x-ct}{R}\right) & \text{when } 0 < \frac{x-ct}{R} < \pi, \\ 0 & \text{otherwise.} \end{cases}$$

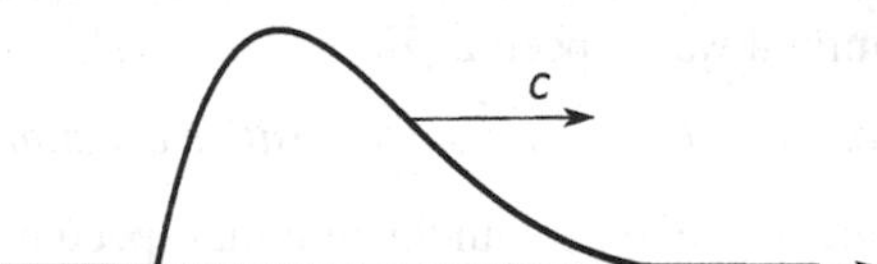

Fig. 6.3: A diagram illustrating $\psi_{\alpha,c,R}(x,t)$.

Lemma 6.3.7 *Suppose $r(x,t) \ge r_0$. For $\eta > 0$ such that $\alpha \ge -r_0 + \frac{c^2}{4} + \frac{1}{R^2} + \eta$, the function $\eta\psi = \eta\psi_{\alpha,c,R}$ is a generalized subsolution of* (6.1), *i.e.*

$$(\eta\psi)_t - (\eta\psi)_{xx} - (\eta\psi)(r(x,t) - \eta\psi) \le 0 \quad \text{in } \mathbb{R} \times (0,\infty)$$

in the generalized sense.

Proof Note that ψ is (locally) the maximum of $\exp\left(-\alpha t - \frac{c}{2}(x-ct)\right)\sin\left(\frac{x-ct}{R}\right)$ and the trivial solution, and the trivial solution is always a subsolution. According to the definition of generalized subsolutions (Definition 1.1.1), it remains to verify that $\exp\left(-\alpha t - \frac{c}{2}(x-ct)\right)\sin\left(\frac{x-ct}{R}\right)$ is a classical subsolution in the region $\left\{(x,t) : 0 < \frac{x-ct}{R} < \pi\right\}$. Denoting $\frac{x-ct}{R}$ by ξ, we have

$$
\begin{aligned}
&\psi_t - \psi_{xx} + \alpha\psi \\
&= \mathrm{e}^{-\alpha t - \frac{c}{2}(x-ct)}\left\{\left[\frac{c^2}{2}\sin\xi - \frac{c}{R}\cos\xi\right] - \left[\frac{c^2}{4}\sin\xi - \frac{c}{R}\cos\xi - \frac{1}{R^2}\sin\xi\right]\right\} \\
&= \psi\left(\frac{c^2}{4} + \frac{1}{R^2}\right).
\end{aligned}
$$

Hence,

$$
(\eta\psi)_t - (\eta\psi)_{xx} - \eta\psi(r(x,t) - \eta\psi) \le \eta\psi\left(-\alpha - r_0 + \frac{c^2}{4} + \frac{1}{R^2} + \eta\right) \le 0,
$$

where we used the fact that $0 \le \psi \le 1$ in the region where $\psi > 0$. □

Definition 6.3.8 Given $\lambda \in (0, \sqrt{r_0})$ and any $\epsilon > 0$ sufficiently small such that

$$
0 < \lambda < \sqrt{r_0 - \epsilon} \quad \text{and} \quad 0 < \epsilon < \frac{r_0 - \epsilon}{\lambda} - \lambda,
$$

let $c = c_\epsilon = \lambda + \frac{r_0 - \epsilon}{\lambda}$, and define

$$
\rho_{r_0,\lambda,\epsilon}(x,t) = \begin{cases} \mathrm{e}^{-\lambda(x-ct)} - \mathrm{e}^{-(\lambda+\epsilon)(x-ct)} & \text{when } x - ct > 0, \\ 0 & \text{otherwise.} \end{cases}
$$

Note that $\epsilon < c - 2\lambda$ and that $0 \le \rho_{r_0,\lambda,\epsilon} \le 1$.

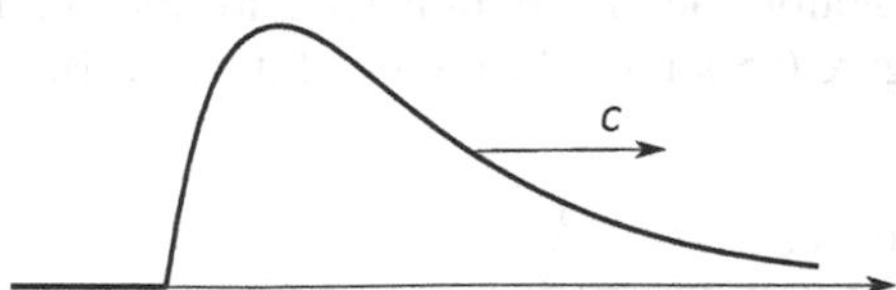

Fig. 6.4: A diagram illustrating $\rho_{r_0,\lambda,\epsilon}(x,t)$.

Lemma 6.3.9 *Suppose $r(x,t) \ge r_0$. For $\eta \in (0,\epsilon]$, the function $\eta\rho$, with $\rho = \rho_{r_0,\lambda,\epsilon}$ being defined above, is a generalized subsolution of* (6.1), *i.e.*

$$
(\eta\rho)_t - (\eta\rho)_{xx} - (\eta\rho)(r_0 - \eta\rho) \le 0 \quad \text{in } \mathbb{R} \times (0,\infty)
$$

in the generalized sense.

Proof Again, it suffices to prove the differential inequality in the region where $\rho > 0$.

$$\begin{aligned}&\rho_t - \rho_{xx} - (r_0 - \epsilon)\rho \\ &= \mathrm{e}^{-\lambda(x-ct)}[c\lambda - \lambda^2 - r_0 + \epsilon] - \mathrm{e}^{-(\lambda+\epsilon)(x-ct)}[c(\lambda+\epsilon) - (\lambda+\epsilon)^2 - r_0 + \epsilon] \\ &= -\mathrm{e}^{-(\lambda+\epsilon)(x-ct)}[c\epsilon - 2\lambda\epsilon - \epsilon^2] \\ &= -\mathrm{e}^{-(\lambda+\epsilon)(x-ct)}\epsilon[c - 2\lambda - \epsilon] \le 0.\end{aligned}$$

Hence,

$$\begin{aligned}(\eta\rho)_t - (\eta\rho)_{xx} - (r_0 - \eta\rho)\eta\rho &\le \eta\rho(-\epsilon + \eta\rho) \\ &\le \eta\rho(-\epsilon + \eta) \le 0,\end{aligned}$$

where we used the facts that $0 \le \rho \le 1$ and $0 < \eta \le \epsilon$. □

We are now in position to prove Theorem 6.3.1.

Proof (of Theorem 6.3.1) Let $r_0 > 0$ be a positive constant, and let $u(x,t)$ be a solution of

$$\begin{cases}\partial_t u - \partial_{xx} u = u(r_0 - u) & \text{for } x \in \mathbb{R},\ t > 0, \\ u(x,0) = u_0(x) & \text{for } x \in \mathbb{R},\end{cases} \tag{6.9}$$

where $u_0 \in C(\mathbb{R})$ is bounded, nonnegative, nontrivial, and compactly supported.

We first prove

$$\lim_{t\to\infty} \sup_{x>ct} |u(x,t)| = 0 \quad \text{for each } c \in (2\sqrt{r_0}, \infty), \tag{6.10}$$

Consider

$$\overline{u}(x,t) = \min\left\{C, \mathrm{e}^{-\sqrt{r_0}(x-2\sqrt{r_0}t)}\right\},$$

with the positive constant $C = \max\{r_0, \sup_{\mathbb{R}} u_0\}$. Since both C and $\mathrm{e}^{\sqrt{r_0}(x-2\sqrt{r_0}t)}$ are (classical) supersolutions, it follows that $\overline{u}$, being the minimum of the two, is a generalized supersolution. See Definition 1.1.1 for the definition of generalized supersolution and Figure 6.5 for an illustration of the construction.

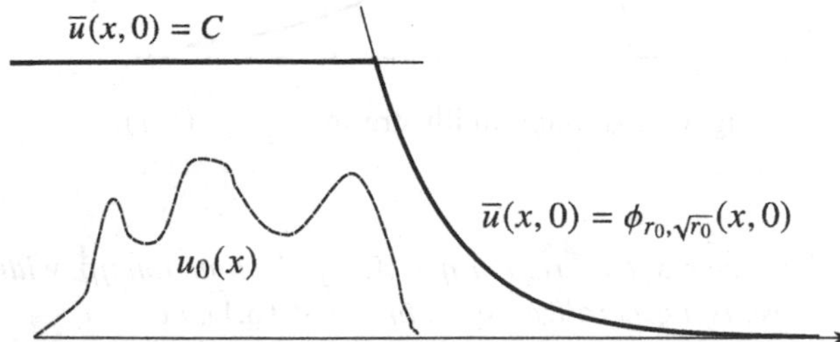

Fig. 6.5: A diagram illustrating the profile of the generalized supersolution $\overline{u}(x,t)$ (solid black line) relative to the compactly supported initial data $u_0(x)$ (dashed line).

Without loss of generality, we may translate u_0 to ensure

$$0 \le u_0(x) \le \overline{u}(x,0) \quad \text{in } \mathbb{R}.$$

By comparison (Corollary 6.2.4), we have

$$0 \le u(x,t) \le \overline{u}(x,t) \quad \text{for } x \in \mathbb{R},\ t > 0.$$

For $c \in (2\sqrt{r_0}, \infty)$, we have

$$\limsup_{t\to\infty} \sup_{x>ct} u(x,t) \le \limsup_{t\to\infty} \sup_{x>ct} \overline{u}(x,t) \le \limsup_{t\to\infty} \mathrm{e}^{-\sqrt{r_0}(c-2\sqrt{r_0})t} = 0.$$

Since $u(x,t) \ge 0$ for all x, t, we have proved (6.10).

It remains to prove a lower bound for the spreading speed, namely,

$$\liminf_{t\to\infty} \inf_{0<x<ct} u(x,t) > 0 \quad \text{for each } c \in (0, 2\sqrt{r_0}). \tag{6.11}$$

Since $c \in (0, 2\sqrt{r_0})$, we may choose $R \gg 1$ and $0 < \eta \ll 1$ (and $\alpha = 0$) such that

$$0 > -r_0 + \frac{c^2}{4} + \frac{1}{R^2} + \eta \quad \text{and} \quad 0 > -r_0 + \frac{1}{R^2} + \eta.$$

By Lemma 6.3.7, $\eta\psi_{0,0,R}$ and $\eta\psi_{0,c,R}$ are subsolutions of (6.9) for all sufficiently small $\eta > 0$. Define

$$\psi(x,t) = \begin{cases} \psi_{0,0,R}(x,t) & \text{for } 0 \le x < R\pi/2, \\ \max\left\{\psi_{0,0,R}(x,t), C, \psi_{0,c,R}(x,t)\right\} & \text{for } \pi/2 \le x < ct + R\pi/2, \\ \psi_{0,c,R}(x,t) & \text{for } ct + R\pi/2 \le x < ct + R\pi, \\ 0 & \text{otherwise,} \end{cases}$$

where the constant C is chosen such that $0 < C < \psi_{0,c,R}(ct + R\pi/2, t) < 1$. (See Figure 6.6 for an illustration of the construction.) Then $\eta\psi(x,t)$ is a generalized subsolution, provided $\eta > 0$ is sufficiently small.

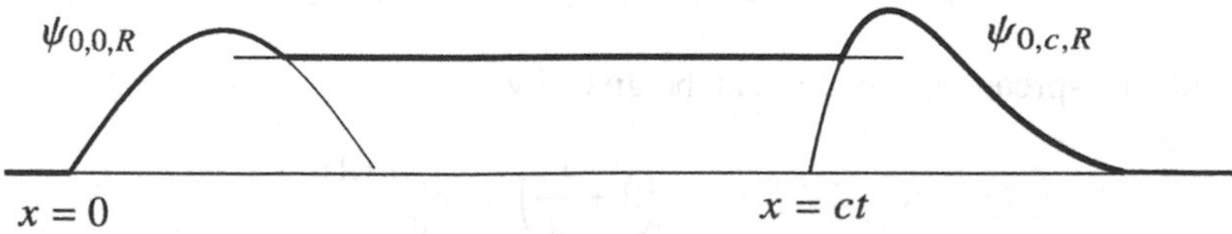

Fig. 6.6: A diagram illustrating the profile of the generalized subsolution $\psi(x,t)$, in solid black line.

Now, given a nonnegative, nontrivial solution $u(x,t)$ of (6.9), the strong maximum principal implies that $u(x,1) > 0$ for all $x \in \mathbb{R}$. Choose $\eta > 0$ small enough such that

$$u(x,1) \geq \eta\psi(x,1) \quad \text{for } x \in \mathbb{R}.$$

This is possible since $u(x,1)$ is bounded from below in the bounded interval on which $\psi(x,1)$ is supported. By the comparison principle, we obtain

$$u(x,t) \geq \eta\psi(x,t) \quad \text{for } x \in \mathbb{R},\ t > 1.$$

This proves (6.11). □

Periodically Varying Environments

Next, we will motivate the formula for spreading speed in spatially periodic environments.

Consider the Fisher–KPP equation where $r(x,t) = \tilde{r}(x)$ for some periodic function $\tilde{r}(x)$. In the case $\tilde{r}(x) \equiv r_0$, we have seen that

$$\tilde{u} = \mathrm{e}^{-\lambda x+\mu t}, \quad \text{with} \quad \mu = \lambda^2 + r_0,$$

satisfies the linearized equation

$$\tilde{u}_t - \tilde{u}_{xx} = r_0\tilde{u} \quad \text{for } x \in \mathbb{R},\ t \in \mathbb{R},$$

and satisfies, for each fixed time $t > 0$, the decay condition

$$\tilde{u}(x,t) \sim \mathrm{e}^{-\lambda x} \quad \text{for } x \gg 1.$$

In particular, $\mu = \lambda^2 + r_0$ is an eigenvalue, admitting an eigenfunction in L^∞, of the linear elliptic problem

$$-v_{xx} + 2\lambda v_x = (\lambda^2 + r_0)v - \mu v \quad \text{for } x \in \mathbb{R}.$$

Note that the spreading speed c_* can be given by

$$c_* = 2\sqrt{r_0} = \inf_{\lambda>0}\left(\lambda + \frac{r_0}{\lambda}\right) = \inf_{\lambda>0}\frac{\mu(\lambda)}{\lambda},$$

where $\mu(\lambda) = \lambda^2 + r_0$.

Now, let $\tilde{r}(x)$ be L-periodic in x. Consider, by analogy, the principal eigenvalue $\mu(\lambda)$ of the periodic eigenvalue problem

$$\begin{cases} -\psi_{xx} + 2\lambda\psi_x = (\lambda^2 + r(x))\psi - \mu(\lambda)\psi & \text{for } x \in \mathbb{R}, \\ \psi(x + L) = \psi(x) & \text{for } x \in \mathbb{R}. \end{cases}$$

The above discussion suggests the following variational formula for the spreading speed

$$c_* = \inf_{\lambda > 0} \frac{\mu(\lambda)}{\lambda}. \tag{6.12}$$

See [7, 22, 24, 64].

6.4 Shifting Environments

Motivated by the poleward movement of isotherms, Potapov and Lewis [51] and subsequently Berestycki et al. [5] considered the effect of climate change on ecosystem dynamics. A list of key problems was given in [5], such as the effect of species mobility and speed of climate shift on the survival/extinction of the species; and to determine the size and form of its population profile if the species survives. Below we briefly discuss one of their results.

Shifting environments with a moving source patch

The following specific problem was considered in [5, 51]:

$$\begin{cases} u_t - u_{xx} = u(1 - u) & \text{for } x \in [c_1 t, c_1 t + L],\ t > 0, \\ u_t - u_{xx} = -\kappa u & \text{for } x \notin [c_1 t, c_1 t + L],\ t > 0, \end{cases} \tag{6.13}$$

where $1, \kappa > 0$ are positive constants. The above model describes a population whose viable habitat is of size L and is moving to the right at the speed c_1. The population has an intrinsic growth rate 1 within the moving patch, and is exposed to a death rate κ outside of that patch. By considering the moving frame $y = x - c_1 t$, we can study the population dynamics of a reaction-diffusion model with temporally stationary coefficients. In fact, it is shown that the long-time dynamics is equivalent to the following bounded domain problem

$$\begin{cases} v_t - v_{yy} - c_1 v_y = v(1 - v) & \text{for } y \in [0, L],\ t > 0, \\ v_y - k_+ v = 0 & \text{for } y = 0,\ t > 0, \\ v_y - k_- v = 0 & \text{for } y = L,\ t > 0, \end{cases}$$

where $k_\pm = \frac{-c_1 \pm \sqrt{c_1^2 + 4\kappa}}{2}$.

By the results in Chapter 5, the population can persist on the moving patch if and only if $\mu_1 < 0$, where $\mu_1 = \mu_1(c_1, L, \kappa)$ is the principal eigenvalue of

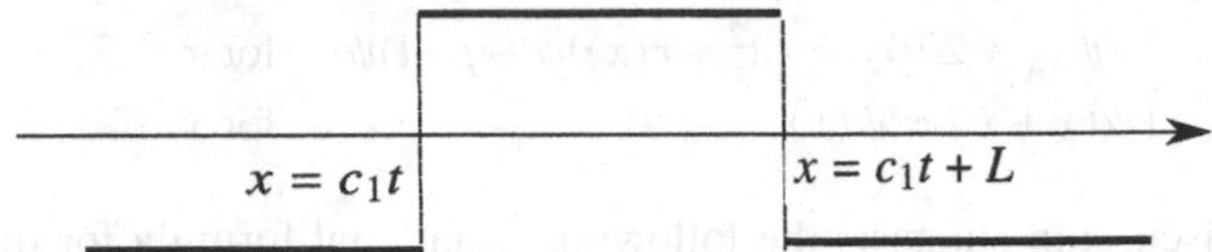

Fig. 6.7: A diagram illustrating the profile of the instrinsic growth rate in a shifting habitat with speed c_1.

$$\begin{cases} -\psi'' - c_1\psi' = \psi + \mu_1\psi & \text{in } 0 < y < L, \\ \psi'(0) - k_+\psi(0) = \psi'(L) - k_-\psi(L) = 0. \end{cases}$$

By a phase plane analysis, it is proved that there exists a critical domain size $L_{crit} > 0$ such that the population persists if and only if

$$L > L_{crit} := \frac{1}{\sqrt{1-(c_1/2)^2}} \arctan\left(\frac{2\sqrt{1-(c_1/2)^2}\sqrt{1/\kappa + (c_1/2)^2}}{1-\kappa - c_1^2/2} \right).$$

See [5] for details.

Shifting boundary connecting an unbounded sink and an unbounded source patch

Consider the situation where the favorable region is an unbounded interval that is retreating at a speed of c_1. This can be modeled by (6.1) by taking $r(x,t) = g(x-c_1 t)$, where $c_1 > 0$ and g is a strictly increasing function. In this case, (6.1) becomes

$$\begin{cases} u_t - u_{xx} = u(g(x-c_1 t) - u) & \text{for } x \in \mathbb{R},\ t > 0, \\ u(x,0) = u_0(x) & \text{for } x \in \mathbb{R}, \end{cases} \tag{6.14}$$

where u_0 is nonnegative, nontrivial, and compactly supported in $\mathbb{R}$.

The authors of [41] first introduced the above problem in the case $g(-\infty) < 0 < g(+\infty)$.

Theorem 6.4.1 *Suppose $g : \mathbb{R} \to \mathbb{R}$ is bounded and increasing, and such that*

$$r_1 := g(-\infty) < 0 \quad \text{and} \quad r_2 := g(+\infty) > 0.$$

Let u be a solution of (6.14) with nonnegative, nontrivial and compactly supported initial data. Then the following statements hold.

(a) *If $c_1 > 2\sqrt{r_2}$, then the population goes to extinction, i.e.*

$$\lim_{t\to\infty} \sup_{x\in\mathbb{R}} u(x,t) = 0.$$

(b) *If* $0 < c_1 < 2\sqrt{r_2}$, *then the population persists, and spreads at* $2\sqrt{r_2}$, *i.e.*

$$\begin{cases} \lim_{t\to\infty} \sup_{x>(2\sqrt{r_2}+\epsilon)t} u(x,t) = 0 & \textit{for each } \epsilon > 0, \\ \liminf_{t\to\infty} \inf_{(c_1+\epsilon)t<x<(2\sqrt{r_2}-\epsilon)t} u(x,t) > 0 & \textit{for each } 0 < \epsilon < (2\sqrt{r_2} - c_1)/2. \end{cases}$$

Remark 6.4.2 Theorem 6.4.1 was proved in [41] under the additional assumption that g is also continuous and piecewise differentiable. Here we give an elementary proof that relaxes these two assumptions. Case (b) of Theorem 6.4.1 is depicted in Figure 6.8.

Remark 6.4.3 We conjecture that the assertion (a) holds also in the critical case when $c_1 = 2\sqrt{r_2}$; see Problem 6.5.1.

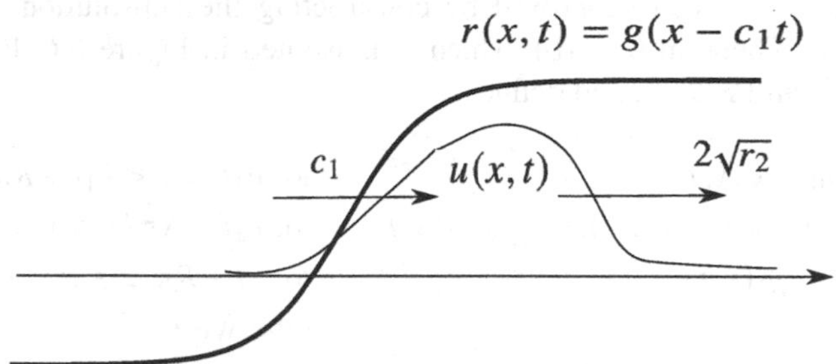

Fig. 6.8: A diagram illustrating the profile of a persisting and spreading population in the case $0 < c_1 < 2\sqrt{r_2}$.

Proof For simplicity, we consider the special case when g is a step function, i.e.

$$r(x,t) = r_2 \quad \text{if } x \geq c_1 t, \quad \text{and} \quad r(x,t) = r_1 \quad \text{if } x < c_1 t.$$

The general case $r(x,t) = g(x - c_1 t)$ can be treated by approximating the profile from above and below by step functions, and taking the limit; see Problem 6.5.3.

First, we prove assertion (a) by constructing the supersolution which is similar as in the proof of Theorem 6.3.1, as illustrated in Figure 6.5. Let

$$\delta = \min\{\sqrt{r_2}(c_1 - 2\sqrt{r_2}), -r_1/2\},$$

and define

$$\begin{aligned} \overline{u}(x,t) &= \min\left\{ e^{-\sqrt{r_2}(x-2\sqrt{r_2}t)}, Ce^{-\delta t} \right\} \\ &= \begin{cases} e^{-\sqrt{r_2}(x-2\sqrt{r_2}t)} & \text{for } x/t \geq 2\sqrt{r_2} + \frac{\delta - \log C}{\sqrt{r_2}}, \\ Ce^{-\delta t} & \text{for } x/t < 2\sqrt{r_2} + \frac{\delta - \log C}{\sqrt{r_2}}, \end{cases} \end{aligned}$$

where $C = \max\{e^{2\delta}, \|u_0\|_{L^\infty(\mathbb{R})}\}$. Using the fact that

$$r(x,t) = r_1 < 0 \quad \text{in } \{(x,t) : \overline{u}(x,t) = Ce^{-\delta t}\} \subseteq \{(x,t) : x < c_1 t\},$$

one can verify that $k\overline{u}$ is a generalized supersolution of (6.14) for any $k \geq 1$. By choosing $k \geq 1$ suitably large, one can arrange that

$$0 \leq u(x,0) \leq k\overline{u}(x,0) \quad \text{for } x \in \mathbb{R}.$$

By comparison (Corollary 6.2.4), we have

$$0 \leq u(x,t) \leq k\overline{u}(x,t) \quad \text{for } x \in \mathbb{R},\ t > 0.$$

Since the function on the right-hand side tends to zero uniformly in $x \in \mathbb{R}$, we obtain assertion (a).

The assertion (b) can be proved by constructing the subsolution similarly as in the proof of Theorem 6.3.1(b), which is illustrated in Figure 6.6. Precisely, fix $c \in (c_1, 2\sqrt{r_2})$ and $R \gg 1$, and define

$$\psi(x,t) = \begin{cases} \psi_{0,c_1,R}(x,t) & \text{for } c_1 t \leq x < c_1 t + R\pi/2, \\ \max\left\{\psi_{0,c_1,R}(x,t), C, \psi_{0,c,R}(x,t)\right\} & \text{for } c_1 t + R\pi/2 \leq x < ct + R\pi/2, \\ \psi_{0,c,R}(x,t) & \text{for } ct + R\pi/2 \leq x < ct + R\pi, \\ 0 & \text{otherwise,} \end{cases}$$

where the constant C is chosen such that

$$0 < C < \min\{\psi_{0,c_1,R}(c_1 t + R\pi/2, t), \psi_{0,c,R}(ct + R\pi/2, t)\} < 1.$$

Then for $\eta > 0$ sufficiently small, $\underline{u}(x,t) = \eta\psi(x,t)$ is a generalized subsolution. One can then argue by the comparison principle to obtain assertion (b). □

Remark 6.4.4 In [68], the above result is considerably generalized to abstract monotone dynamical systems that are "sandwiched" between two limiting homogeneous KPP systems as $x \to \pm\infty$, where the one at $+\infty$ has positive growth rate, and the other one has negative growth rate. This work, which is based on the method of recursion [44, 63], includes Theorem 6.4.1 as a special case.

Shifting boundary connecting two unbounded source patches and nonlocally pulling

In this section, we consider the case $r(x,t) = g(x - c_1 t)$, where g is increasing, and

$$0 < r_1 := g(-\infty) < r_2 := g(+\infty). \tag{6.15}$$

It turns out that the determination of spreading speed is more delicate in this case. We first make several observations:

- Since $r(x,t)$ is bounded from above and below by positive constants, i.e, $r_1 \le r(x,t) \le r_2$, it is clear that the spreading speed c_* of (6.14), if it exists, must satisfy
$$2\sqrt{r_1} \le c_* \le 2\sqrt{r_2}.$$
- Suppose $c_1 < 2\sqrt{r_2}$, then the proof of Theorem 6.4.1(b) can be repeated to deduce that the population will spread at speed $2\sqrt{r_2}$.
- Suppose $c_1 > 2\sqrt{r_2}$. Since the population cannot migrate faster than $2\sqrt{r_2}$, we have
$$c_* \le 2\sqrt{r_2} < c_1,$$
i.e. it will fail to keep up with the good patch with larger growth rate r_2. See Figure 6.9 for an illustration.
- If c_1 is sufficiently large, i.e. the better patch retreats at a very fast speed, it is then expected that the species cannot take advantage of the better patch and will spread at speed $2\sqrt{r_1}$.
- Suppose the spreading speed c_* exists for all $c_1 \ge 0$, then one can prove by standard comparison that c_* is a nonincreasing function of c_1. In particular, there must be a transition between the minimum value of $c_* = 2\sqrt{r_1}$ and the maximum value of $c_* = 2\sqrt{r_2}$.

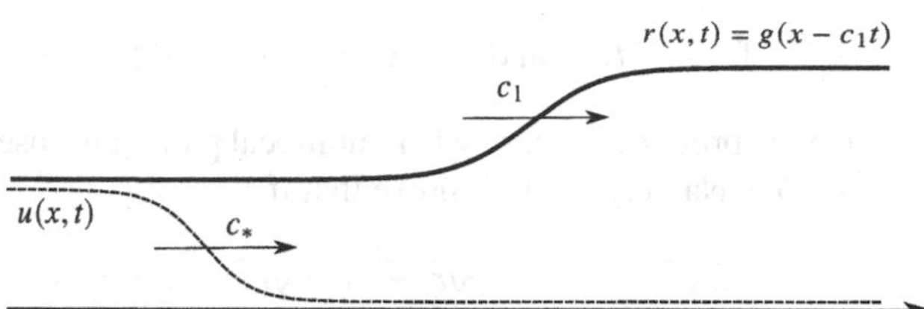

Fig. 6.9: A diagram illustrating the profile of a persisting and spreading population in the case $0 < c_* < c_1$. The sold line is the growth rate of the habitat moving at speed c_1. The dashed line is the population density of the species which is invading at speed c_*.

If $c_* < c_1$, then the population cannot keep up with the good patch. It turns out that, in some parameter regime, the presence of a good habitat at the large space horizon can enhance the invasion speed of the population. This phenomenon was first discovered by Holzer and Scheel [29], and is coined as "nonlocal pulling" in [25]. Roughly speaking, the presence of the better patch at the large space horizon can modify the decay rate of the population. In some cases, this allows the selection of an invasion speed that corresponds to a supercritical wave speed, which is strictly greater than $2\sqrt{r_1}$. The speed is determined as follows, based on the results in [39].

Theorem 6.4.5 *Let $g : \mathbb{R} \to \mathbb{R}$ be bounded and increasing, such that* (6.15) *holds. Then the model* (6.14) *admits a spreading speed c_* which is given by*

$$c_* = \begin{cases} 2\sqrt{r_2} & \text{if } 0 \le c_1 \le 2\sqrt{r_2}, \\ \frac{c_1}{2} - \sqrt{r_2 - r_1} + \frac{r_1}{\frac{c_1}{2} - \sqrt{r_2 - r_1}} & \text{if } 2\sqrt{r_2} < c_1 < 2(\sqrt{r_2 - r_1} + \sqrt{r_1}), \\ 2\sqrt{r_1} & \text{if } c_1 \ge 2(\sqrt{r_2 - r_1} + \sqrt{r_1}). \end{cases}$$

Remark 6.4.6 If $c_1 \notin (2\sqrt{r_2}, 2(\sqrt{r_2 - r_1} + \sqrt{r_1}))$, then the spreading speed $c_* = 2\sqrt{r_i}$ and is determined by the environment where the invasion front is located, i.e. the invasion is *locally pulled*. If $c_1 \in (2\sqrt{r_2}, 2(\sqrt{r_2 - r_1} + \sqrt{r_1}))$, then the invasion front is located at the patch with growth rate r_1, but $c_* > 2\sqrt{r_1}$. In other words, the spreading speed is enhanced by the better patch which is at a distance of $O(t)$ ahead of the invasion front, i.e. the invasion front is *nonlocally pulled*. We refer to [23] for a definition of pulled versus pushed fronts in terms of "inside-dynamics". See [25] for a discussion of nonlocally pulled fronts in competition models and also [15] for a related discussion in predator-prey models.

Proof (of Theorem 6.4.5) We will present a proof based on constructing generalized super- and subsolutions by adapting the ideas in [25]. The results in [39] hold under more general assumptions, but are based on a different approach involving the WKB-ansatz and Hamilton–Jacobi equations [17]. We leave to the reader (Problem 6.5.3) to show that the general case can be reduced to the case when g is piecewise constant. Namely,

$$r(x,t) = r_1 \quad \text{if } x < c_1 t, \quad \text{and} \quad r(x,t) = r_2 \quad \text{if } x \ge c_1 t. \tag{6.16}$$

Also, we only sketch the proof of the case where nonlocal pulling is observed, which is depicted in Figure 6.9. Namely, we will prove that if

$$2\sqrt{r_2} < c_1 < 2(\sqrt{r_2 - r_1} + \sqrt{r_1}),$$

then the spreading speed c_* is given by

$$c_* = \lambda_* + \frac{r_1}{\lambda_*}, \quad \text{where} \quad \lambda_* = \frac{c_1}{2} - \sqrt{r_2 - r_1} \in (0, \sqrt{r_1}).$$

First, we construct the generalized supersolution. Define

$$\begin{aligned} \overline{u}(x,t) &= \min\left\{r_1, \phi_{r_1,\lambda_*}(x,t), \phi_{r_2,c_1/2}(x,t)\right\} \\ &= \begin{cases} r_1 & \text{for } x \le \left(\lambda_* + \frac{r_1}{\lambda_*}\right)t - \frac{1}{\lambda_*}\log r_1, \\ \exp\left(-\lambda_* x + (\lambda_*^2 + r_1)t\right) & \text{for } \left(\lambda_* + \frac{r_1}{\lambda_*}\right)t - \frac{1}{\lambda_*}\log r_1 < x \le c_1 t, \\ \exp\left(-\frac{c_1}{2}x + \left(\frac{c_1^2}{4} + r_2\right)t\right) & \text{for } x > c_1 t. \end{cases} \end{aligned} \tag{6.17}$$

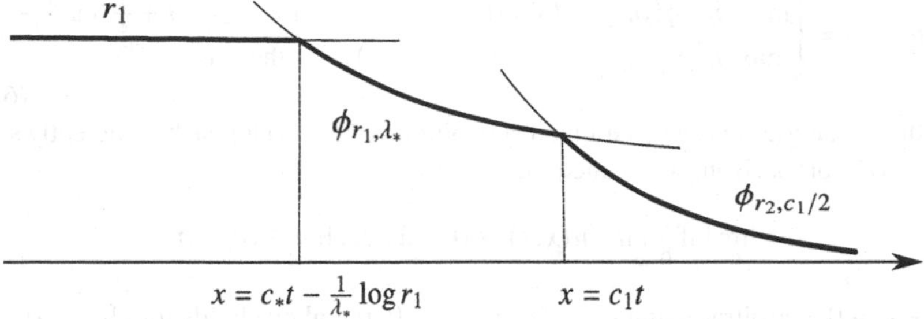

Fig. 6.10: A diagram illustrating the profile of the generalized supersolution $\overline{u}(x,t)$ (solid black line) in the proof of Theorem 6.4.5. Here $\lambda_* = \frac{c_1}{2} - \sqrt{r_2 - r_1}$ and $c_* = \lambda_* + \frac{r_1}{\lambda_*}$.

Since $\overline{u}$ is defined as the minimum among three smooth functions, it suffices to check that each of the smooth functions is a classical supersolution in the respective regions in the (x,t) plane in which they coincide with $\overline{u}(x,t)$. We omit the details, except to note that the choice of exponents $(\lambda_1, \lambda_2) = (\lambda_*, c_1/2)$ is the optimum among all admissible choices of $(\lambda_1, \lambda_2) \in (0, \sqrt{r_1}] \times (0, \sqrt{r_2}]$ to produce the slowest possible supersolution. By comparison, we deduce that

$$\lim_{t\to\infty} \sup_{x>ct} u(x,t) = 0 \quad \text{for each } c > c_*,$$

i.e. c_* is an upper bound of the spreading speed.

Next, we construct a generalized subsolution to show that the population spreads at least as fast as c_*. Recall that $\lambda_* = \frac{c_1}{2} - \sqrt{r_2 - r_1}$. Fix an arbitrarily small $\epsilon > 0$, define

$$c_\epsilon = \lambda_* + \frac{r_0 - \epsilon}{\lambda_*}, \quad \text{and} \quad \alpha = \lambda_*(c_1 - c_\epsilon). \tag{6.18}$$

We leave it for the reader to verify that

$$\alpha > -r_2 + \frac{c_1^2}{4}.$$

Granted, then we may take $R \gg 1$ such that

$$\alpha > -r_2 + \frac{c_1^2}{4} + \frac{1}{R^2}. \tag{6.19}$$

By our choices of parameters, it follows from Lemmas 6.3.7 and 6.3.9 that for $\epsilon, \eta > 0$ small, $\eta^3 C$, $\eta^2 \rho_{r_1,\lambda_*,\epsilon}$, and $\eta\psi_{\alpha,c_1,R}$ are subsolutions. Now, we glue them together in the following way, as illustrated in Figure 6.11:

$$\underline{u}(x,t) = \begin{cases} \max\{\eta^3, \eta^2 \rho_{r_1,\lambda_*,\epsilon}(x,t)\} & \text{for } x \le c_\epsilon t + \frac{1}{\epsilon}\log\frac{\lambda+\epsilon}{\lambda}, \\ \max\{\eta^2 \rho_{r_1,\lambda_*,\epsilon}(x,t), \eta\psi_{\alpha,c_1,R}(x,t)\} & \text{otherwise.} \end{cases} \tag{6.20}$$

It follows that $\underline{u}(x,t)$ is a generalized subsolution of (6.14) for each sufficiently small $\eta > 0$. By comparison, we deduce that

$$\liminf_{t\to\infty} \inf_{0<x<ct} u(x,t) > 0 \quad \text{for each } c \in (0, c_\epsilon).$$

Since $\epsilon > 0$ is arbitrary, and $c_\epsilon \nearrow c_*$ as $\epsilon \searrow 0$, the above holds for all $c \in (0, c_*)$. The proof of the theorem is finished. □

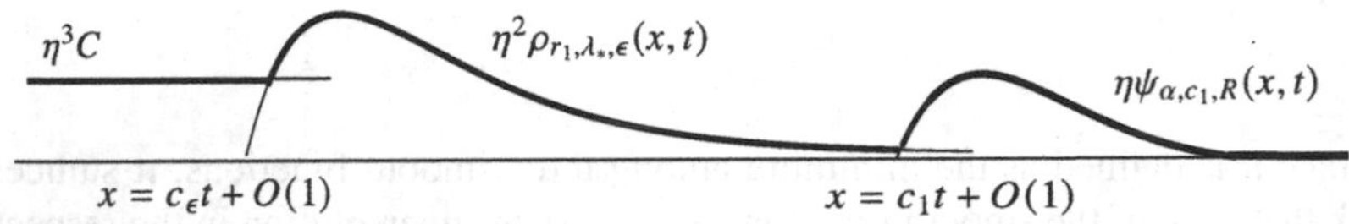

Fig. 6.11: A diagram illustrating the profile of the generalized subsolution $\underline{u}(x,t)$ (solid black line) in the proof of Theorem 6.4.5. Here $\lambda_* = \frac{c_1}{2} - \sqrt{r_2 - r_1}$ and $c_\epsilon = \lambda_* + \frac{r_1 - \epsilon}{\lambda_*}$.

6.5 Further Reading

The pioneering works of [21, 37] have inspired extensive further research in the asymptotic spreading of populations, including extension to heterogeneous environments, and more recently, to shifting environments. The definition of spreading speed, or the asymptotic speed of spread, was originally introduced in [2, 3] for reaction-diffusion equations, and in [1] for integral-differential equations. Reaction-diffusion models with various kinds of delays [34, 35, 36, 39, 58, 65] can also be treated by transforming them into integral equations [14, 53, 57], where the definition also applies.

For spatially periodic environments, Weinberger [63, 64] introduced an elaborate method and showed the existence of spreading speeds for discrete-time recursions. Subsequently, the general theory on the existence of spreading speeds and its coincidence with the minimal wave speed was developed in [44] for monotone dynamical systems and [19] for the time-space periodic monotone systems with weak compactness. Using the super- and subsolutions method and the principal eigenvalue of time-space periodic parabolic problems, spreading properties in time-space periodic and more general environments are investigated in [7], as well as in [55]. See also [26, 62] and references therein for spreading results concerning models with nonlocal diffusion and/or delays.

The Hamilton–Jacobi approach for analyzing asymptotic spreading was introduced by Freidlin and Gärtner [22, 24], who derived the formula (6.12) for the spreading speed in periodic environments based on probabilistic arguments. Later, Evans and Souganidis [17] provided a proof by PDE arguments. Based on this approach and homogenization ideas, Berestycki and Nadin [8, 9] showed the existence of spreading speeds for spatially almost periodic, random stationary ergodic, and other general heterogeneous environments. In these works, the spreading speed is expressed as a minimax formula in terms of suitable notions of generalized principal eigenvalues in unbounded domains. See also [45] for the related result on KPP models with nonlocal diffusion.

The mathematical work on shifting environments started with [5, 51], and has stimulated significant further research; see the survey article [59]. We also mention [6, 11] for works concerning the existence of forced traveling waves; and [43, 60, 61] for works concerning nonlocal diffusion models in shifting environments.

The authors in [41] considered monotone habitats whose favorable part shrinks through time. Specifically, they considered the model (6.14) with growth rate $r(x,t) = g(x - c_1 t)$ with g being an increasing function satisfying $g(-\infty) < 0 < g(+\infty)$. If $c_1 = 0$, then $g(x)$ can be seen as the spatially varying baseline for population growth. For $c_1 > 0$, the source region $\{x \in \mathbb{R} : \ g(x - c_1 t) > 0\}$ shrinks relative to the sink region $\{x \in \mathbb{R} : \ g(x - c_1 t) < 0\}$ as t increases. The existence of forced traveling wave solutions was considered in [18, 32]. Similar results are obtained in [31] for the degenerate case when $g(-\infty) = 0 < g(+\infty)$.

Based on the fact that habitat degradation due to climate change may not always drive the species to extinction, one is led to consider (6.14) in the case $0 < g(-\infty) < g(+\infty)$. See [6, 10, 30, 39, 67]. In particular, it is proved in [39] that, when $r(x,t) = \sum_{i=1}^{N} g_i(x - c_i t)$, where g_i are bounded, positive and increasing functions, the spreading speed can be characterized by

$$c_* = \sup\{s > 0 : \hat{\rho}(s) = 0\},$$

where $\hat{\rho}(s)$ is the unique viscosity solution of the following Hamilton–Jacobi equation

$$\begin{cases} \min\{\rho, \rho - s\rho' + |\rho'|^2 + R(s)\} = 0 \quad \text{in } (0,\infty), \\ \rho(0) = 0, \quad \text{and} \quad \rho(s)/s \to \infty \ \text{ as } s \to \infty, \end{cases}$$

where $R\left(\frac{x}{t}\right) = \lim_{\epsilon \to 0} r\left(\frac{x}{\epsilon}, \frac{t}{\epsilon}\right)$. For an introduction to the basic theory of viscosity solutions, see [4].

It turns out that in a shifting environment connecting two favorable habitats, i.e. $0 < g(-\infty) < g(+\infty)$, the spreading speed may be *nonlocally pulled* (see Theorem 6.4.5). This phenomenon was first found in [29], in a slightly different setting of staged invasion waves. The name *nonlocal pulling* was introduced in [25]. See also [20] for the case of shifting diffusivity.

Finally, we mention some works on multiple species problems. The spreading property of two competing species was first established in [40, 42], concerning the invasion speed of a novel species into the territory of a well-established species.

See also [33] for the existence of the stacked invasion fronts in monotone reaction-diffusion systems when all components have the same diffusion rate.

Motivated by the co-invasion of tree species into the North American continent at the end of last ice age, Shigesada et al. [56] conjectured, with numerical evidence, the existence of stacked invasion fronts when there are N competing species with compactly supported initial data. For the resolution of this question when $N = 2$, see [13] for the bistable case, and [25, 46, 47] for the monostable case. Recently, progress has also been obtained for multiple-species systems without monotonicity structure [15, 16, 48]. It turns out that nonlocal pulling also takes place in the competition of two or three-species having different motility rate, even when the environment is homogeneous. This is because the faster invading species modifies and creates a virtually shifting environment for the slower species. For the existence and qualitative properties of entire solutions consisting of co-invasion waves, see [27, 38, 49, 54].

Problems

6.5.1 Prove that the conclusion of assertion (a) of Theorem 6.4.1 holds also when $c_1 = 2\sqrt{r_2}$. Namely, the solution u of (6.14) with nonnegative, nontrivial and compactly supported initial data satisfies

$$\lim_{t\to\infty} \sup_{x\in\mathbb{R}} u(x,t) = 0.$$

6.5.2 Show the remaining two cases of Theorem 6.4.5, where $c_* = 2\sqrt{r_1}$ or $c_* = 2\sqrt{r_2}$.

6.5.3 In Section 6.4, we have proved Theorem 6.4.5 under the assumption that $r(x,t)$ satisfies (6.16), i.e. it is piecewise constant. Derive Theorem 6.4.5 for $r(x,t) = g(x - c_1 t)$ with arbitrary increasing function $g : \mathbb{R} \to \mathbb{R}$ such that $0 < \inf g < \sup g$. [Hint: For each $\delta > 0$, there exist x_δ and y_δ such that

$$(\inf g - \delta) + \chi_{(x_\delta,\infty)}(\sup g - \inf g) \le g(s) \le (\inf g + \delta) + \chi_{(y_\delta,\infty)}(\sup g - \inf g)$$

for all $s \in \mathbb{R}$.]

References

1. D. G. Aronson, *The asymptotic speed of propagation of a simple epidemic*, in Nonlinear diffusion (NSF-CBMS Regional Conf. Nonlinear Diffusion Equations, Univ. Houston, Houston, Tex., 1976), 1977, pp. 1–23. Res. Notes Math., No. 14.
2. D. G. Aronson and H. F. Weinberger, *Nonlinear diffusion in population genetics, combustion, and nerve pulse propagation*, in Partial differential equations and related topics (Program, Tulane Univ., New Orleans, La., 1974), 1975, pp. 5–49. Lecture Notes in Math., Vol. 446.

3. ———, *Multidimensional nonlinear diffusion arising in population genetics*, Adv. in Math., 30 (1978), pp. 33–76.
4. G. Barles, *An introduction to the theory of viscosity solutions for first-order Hamilton-Jacobi equations and applications*, in Hamilton-Jacobi equations: approximations, numerical analysis and applications, vol. 2074 of Lecture Notes in Math., Springer, Heidelberg, 2013, pp. 49–109.
5. H. Berestycki, O. Diekmann, C. J. Nagelkerke, and P. A. Zegeling, *Can a species keep pace with a shifting climate?*, Bull. Math. Biol., 71 (2009), pp. 399–429.
6. H. Berestycki and J. Fang, *Forced waves of the Fisher-KPP equation in a shifting environment*, J. Differential Equations, 264 (2018), pp. 2157–2183.
7. H. Berestycki, F. Hamel, and G. Nadin, *Asymptotic spreading in heterogeneous diffusive excitable media*, J. Funct. Anal., 255 (2008), pp. 2146–2189.
8. H. Berestycki and G. Nadin, *Spreading speeds for one-dimensional monostable reaction-diffusion equations*, J. Math. Phys., 53 (2012), pp. 115619, 23.
9. H. Berestycki and G. Nadin, *Asymptotic spreading for general heterogeneous Fisher-KPP type equations*, Mem. Amer. Math. Soc., (2022). (in press).
10. J. Bouhours and T. Giletti, *Spreading and vanishing for a monostable reaction-diffusion equation with forced speed*, J. Dynam. Differential Equations, 31 (2019), pp. 247–286.
11. J. Bouhours and G. Nadin, *A variational approach to reaction-diffusion equations with forced speed in dimension 1*, Discrete Contin. Dyn. Syst., 35 (2015), pp. 1843–1872.
12. M. Bramson, *Convergence of solutions of the Kolmogorov equation to travelling waves*, Mem. Amer. Math. Soc., 44 (1983), pp. iv+190.
13. C. Carrère, *Spreading speeds for a two-species competition-diffusion system*, J. Differential Equations, 264 (2018), pp. 2133–2156.
14. O. Diekmann, *Run for your life. A note on the asymptotic speed of propagation of an epidemic*, J. Differential Equations, 33 (1979), pp. 58–73.
15. A. Ducrot, T. Giletti, J.-S. Guo, and M. Shimojo, *Asymptotic spreading speeds for a predator-prey system with two predators and one prey*, Nonlinearity, 34 (2021), pp. 669–704.
16. A. Ducrot, T. Giletti, and H. Matano, *Spreading speeds for multidimensional reaction-diffusion systems of the prey-predator type*, Calc. Var. Partial Differential Equations, 58 (2019), p. 34. Paper No. 137.
17. L. C. Evans and P. E. Souganidis, *A PDE approach to geometric optics for certain semilinear parabolic equations*, Indiana Univ. Math. J., 38 (1989), pp. 141–172.
18. J. Fang, Y. Lou, and J. Wu, *Can pathogen spread keep pace with its host invasion?*, SIAM J. Appl. Math., 76 (2016), pp. 1633–1657.
19. J. Fang, X. Yu, and X.-Q. Zhao, *Traveling waves and spreading speeds for time-space periodic monotone systems*, J. Funct. Anal., 272 (2017), pp. 4222–4262.
20. G. Faye, T. Giletti, and M. Holzer, *Asymptotic spreading for Fisher-KPP reaction-diffusion equations with heterogeneous shifting diffusivity*, Discrete Contin. Dyn. Syst. Ser. S, 15 (2022), pp. 2467–2496.
21. R. A. Fisher, *The wave of advance of advantageous genes*, Annals of Eugenics, 7 (1937), pp. 355–369.
22. M. I. Freidlin, *On wavefront propagation in periodic media*, in Stochastic analysis and applications, vol. 7 of Adv. Probab. Related Topics, Dekker, New York, 1984, pp. 147–166.
23. J. Garnier, T. Giletti, F. Hamel, and L. Roques, *Inside dynamics of pulled and pushed fronts*, J. Math. Pures Appl. (9), 98 (2012), pp. 428–449.
24. J. Gertner and M. I. Freĭdlin, *The propagation of concentration waves in periodic and random media*, Dokl. Akad. Nauk SSSR, 249 (1979), pp. 521–525.
25. L. Girardin and K.-Y. Lam, *Invasion of open space by two competitors: spreading properties of monostable two-species competition-diffusion systems*, Proc. Lond. Math. Soc. (3), 119 (2019), pp. 1279–1335.
26. S. A. Gourley and J. Wu, *Delayed non-local diffusive systems in biological invasion and disease spread*, in Nonlinear dynamics and evolution equations, vol. 48 of Fields Inst. Commun., Amer. Math. Soc., Providence, RI, 2006, pp. 137–200.
27. J.-S. Guo and C.-H. Wu, *Entire solutions originating from traveling fronts for a two-species competition-diffusion system*, Nonlinearity, 32 (2019), pp. 3234–3268.

28. F. Hamel, J. Nolen, J.-M. Roquejoffre, and L. Ryzhik, *A short proof of the logarithmic Bramson correction in Fisher-KPP equations*, Netw. Heterog. Media, 8 (2013), pp. 275–289.
29. M. Holzer and A. Scheel, *Accelerated fronts in a two-stage invasion process*, SIAM J. Math. Anal., 46 (2014), pp. 397–427.
30. C. Hu, J. Shang, and B. Li, *Spreading speeds for reaction-diffusion equations with a shifting habitat*, J. Dynam. Differential Equations, 32 (2020), pp. 1941–1964.
31. H. Hu, T. Yi, and X. Zou, *On spatial-temporal dynamics of a Fisher-KPP equation with a shifting environment*, Proc. Amer. Math. Soc., 148 (2020), pp. 213–221.
32. H. Hu and X. Zou, *Existence of an extinction wave in the Fisher equation with a shifting habitat*, Proc. Amer. Math. Soc., 145 (2017), pp. 4763–4771.
33. M. Iida, R. Lui, and H. Ninomiya, *Stacked fronts for cooperative systems with equal diffusion coefficients*, SIAM J. Math. Anal., 43 (2011), pp. 1369–1389.
34. D. A. Jones, H. L. Smith, and H. R. Thieme, *Spread of viral infection of immobilized bacteria*, Netw. Heterog. Media, 8 (2013), pp. 327–342.
35. ———, *Spread of phage infection of bacteria in a petri dish*, Discrete Contin. Dyn. Syst. Ser. B, 21 (2016), pp. 471–496.
36. D. A. Jones, H. L. Smith, H. R. Thieme, and G. Röst, *On spread of phage infection of bacteria in a Petri dish*, SIAM J. Appl. Math., 72 (2012), pp. 670–688.
37. A. N. Kolmogorov, I. G. Petrovskii, and N. S. Piskunov, *Etude de l'équation de diffusion avec accroissement de la quantité de matière, et son application à un problème biologique*, Bjul. Moskowskogo Gos. Univ., 17 (1937), pp. 1–26.
38. K.-Y. Lam, R. B. Salako, and Q. Wu, *Entire solutions of diffusive Lotka–Volterra system*, J. Differential Equations, 269 (2020), pp. 10758–10791.
39. K.-Y. Lam and X. Yu, *Asymptotic spreading of KPP reactive fronts in heterogeneous shifting environments*, 2021. To appear in J. Math. Pures Appl. arXiv:2101.06698 [math.AP].
40. M. A. Lewis, B. Li, and H. F. Weinberger, *Spreading speed and linear determinacy for two-species competition models*, J. Math. Biol., 45 (2002), pp. 219–233.
41. B. Li, S. Bewick, J. Shang, and W. F. Fagan, *Persistence and spread of a species with a shifting habitat edge*, SIAM J. Appl. Math., 74 (2014), pp. 1397–1417.
42. B. Li, H. F. Weinberger, and M. A. Lewis, *Spreading speeds as slowest wave speeds for cooperative systems*, Math. Biosci., 196 (2005), pp. 82–98.
43. W.-T. Li, J.-B. Wang, and X.-Q. Zhao, *Spatial dynamics of a nonlocal dispersal population model in a shifting environment*, J. Nonlinear Sci., 28 (2018), pp. 1189–1219.
44. X. Liang and X.-Q. Zhao, *Asymptotic speeds of spread and traveling waves for monotone semiflows with applications*, Comm. Pure Appl. Math., 60 (2007), pp. 1–40.
45. X. Liang and T. Zhou, *Spreading speeds of nonlocal KPP equations in almost periodic media*, J. Funct. Anal., 279 (2020), pp. 108723, 58.
46. G. Lin and W.-T. Li, *Asymptotic spreading of competition diffusion systems: the role of interspecific competitions*, European J. Appl. Math., 23 (2012), pp. 669–689.
47. Q. Liu, S. Liu, and K.-Y. Lam, *Asymptotic spreading of interacting species with multiple fronts I: a geometric optics approach*, Discrete Contin. Dyn. Syst., 40 (2020), pp. 3683–3714.
48. ———, *Stacked invasion waves in a competition-diffusion model with three species*, J. Differential Equations, 271 (2021), pp. 665–718.
49. Y. Morita and K. Tachibana, *An entire solution to the Lotka–Volterra competition-diffusion equations*, SIAM J. Math. Anal., 40 (2009), pp. 2217–2240.
50. J. Nolen, J.-M. Roquejoffre, and L. Ryzhik, *Convergence to a single wave in the Fisher-KPP equation*, Chinese Ann. Math. Ser. B, 38 (2017), pp. 629–646.
51. A. B. Potapov and M. A. Lewis, *Climate and competition: the effect of moving range boundaries on habitat invasibility*, Bull. Math. Biol., 66 (2004), pp. 975–1008.
52. M. H. Protter and H. F. Weinberger, *Maximum principles in differential equations*, Springer-Verlag, New York, 1984. Corrected reprint of the 1967 original.
53. L. Rass and J. Radcliffe, *Spatial deterministic epidemics*, vol. 102 of Mathematical Surveys and Monographs, American Mathematical Society, Providence, RI, 2003.
54. R. B. Salako, *Invasion entire solutions for two-species diffusive monostable competitive systems*, Nonlinear Anal. Real World Appl., 59 (2021), p. 37. Paper No. 103264.

55. W. Shen, *Variational principle for spreading speeds and generalized propagating speeds in time almost periodic and space periodic KPP models*, Trans. Amer. Math. Soc., 362 (2010), pp. 5125–5168.
56. N. Shigesada and K. Kawasaki, *Biological invasions: theory and practice*, Oxford University Press, UK, 1997.
57. H. R. Thieme, *Asymptotic estimates of the solutions of nonlinear integral equations and asymptotic speeds for the spread of populations*, J. Reine Angew. Math., 306 (1979), pp. 94–121.
58. H. R. Thieme and X.-Q. Zhao, *Asymptotic speeds of spread and traveling waves for integral equations and delayed reaction-diffusion models*, J. Differential Equations, 195 (2003), pp. 430–470.
59. J.-B. Wang, W.-T. Li, F.-D. Dong, and S.-X. Qiao, *Recent developments on spatial propagation for diffusion equations in shifting environments*, Discrete Contin. Dyn. Syst. Ser. B, 27 (2022), pp. 5101–.
60. J.-B. Wang and C. Wu, *Forced waves and gap formations for a Lotka–Volterra competition model with nonlocal dispersal and shifting habitats*, Nonlinear Anal. Real World Appl., 58 (2021), p. 19. Paper No. 103208.
61. J.-B. Wang and X.-Q. Zhao, *Uniqueness and global stability of forced waves in a shifting environment*, Proc. Amer. Math. Soc., 147 (2019), pp. 1467–1481.
62. Z.-C. Wang, W.-T. Li, and S. Ruan, *Travelling wave fronts in reaction-diffusion systems with spatio-temporal delays*, J. Differential Equations, 222 (2006), pp. 185–232.
63. H. F. Weinberger, *Long-time behavior of a class of biological models*, SIAM J. Math. Anal., 13 (1982), pp. 353–396.
64. ———, *On spreading speeds and traveling waves for growth and migration models in a periodic habitat*, J. Math. Biol., 45 (2002), pp. 511–548.
65. Z. Xu and D. Xiao, *Spreading speeds and uniqueness of traveling waves for a reaction diffusion equation with spatio-temporal delays*, J. Differential Equations, 260 (2016), pp. 268–303.
66. Q. X. Ye, Z. Y. Li, M. Wang, and Y. Wu, *Introduction to Reaction-Diffusion Equations (Chinese)*, Foundations of Modern Mathematics Series, Science Press, Beijing, second ed., 2011.
67. T. Yi, Y. Chen, and J. Wu, *Asymptotic propagations of asymptotical monostable type equations with shifting habitats*, J. Differential Equations, 269 (2020), pp. 5900–5930.
68. T. Yi and X.-Q. Zhao, *Propagation dynamics for monotone evolution systems without spatial translation invariance*, J. Funct. Anal., 279 (2020), pp. 108722, 50.

Chapter 7
The Lotka–Volterra Competition-Diffusion Systems for Two Species

Abstract In this chapter we study the Lotka–Volterra competition-diffusion models for two species in bounded domains. As these reaction-diffusion systems are order-preserving, we first cast them into the setting of strongly monotone dynamical systems and derive some general conclusions that hold for such systems. Next, the global dynamics of the Lotka–Volterra system with constant coefficients are studied, via the Lyapunov functions and LaSalle's invariance principle. For the Lotka–Volterra system with heterogeneous coefficients, the global dynamics of the system with weak intraspecific competitions are characterized when the diffusion rates for two species are both small or large. When two species are identical except for their diffusion rates, it is shown that the species with the slower diffusion drives its competitor to extinction; this phenomenon is termed "evolution of slow dispersal" in the literature. For some weak competition cases with general diffusion rates, the full dynamics of the system is completely characterized. In particular, it is proved that one of the species is able to drive the other species to extinction for arbitrary initial conditions, even though without diffusion two species could always coexist. Finally, we consider the two-species competition model in rivers and prove that the species with the faster diffusion drives its competitor to extinction, in strong contrast with the "evolution of slow dispersal" phenomenon.

7.1 Elements from the Theory of Monotone Dynamical Systems

Consider

$$\begin{cases} u_t + \mathcal{L}_1 u = u(a_1(x) - b_1(x)u - c_1(x)v) & \text{in } \Omega \times (0,\infty), \\ v_t + \mathcal{L}_2 v = v(a_2(x) - b_2(x)u - c_2(x)v) & \text{in } \Omega \times (0,\infty), \\ \mathcal{B}_1 u = \mathcal{B}_2 v = 0 & \text{on } \partial\Omega \times (0,\infty), \\ u(x,0) = u_0(x) \quad \text{and} \quad v(x,0) = v_0(x) & \text{in } \Omega, \end{cases} \tag{7.1}$$

K. -Y. Lam, Y. Lou, *Introduction to Reaction-Diffusion Equations*, Lecture Notes on Mathematical Modelling in the Life Sciences, https://doi.org/10.1007/978-3-031-20422-7_7

where $\mathcal{L}_k = -\tilde{a}_k^{ij}(x)D_{ij} + \tilde{b}_k^j(x)D_j$ is a non-divergence form uniformly elliptic operator with Hölder continuous coefficients, and $\mathcal{B}_k = p_k^j D_j + p_k^0$ is an oblique derivative operator on $\partial\Omega$, such that

$$n_j p_k^j > 0, \quad \text{and} \quad p_k^0 \geq 0 \quad \text{on } \partial\Omega,$$

where $\mathbf{n}(x) = (n_j(x))_{j=1}^N$ is the outer unit normal vector on $\partial\Omega$. All other coefficients are assumed to be smooth and satisfy the competition hypotheses

$$b_k > 0, \quad c_k > 0 \quad \text{in } \overline{\Omega}.$$

The functions u and v represent the population densities of two competing species, a_i denotes their respective carrying capacities, b_1, c_2 account for self-regulation of each species, and b_2, c_1 account for interspecific competition.

With the special case $\mathcal{L}_i = -d_i\Delta$ in mind, the model (7.1) is referred to in the literature as the diffusive Lotka–Volterra model for two competing species [7].

For $i = 1, 2$, let $\tilde{\mu}_i$ and $\tilde{\phi}_i > 0$ be the principal eigenvalue and eigenfunction of

$$\mathcal{L}_i\tilde{\phi} = a_i\tilde{\phi} + \tilde{\mu}_i\tilde{\phi} \quad \text{in } \Omega, \quad \mathcal{B}_i\tilde{\phi} = 0 \quad \text{on } \partial\Omega. \tag{7.2}$$

We always assume in this chapter that $\tilde{\mu}_i < 0$ for $i = 1, 2$. In this case, Theorem 5.2.2 guarantees that the system (7.1) has a trivial equilibrium $E_0 = (0, 0)$, and two semi-trivial equilibria $E_1 = (\tilde{u}, 0)$ and $E_2 = (0, \tilde{v})$, where $\tilde{u}$ and $\tilde{v}$ are, respectively, the unique positive solution to

$$\mathcal{L}_1\tilde{u} = \tilde{u}(a_1 - b_1\tilde{u}) \text{ in } \Omega, \quad \mathcal{B}_1\tilde{u} = 0 \text{ on } \partial\Omega, \tag{7.3}$$

and

$$\mathcal{L}_2\tilde{v} = \tilde{v}(a_2 - c_2\tilde{v}) \text{ in } \Omega, \quad \mathcal{B}_2\tilde{v} = 0 \text{ on } \partial\Omega. \tag{7.4}$$

Also, we call $E^* = (u^*, v^*)$ a positive equilibrium if $u^* > 0$ and $v^* > 0$ in $\overline{\Omega}$.

Lemma 7.1.1 *Let $E_1 = (\tilde{u}, 0)$ and $E_2 = (0, \tilde{v})$.*

1. *The equilibrium E_1, if it exists, is linearly unstable (resp. linearly stable) if and only if $\mu_{E_1} < 0$ (resp. $\mu_{E_1} > 0$), where μ_{E_1} is the principal eigenvalue of*

$$\mathcal{L}_2\phi = \phi(a_2 - b_2\tilde{u}) + \mu\phi \text{ in } \Omega, \quad \mathcal{B}_2\phi = 0 \text{ on } \partial\Omega. \tag{7.5}$$

2. *Symmetrically, the equilibrium E_2, if it exists, is linearly unstable (resp. linearly stable) if and only if $\mu_{E_2} < 0$ (resp. $\mu_{E_2} > 0$), where μ_{E_2} is the principal eigenvalue of*

$$\mathcal{L}_1\phi = \phi(a_1 - c_1\tilde{v}) + \mu\phi \text{ in } \Omega, \quad \mathcal{B}_1\phi = 0 \text{ on } \partial\Omega. \tag{7.6}$$

Proof We will only prove assertion 1, as assertion 2 follows from a symmetric argument. The linear stability of $(\tilde{u}, 0)$ is determined by the following linearized eigenvalue problem:

$$\begin{cases} \mathcal{L}_1\varphi = (a_1 - 2b_1\tilde{u})\varphi - c_1\tilde{u}\psi + \lambda\varphi & \text{in } \Omega, \\ \mathcal{L}_2\psi = (a_2 - b_2\tilde{u})\psi + \lambda\psi & \text{in } \Omega, \\ \mathcal{B}_1\varphi = \mathcal{B}_2\psi = 0 & \text{on } \partial\Omega. \end{cases} \tag{7.7}$$

By Theorem 3.3.2, the system (7.7) has a principal eigenvalue $\lambda_1 \in \mathbb{R}$ such that $(\tilde{u}, 0)$ is linearly unstable (resp. stable) if $\lambda_1 < 0$ (resp. $\lambda_1 > 0$). It remains to show that $\text{sign}\,(\lambda_1) = \text{sign}\,(\mu_{E_1})$.

Step 1. We claim that the principal eigenvalue $\tilde{\mu}$ of

$$\begin{cases} \mathcal{L}_1\tilde{\phi} = (a_1 - 2b_1\tilde{u})\tilde{\phi} + \mu\tilde{\phi} & \text{in } \Omega, \\ \mathcal{B}_1\tilde{\phi} = p_1^j D_j\tilde{\phi} + p_1^0\tilde{\phi} = 0 & \text{on } \partial\Omega \end{cases} \tag{7.8}$$

is positive. Indeed, $\tilde{\mu}$ exists thanks to Theorem 1.3.6. Moreover, $\tilde{\mu} > 0$ follows by applying Lemma 1.3.12 with $(\underline{\mu}, w) = (0, \tilde{u})$, where $\tilde{u}$ is the unique positive solution of (7.3).

Step 2. Suppose that $\mu_{E_1} > 0$. Let λ_1 be the principal eigenvalue of (7.7) with eigenfunction $\varphi_1 \geq 0 \geq \psi_1$ where at least one of them is nontrivial. If $\psi_1 \not\equiv 0$, then λ_1 is an eigenvalue of (7.5) and we automatically have $\lambda_1 \geq \mu_{E_1} > 0$. If $\psi_1 \equiv 0$, then λ_1 is an eigenvalue of (7.8). Since the principal eigenvalue $\tilde{\mu}$ of (7.8) is positive, we thus have $\lambda_1 \geq \tilde{\mu} > 0$. This proves the case $\mu_{E_1} > 0$.

Step 3. Consider the inhomogeneous problem

$$\begin{cases} \mathcal{L}_1 w = (a_1 - 2b_1\tilde{u})w + \mu w + f & \text{in } \Omega, \\ \mathcal{B}_1 w = p_1^j D_j w + p_1^0 w = 0 & \text{on } \partial\Omega. \end{cases} \tag{7.9}$$

We claim that for each $\mu \leq 0$ and $f \in C^\alpha(\overline{\Omega})$, there exists a unique $w = T_\mu[f] \in C^{2+\alpha}(\overline{\Omega})$ which satisfies (7.9). Furthermore, $w > 0$ in $\overline{\Omega}$ if $f \geq, \not\equiv 0$ in Ω.

Indeed, (7.8) has no nonpositive eigenvalue, so $w = T_\mu[f]$ is well-defined for $\mu \leq 0$ by the Fredholm alternative. Moreover, $w \in C^{2+\alpha}(\overline{\Omega})$ thanks to Schauder estimates [21, Chapter 6]. To see that $w > 0$ in $\overline{\Omega}$, we write $w = \phi z$, where ϕ is a positive eigenfunction of (7.8). Then z satisfies

$$\begin{cases} \mathcal{L}_2 z - \frac{1}{\phi}(a_1^{ij} D_i\phi)D_j z + (\tilde{\mu} - \mu)z = f \geq 0 & \text{in } \Omega, \\ p_1^j D_j z = 0 & \text{on } \partial\Omega. \end{cases}$$

Since $\tilde{\mu} - \mu > 0$, it follows from the classical maximum principle that $z \geq 0$ in Ω if $f \geq 0$ in Ω, and $z > 0$ in $\overline{\Omega}$ if $f \geq, \not\equiv 0$.

Step 4. Now, suppose the principal eigenvalue μ_{E_1} of (7.5) is nonpositive and denote by ϕ_1 its associated positive eigenfunction. We can now observe that μ_{E_1} is an eigenvalue of (7.7) with eigenfunction $(\varphi_1, \psi_1) = (T_{\mu_{E_1}}[c_1\tilde{u}\phi_1], -\phi_1)$. We claim that μ_{E_1} is the eigenvalue with the least real part. Indeed, if λ' is an eigenvalue of (7.7) with eigenfunction (φ, ψ). If $\psi \not\equiv 0$, then λ' is an eigenvalue of (7.5), and hence

$\mathrm{Re}\,\lambda' \geq \mu_{E_1}$. If $\psi \equiv 0$, then λ' is an eigenvalue of (7.8), whence $\mathrm{Re}\,\lambda' \geq \tilde{\mu} > 0$. In conclusion, when the principal eigenvalue μ_{E_1} of (7.5) is nonpositive, it is also the eigenvalue of (7.7) with the least real part. In conclusion, the principal eigenvalue λ_1 of (7.7) is given by $\lambda_1 = \mu_{E_1} \leq 0$. □

Theorem 7.1.2 *Suppose that* (7.1) *has no positive equilibria.*

1. *If* $E_1 = (\tilde{u}, 0)$ *is linearly unstable, then* $(u, v) \to (0, \tilde{v})$ *whenever* $(u_0, v_0) \in C(\overline{\Omega}; \mathbb{R}^2_+)$ *and* $v_0 \not\equiv 0$.
2. *If* $E_2 = (0, \tilde{v})$ *is linearly unstable, then* $(u, v) \to (\tilde{u}, 0)$ *whenever* $(u_0, v_0) \in C(\overline{\Omega}; \mathbb{R}^2_+)$ *and* $u_0 \not\equiv 0$.

Proof We present a direct proof of assertion 1 due to [39]. The proof of assertion 2 is similar and is omitted. Recall that $E_1 = (\tilde{u}, 0)$ and let ϕ be the positive eigenfunction of (7.5). Define

$$(\overline{u}_0, \underline{v}_0) = ((1+\epsilon)\tilde{u}, \delta\phi),$$

and write $f_i(x, u, v) = a_i(x) - b_i(x)u - c_i(x)v$.

Step 1. There exists an $\epsilon_0 > 0$ such that for $\delta, \epsilon \in (0, \epsilon_0]$, $(\overline{u}_0, \underline{v}_0)$ is a supersolution.

Defining $f_i(x, u, v) = a_i(x) - b_i(x)u - c_i(x)v$, we proceed by a direct computation.

$$\begin{aligned}
&-\mathcal{L}_1[(1+\epsilon)\tilde{u}] + (1+\epsilon)\tilde{u} f_1(x, (1+\epsilon)\tilde{u}, \delta\phi)\\
&= (1+\epsilon)[-\mathcal{L}_1\tilde{u} + \tilde{u} f_1(x, (1+\epsilon)\tilde{u}, \delta\phi)]\\
&= (1+\epsilon)\tilde{u}[-f_1(x, \tilde{u}, 0) + f_1(x, (1+\epsilon)\tilde{u}, \delta\phi)] \leq 0,
\end{aligned}$$

where the last inequality is due to f_1 being decreasing in u and v. Also,

$$\begin{aligned}
&-\mathcal{L}_2[\delta\phi] + \delta\phi f_2(x, (1+\epsilon)\tilde{u}, \delta\phi)\\
&= \delta[-\mathcal{L}_2\phi + \phi f_2(x, (1+\epsilon)\tilde{u}, \delta\phi)]\\
&= \delta\phi[-f_2(x, \tilde{u}, 0) - \mu_{E_1} + f_2(x, (1+\epsilon)\tilde{u}, \delta\phi)]\\
&= \delta\phi[-\mu_{E_1} + O(|\delta| + |\epsilon|)] \geq 0
\end{aligned}$$

provided that δ and ϵ are both sufficiently small.

Step 2. Let $(\overline{u}, \underline{v})$ be the solution to (7.1) with initial data $((1+\epsilon)\tilde{u}, \delta\phi)$ for some $\epsilon, \delta \in (0, \epsilon_0]$, then $(\overline{u}, \underline{v}) \to (0, \tilde{v})$ as $t \to \infty$.

Indeed, since $\overline{u}$ and $\underline{v}$ are monotone in t, we have $(\overline{u}, \underline{v}) \to (\overline{u}_\infty, \underline{v}_\infty)$, where $(\overline{u}_\infty, \underline{v}_\infty)$ is a nonnegative equilibrium of (7.1) satisfying

$$\overline{u}_\infty(x) \leq (1+\epsilon)\tilde{u}(x) \quad \text{and} \quad \underline{v}_\infty(x) \geq \delta\phi(x) > 0.$$

The nonexistence of positive equilibrium implies that $(\overline{u}_\infty, \underline{v}_\infty) = (0, \tilde{v})$.

Step 3. Let (u, v) be a solution of (7.1) such that $v_0 \geq, \equiv 0$. We derive a rough estimate of (u, v).

$$\limsup_{t\to\infty} \sup_{\Omega}(u(x,t) - \tilde{u}(x)) \leq 0 \quad \text{and} \quad \limsup_{t\to\infty} \sup_{\Omega}(v(x,t) - \tilde{v}(x)) \leq 0. \tag{7.10}$$

To this end, choose $M > 0$ such that $u(x,0) \le (1+M)\tilde{u}(x)$ in $\overline{\Omega}$. This is possible since $\inf_\Omega \tilde{u} > 0$. Next, observe that the pair of functions

$$u(x,t) \quad \text{and} \quad \overline{u}(x,t) := \frac{M}{1+tM\inf_\Omega(b_1\tilde{u})}\tilde{u}(x) + \tilde{u}(x)$$

form a pair of sub- and superequilibria of

$$\begin{cases} \vartheta_t + \mathcal{L}_1\vartheta = \vartheta(a_1 - b_1\vartheta) & \text{in } \Omega\times(0,\infty), \\ \mathcal{B}_1\vartheta = 0 & \text{on } \partial\Omega\times(0,\infty), \end{cases}$$

and that $u(x,0) \le \overline{u}(x,0)$ by our choice of M. By comparison, we deduce the first part of (7.10). The second part of (7.10) is analogous.

Step 4. Let (u,v) be a solution of (7.1) such that $v_0 \ge,\equiv 0$. If $u_0 \equiv 0$, then $u \equiv 0$ in $\overline{\Omega}\times(0,\infty)$ and it follows from Theorem 5.2.2 that $(u,v) \to (0,\tilde{v})$. Suppose $u_0 \ge,\equiv 0$, $v_0 \ge,\not\equiv 0$. By the strong maximum principle, $u(x,t) > 0$ and $v(x,t) > 0$ in $\overline{\Omega}\times(0,\infty)$. Moreover, Step 2 implies that there is a $T_0 > 0$ such that $0 \le u(x,T_0) \le (1+\epsilon_0)\tilde{u}$ for $x\in\Omega$. Since $v(x,T_0) > 0$ in $\overline{\Omega}$, we may take $\delta\in(0,\epsilon_0]$ such that $v(x,T_0) \ge \delta\phi(x)$ in Ω. Consider the two sets of solutions $(u,v)(x,t+T_0)$ and $(\overline{u},\underline{v})$ of (7.1) with initial data (u_0,v_0) and $((1+\epsilon_0)\tilde{u},\delta\phi)$ respectively. Since they are ordered at $t=0$, we can apply the comparison principle to $(u,v)(x,t+T_0)$ and $(\overline{u},\underline{v})(x,t)$ to get

$$\limsup_{t\to\infty}\sup_\Omega u(x,t) \le 0, \quad \text{and} \quad \liminf_{t\to\infty}\inf_\Omega(v(x,t)-\tilde{v}(x)) \ge 0,$$

where we used $(\overline{u},\underline{v}) \to (0,\tilde{v})$ by Step 3. Combining with (7.10), we deduce that $(u,v)\to(0,\tilde{v})$. This completes the proof. □

In the preceding result, we see that for Lotka–Volterra systems, the nonexistence of positive equilibria combined with the instability of E_1 implies that E_2 attracts all nontrivial solutions. This is true for a general class of abstract competitive systems; see Proposition E.2.7.

Can we weaken the assumption further? For more general competitive systems such as

$$\begin{cases} u_t + \mathcal{L}_1 u = u f_1(x,u,v) & \text{in } \Omega\times(0,\infty), \\ v_t + \mathcal{L}_2 v = v f_2(x,u,v) & \text{in } \Omega\times(0,\infty), \\ \mathcal{B}_1 u = \mathcal{B}_2 v = 0 & \text{on } \partial\Omega\times(0,\infty), \\ u(x,0) = u_0(x) \quad \text{and} \quad v(x,0) = v_0(x) & \text{in } \Omega, \end{cases} \tag{7.11}$$

where

$$\frac{\partial f_i}{\partial u}(x,u,v) < 0 \quad \text{and} \quad \frac{\partial f_i}{\partial v}(x,u,v) < 0 \quad \text{for } x\in\overline{\Omega},\ u,v\in(0,\infty),$$

the associated semiflow preserves the competitive order (see Theorem 3.3.4), and yet the nonexistence of positive equilibrium alone does not imply the global attractivity of E_1 or E_2 among all nonnegative, nontrivial solutions. Indeed, for competitive systems satisfying (H1')–(H4') (which is recalled below), a general theorem due to Hsu et al. [37] (see Theorem E.2.5) asserts that, in the absence of positive equilibria, every trajectory initiating in $I = \{(u_0, v_0) : 0 < u_0 \le \tilde{u},\ 0 < v_0 \le \tilde{v}\}$ converges to the same boundary equilibrium, say E_1. However, the same boundary equilibrium may fail to be globally attractive with respect to all nonnegative, nontrivial data. See Problem 7.5.1 for a counterexample. Fortunately, such a pathological situation does not happen for Lotka–Volterra systems. In Appendix E, we slightly strengthen the trichotomy result in [37] by proposing the condition (H5'). As we will see below, the Lotka–Volterra systems satisfy all conditions (H1')–(H5') and thus enjoy a stronger trichotomy result; see Theorem 7.1.6, which is based on Theorem E.2.13 of Appendix E. (The readers who are not interested in the theory of monotone dynamical systems may wish to skip to Section 7.2.) For this purpose, let

$$\begin{cases} X = X_1 \times X_2 = C(\overline{\Omega}) \times C(\overline{\Omega}), \quad C = C_1 \times C_2 = C(\overline{\Omega}; \mathbb{R}_+) \times C(\overline{\Omega}; \mathbb{R}_+), \\ K = X_1^+ \times (-X_2^+) = C(\overline{\Omega}; \mathbb{R}_+) \times (-C(\overline{\Omega}; \mathbb{R}_+)), \end{cases}$$

so that X is a Banach space with an order induced by K, and consider the semiflow $\Phi_t = (\Phi_t^{(1)}, \Phi_t^{(2)}) : C \to C$ generated by (7.1). We now recall (H1')–(H5') from Appendix E.

(H1') Φ_t is order compact[1] for each $t > 0$ and strictly monotone with respect to $<_K$. That is, $(u_0, v_0) <_K (\bar{u}_0, \bar{v}_0)$ implies $\Phi_t(u_0, v_0) <_K \Phi_t(\bar{u}_0, \bar{v}_0)$.

(H2') For each $t > 0$, there exist maps $\eta : C \to X$ and $f_i : C_i \to C_i$ $(i = 1, 2)$ such that

$$\Phi_t(u_0, v_0) = (f_1(u_0), f_2(v_0)) + \eta(u_0, v_0),$$

where $\|\eta(x)\|/\|x\| \to 0$ as $\|x\| \to 0$ and, for $i = 1, 2$, f_i is compact, positively homogeneous of degree one, strongly monotone with respect to the order generated by C_i, and its Bonsall cone spectral radius[2] satisfies $\tilde{r}_{C_i}(f_i) > 1$.

(H3') $\Phi_t(C_1 \times \{0\}) \subset C_1 \times \{0\}$ for $t \ge 0$. There exists a $\tilde{u} \in \operatorname{Int} C_1$ such that $E_1 = (\tilde{u}, 0)$ is an equilibrium of Φ_t, and $\Phi_t(u_0, 0) \to (\tilde{u}, 0)$ as $t \to \infty$ for every $u_0 \in C_1 \setminus \{0\}$. The symmetric conditions hold for Φ_t on $\{0\} \times C_2$, where the equilibrium is denoted by $E_2 = (0, \tilde{v})$.

(H4') If $(u_0, v_0), (\bar{u}_0, \bar{v}_0) \in C$ satisfy $(u_0, v_0) <_K (\bar{u}_0, \bar{v}_0)$ and at least one of them belongs to $\operatorname{Int} C$, then $\Phi_t(u_0, v_0) \ll_K \Phi_t(\bar{u}_0, \bar{v}_0)$ for $t > 0$. If $(u_0, v_0) \in C$ satisfies $u_0 \ne 0$ and $v_0 \ne 0$, then $\Phi_t(u_0, v_0) \in \operatorname{Int} C$ for $t > 0$.

(H5') For each $t > 0$, Φ_t is continuously differentiable, and the map $D\Phi_t(E_2) : X \to X$ is compact and $r(D_2\Phi_t^{(2)}(E_2)) < 1$ for all $t > 0$. If $r(D\Phi_\tau(E_2)) \ge 1$ for some τ, then we also have

[1] For the definition of order compactness, see Definition D.1.1.

[2] For the definition of the Bonsall cone spectral radius, see Definition B.2.5. If $X = C(\overline{\Omega})$, $C_i = C(\overline{\Omega}; \mathbb{R}_+)$ and f is linear, then the Bonsall cone spectral radius coincides with the spectral radius. See Lemma B.4.9.

$$D\Phi_T(E_2)[\phi,\psi] \geq_K (\phi,\psi) \quad \text{for some } \phi,\psi \in \operatorname{Int} C_1 \times (-\operatorname{Int} X_2^+).$$

A symmetric condition holds for $D\Phi_t(E_1)$.

Here $\Phi_t(u_0,v_0) = (\Phi_t^{(1)}(u_0,v_0), \Phi_t^{(2)}(u_0,v_0)) \in X_1 \times X_2$ and $D_2\Phi_t^{(2)}$ is the Fréchet derivative of $\Phi_t^{(2)}$ with respect to the second component of the input variables.

Lemma 7.1.3 *The semiflow $\Phi_t : C \to C$ generated by* (7.1) *is continuously differentiable, and satisfies the conditions* (H1')–(H5').

Remark 7.1.4 Another example of a competitive system that satisfies (H1')–(H5') is the two-species phytoplankton model, which is a system of integro-PDEs modeling phytoplankton populations engaging in competition for light via mutual shading; see Chapter 8.

Proof (of Lemma 7.1.3) Now, (H1') and (H4') follow directly from the strong comparison principle of (7.1).

Next, notice that the semiflow operator is continuously differentiable. Furthermore, if $(\tilde{u},\tilde{v})$ is an equilibrium of Φ_t, then $D\Phi_t(\tilde{u},\tilde{v}) : X \to X$ is given by $D\Phi_t(\tilde{u},\tilde{v})[p_0,q_0] = (p(\cdot,t), q(\cdot,t))$, where (p,q) is the unique solution of the linear parabolic system

$$\begin{cases} p_t + \mathcal{L}_1 p = (a_1 - 2b_1\tilde{u} - c_1\tilde{v})p - c_1\tilde{u}q & \text{in } \Omega \times (0,\infty), \\ q_t + \mathcal{L}_2 q = -b_2\tilde{v}p + (a_2 - b_2\tilde{u} - 2c_2\tilde{v})q & \text{in } \Omega \times (0,\infty), \\ \mathcal{B}_1 p = \mathcal{B}_2 q = 0 & \text{on } \partial\Omega \times (0,\infty), \\ p(x,0) = p_0(x), \quad q(x,0) = q_0(x) & \text{in } \Omega. \end{cases} \tag{7.12}$$

To verify condition (H2'), set $(\tilde{u},\tilde{v}) = (0,0)$ in (7.12) so that the system decouples. Next, observe that

$$D\Phi_t(0,0)[p_0,q_0] = (f_1(p_0), f_2(q_0)) = (e^{t(-\mathcal{L}_1+a_1)}[p_0], e^{t(-\mathcal{L}_2+a_2)}[q_0]),$$

where $e^{t(-\mathcal{L}_i+a_i)}$ denotes the analytic semigroup generated by $\psi_t + \mathcal{L}_i\psi = a_i\psi$ in the space $C(\overline{\Omega})$, such that the solution respects the appropriate boundary condition in positive time. For $i = 1,2$, let $\tilde{\mu}_i$ and $\tilde{\phi}_i(x) > 0$ be the principal eigenvalues and eigenfunctions of (7.2). Then $f_i(\tilde{\phi}_i) = e^{-\tilde{\mu}_i t}\tilde{\phi}_i$. Since f_i is linear and strictly positive in $C(\overline{\Omega})$ and since we enforce the condition $\mu_i < 0$ in this chapter, it follows from the strong form of the Krein–Rutman Theorem (Theorem B.3.2) that $r(f_i) = e^{-\tilde{\mu}_i t} > 1$. This verifies (H2'), in view of Remark E.2.1. (H3') follows from Theorem 5.2.2 and the fact that $\mu_i < 0$.

We will verify the first part of (H5') using Lemma E.2.10. For that purpose, observe that for $i = 1,2$, $D_i\Phi_t^{(i)}(E_2)$ is again an analytic semigroup generated by certain linear elliptic operators, so they are automatically compact and strongly monotone with respect to $X_i^+ = C_i$. Moreover, $r(D_2\Phi_t^{(2)}(E_2)) = e^{-\tilde{\mu}t}$, where $\tilde{\mu}$ is the principal eigenvalue of

$$\mathcal{L}_2\tilde{\phi} = (a_2 - 2c_2\tilde{v})\tilde{\phi} + \tilde{\mu}\tilde{\phi} \quad \text{in } \Omega, \quad \mathcal{B}_2\tilde{\phi} = 0 \quad \text{on } \partial\Omega.$$

Arguing as in Step 1 in the proof of Lemma 7.1.1, $\tilde{\mu} > 0$, so that $r(D_1\Phi_t^{(1)}(E_1)) = \mathrm{e}^{-\tilde{\mu}t} < 1$. This verifies the first two hypotheses of Lemma E.2.10. The last hypothesis, which translates to $D_1\Phi_t^{(2)}(E_2)[p_0] \neq 0$ for $p_0 \in C(\overline{\Omega})$, $p_0 > 0$ in $\overline{\Omega}$, holds since $c_1, b_2 > 0$ in $\overline{\Omega}$. Now, we can invoke Lemma E.2.10 to conclude that the first part of (H5') concerning $D\Phi_t(E_2)$ holds. The second part of (H5') concerning $D\Phi_t(E_1)$ holds by a symmetric argument. □

Denote $[0, \infty)$ by $\mathbb{R}_+$, and let

$$K = C(\overline{\Omega}; \mathbb{R}_+) \times [-C(\overline{\Omega}; \mathbb{R}_+)].$$

In the following, the partial order $(u_0, v_0) \leq_K (u_0', v_0')$ (resp. $<_K$, $\ll_K$) holds if and only if $(u_0' - u_0, v_0' - v_0) \in K$ (resp. $\in K \setminus \{(0,0)\}$, $\in \operatorname{Int} K$). For example, $(u_0, v_0) \leq_K (u_0', v_0')$ if $u_0 \leq u_0'$ and $v_0 \geq v_0'$ in Ω.

Theorem 7.1.5 (Compression result) *If the semitrivial equilibria E_1 and E_2 are linearly unstable (i.e. the principal eigenvalues μ_{E_1}, μ_{E_2} of (7.5) and (7.6) are both negative), then (7.1) has positive equilibria $E_* = (u_*, v_*)$ and $E^* = (u^*, v^*)$, such that $u_* \leq u^*$ and $v_* \geq v^*$ and the following statements hold:*

1. $(u, v) \to E_*$ *if* $E_2 \ll_K (u_0, v_0) \leq_K E_*$, $u_0, v_0 \not\equiv 0$;
2. $(u, v) \to E^*$ *if* $E^* \leq_K (u_0, v_0) \ll_K E_1$, $u_0, v_0 \not\equiv 0$;
3. *If u_0, v_0 are nonnegative and nontrivial, then*

$$\begin{cases} u_*(x) \leq \liminf\limits_{t\to\infty} u(x,t) \leq \limsup\limits_{t\to\infty} u(x,t) \leq u^*(x), \\ v^*(x) \leq \liminf\limits_{t\to\infty} v(x,t) \leq \limsup\limits_{t\to\infty} v(x,t) \leq v_*(x). \end{cases}$$

If, in addition, $E_* = E^*$, then $(u, v) \to E^*$ if $u_0 \geq, \not\equiv 0$ and $v_0 \geq, \not\equiv 0$.

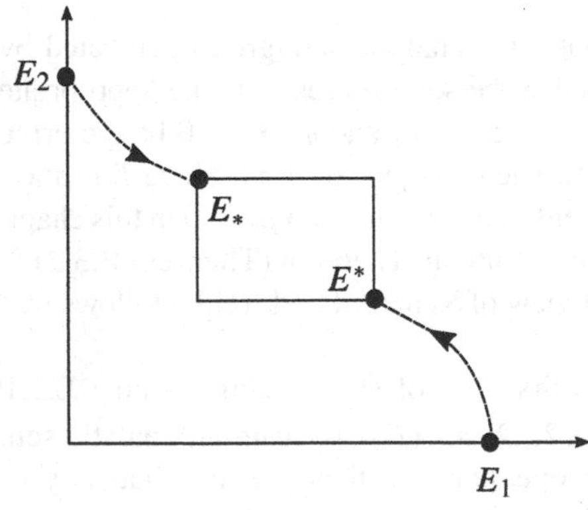

Fig. 7.1: A diagram illustrating the compression result (Theorem 7.1.5).

Proof We apply Theorem E.2.15. Having already verified (H1')–(H4'), we observe that $D_2\Phi_t^{(2)}(E_1) : C(\overline{\Omega}) \to C(\overline{\Omega})$ and $D_1\Phi_t^{(1)}(E_2) : C(\overline{\Omega}) \to C(\overline{\Omega})$ are compact and strongly positive with respect to $C(\overline{\Omega}; \mathbb{R}^2_+)$, and that

$$r(D_2\Phi_t^{(2)}(E_1)) = \mathrm{e}^{-\mu_{E_1}t} > 1, \quad \text{and} \quad r(D_1\Phi_t^{(1)}(E_2)) = \mathrm{e}^{-\mu_{E_2}t} > 1,$$

where μ_{E_1} and μ_{E_2} are, respectively, the principal eigenvalues of (7.5) and (7.6). The result directly follows from an application of Theorem E.2.15. □

Theorem 7.1.6 *Consider the semiflow generated by* (7.1). *Exactly one of the following statements hold.*

1. (7.1) *has at least one positive equilibrium* (u^*, v^*) *such that* $u^* > 0$ *and* $v^* > 0$ *in* $\overline{\Omega}$;
2. E_1 *is globally asymptotically stable among all solutions of* (7.1) *with nonnegative, nontrivial initial data;*
3. E_2 *is globally asymptotically stable among all solutions of* (7.1) *with nonnegative, nontrivial initial data.*

Proof Since we have already verified (H1')–(H5'), the result is a direct consequence of Theorem E.2.13. □

Theorem 7.1.7 *If* (7.1) *has at least one positive equilibrium, and if every positive equilibrium is locally asymptotically stable, then there exists a unique positive equilibrium* E^*. *Furthermore,* $(u, v) \to E^*$ *for every solution* (u, v) *of* (7.1) *with nonnegative initial data* (u_0, v_0) *satisfying* $u_0 \not\equiv 0$ *and* $v_0 \not\equiv 0$.

Proof This is a direct consequence of Theorem E.2.14. □

7.2 Lotka–Volterra Systems with Constant Coefficients

Consider

$$\begin{cases} u_t - d_1\Delta u = u(a_1 - b_1u - c_1v) & \text{in } \Omega \times (0, \infty), \\ v_t - d_2\Delta v = v(a_2 - b_2u - c_2v) & \text{in } \Omega \times (0, \infty), \\ \mathbf{n} \cdot \nabla u = \mathbf{n} \cdot \nabla v = 0 & \text{on } \partial\Omega \times (0, \infty), \end{cases} \tag{7.13}$$

with nonnegative and nontrivial initial data. The following theorem was first proved in [4].

Theorem 7.2.1 *Suppose that* a_i, b_i, c_i *are positive constants such that* $\frac{b_1}{b_2} > \frac{a_1}{a_2} > \frac{c_1}{c_2}$. *Then for any solution* (u, v) *of* (7.13) *with nonnegative, nontrivial initial value, we have* $(u, v) \to (u^*, v^*)$, *where*

$$(u^*, v^*) = \left(\frac{a_2c_1 - a_1c_2}{b_2c_1 - b_1c_2}, \frac{a_2b_1 - a_1b_2}{b_2c_1 - b_1c_2} \right).$$

We prove Theorem 7.2.1 by constructing a Lyapunov function and invoking LaSalle's invariance principle.

Definition 7.2.2 1. $W : [C(\overline{\Omega})]^2 \to \mathbb{R}$ is a Lyapunov function with respect to the semiflow generated by (7.13) in $[C(\overline{\Omega})]^2$ if W is differentiable and

$$\frac{\mathrm{d}}{\mathrm{d}t} W(u(\cdot,t), v(\cdot,t)) \le 0 \quad \text{for } t > 0,$$

for each positive solution (u, v) of (7.13).

2. For $(u_0, v_0) \in [C(\overline{\Omega})]^2$, define

$$\dot{W}(u_0, v_0) = \left.\frac{\mathrm{d}}{\mathrm{d}t} W(u(\cdot,t), v(\cdot,t))\right|_{t=0},$$

where (u, v) is the solution of (7.13) with initial data (u_0, v_0).

3. Define

$$E = \{(u_0, v_0) \in [C(\overline{\Omega})]^2 : \dot{W}(u_0, v_0) = 0\}.$$

4. We say that $\mathfrak{M} \subset E$ is the maximal invariant set of E if

$$\mathfrak{M} = \cup_{(\hat{u},\hat{v})} \{(\hat{u}(\cdot,t), \hat{v}(\cdot,t)) : t \in \mathbb{R}\}$$

where the union is taken over all bounded entire solutions[3] $(\hat{u}, \hat{v})$ of (7.13) such that

$$(\hat{u}(\cdot,t), \hat{v}(\cdot,t)) \in E \quad \text{for each } t \in \mathbb{R}.$$

Theorem 7.2.3 (LaSalle's invariance principle) *Suppose that* $W : [C(\overline{\Omega})]^2 \to \mathbb{R}$ *is a Lyapunov function with respect to the semiflow generated by* (7.13) *in* $[C(\overline{\Omega})]^2$. *Then for each solution* (u, v) *of* (7.13) *with pre-compact trajectory (i.e.* $\{(u(\cdot,t), v(\cdot,t) : t \ge 0\}$ *is precompact in* $[C(\overline{\Omega})]^2$*), it holds that*

$$\mathrm{dist}((u(\cdot,t), v(\cdot,t)), \mathfrak{M}) \to 0 \quad \text{as } t \to \infty.$$

Proof See, e.g. [23, 31]. □

Proof (of Theorem 7.2.1) Fix a solution (u, v) of (7.13) with positive initial data in $[C(\overline{\Omega})]^2$. Thanks to parabolic estimates, $t \mapsto (u(\cdot,t), v(\cdot,t))$ is continuous in $C(\overline{\Omega}; \mathbb{R}^2)$ and

$$\sup_{t \ge 1} \|(u(\cdot,t), v(\cdot,t)\|_{C^2(\overline{\Omega})} < +\infty,$$

so it has a precompact trajectory. Moreover, $u > 0$ and $v > 0$ in $\overline{\Omega} \times [0, \infty)$.

[3] We say that $(\hat{u}, \hat{v})$ is an entire solution if $(\hat{u}, \hat{v}) \in C^{2,1}(\overline{\Omega} \times (-\infty, +\infty))$ is a classical solution of the competition system.

By replacing (u, v) with $\left(\frac{b_1}{a_1}u, \frac{c_2}{a_2}v\right)$, we may reduce to the case that

$$a_1 = b_1, a_2 = c_2, c_1 < a_1, b_2 < a_2. \tag{7.14}$$

In this case,

$$(u^*, v^*) := \left(\frac{1 - c_0}{1 - b_0 c_0}, \frac{1 - b_0}{1 - b_0 c_0}\right), \quad \text{where } (b_0, c_0) = \left(\frac{a_1 b_2}{a_2 b_1}, \frac{a_2 c_1}{a_1 c_2}\right).$$

We will apply LaSalle's invariance principle to show that $(u(\cdot, t), v(\cdot, t)) \to (u^*, v^*)$. To this end, we set

$$W(u, v) = \int_\Omega \left\{\frac{1}{a_1}\left[u - u^* - u^* \log\frac{u}{u^*}\right] + \frac{1}{a_2}\left[v - v^* - v^* \log\frac{v}{v^*}\right]\right\} \mathrm{d}x,$$

to obtain

$$\begin{aligned}\frac{\mathrm{d}}{\mathrm{d}t} W(u(\cdot, t), v(\cdot, t)) = &- \int_\Omega \left[|u - u^*|^2 + (b_0 + c_0)(u - u^*)(v - v^*) + |v - v^*|^2\right] \\ &- \frac{d_1 u^*}{a_1} \int_\Omega \frac{|\nabla u|^2}{u^2} - \frac{d_2 v^*}{a_2} \int_\Omega \frac{|\nabla v|^2}{v^2}.\end{aligned}$$

Using $0 < b_0 + c_0 < 2$ it is not hard to see that there is a $\delta > 0$ such that

$$\frac{\mathrm{d}}{\mathrm{d}t} W(u(\cdot, t), v(\cdot, t)) \le -\delta \int_\Omega (|u(x, t) - u^*|^2 + |v(x, t) - v^*|^2) \le 0.$$

This shows that W is a Lyapunov function. From this, we infer that the set E and its maximal invariance set $\mathfrak{M}$ (see Definition 7.2.2) coincides with the singleton set

$$E = \mathfrak{M} = \{(u^*, v^*)\}.$$

By LaSalle's invariance principle (Theorem 7.2.3), it follows that $(u, v) \to (u^*, v^*)$ as $t \to \infty$. This shows that the equilibrium (u^*, v^*) attracts all positive solutions.

Finally, if the initial condition of (u, v) is nonnegative and nontrivial, then it follows from the strong maximum principle that $u(x, t) > 0$ and $v(x, t) > 0$ in $\overline{\Omega} \times (0, \infty)$, so it converges to (u^*, v^*) as well. □

Theorem 7.2.4 *Let* (u, v) *be a solution to* (7.13). *If* $a_i = b_i = c_i = 1$, *then* $(u, v) \to (s, 1 - s)$ *for some* $s \in [0, 1]$.

Proof By considering the Lyapunov function $W : [C(\overline{\Omega})]^2 \to \mathbb{R}$,

$$W(u_0, v_0) = \int_\Omega \left[u_0(x) - \frac{1}{2} - \frac{1}{2}\log\left(\frac{u_0(x)}{1/2}\right) + v_0(x) - \frac{1}{2} - \frac{1}{2}\log\left(\frac{v_0(x)}{1/2}\right)\right],$$

then a similar calculation as in the proof of Theorem 7.2.1 gives

$$\frac{\mathrm{d}}{\mathrm{d}t}W(u(\cdot,t),v(\cdot,t)) = -\int_\Omega (u+v-1)^2 - \frac{d_1}{2}\int_\Omega \frac{|\nabla u|^2}{u^2} - \frac{d_2}{2}\int_\Omega \frac{|\nabla v|^2}{v^2}. \quad (7.15)$$

To determine the maximal invariance subset $\mathfrak{M}$ of $E = \{\dot{W} = 0\}$, let $(\hat{u},\hat{v})$ be an entire solution of (7.13) (with $a_i = b_i = c_i = 1$) such that $\frac{\mathrm{d}}{\mathrm{d}t}W(\hat{u},\hat{v}) = 0$ for all $t \in \mathbb{R}$. The identity (7.15) then implies that $\nabla\hat{u} \equiv \nabla\hat{v} \equiv 0$, and $\hat{u} = 1-\hat{v}$. This implies that $\mathfrak{M} = \{(s, 1-s) : 0 \le s \le 1\}$. It follows from LaSalle's invariance principle that

$$\mathrm{dist}((u(\cdot,t),v(\cdot,t)), \{(s,1-s) : 0 \le s \le 1\}) \to 0 \quad \text{as } t \to \infty.$$

To strengthen the conclusion to $(u,v) \to (s_0, 1-s_0)$ for some $s_0 \in [0,1]$, it suffices to note that the omega limit set is non-ordered and hence can only contain a single element of the ordered set $\{(s,1-s) : 0 \le s \le 1\}$. More precisely, fix a $s_0 \in [0,1]$ such that

$$(u(\cdot,t_k),v(\cdot,t_k)) \to (s_0, 1-s_0) \quad \text{for some } t_k \to \infty.$$

Then for $\epsilon > 0$,

$$s_0 - \epsilon \le u(x,t) \le s_0 + \epsilon, \quad 1 - s_0 - \epsilon \le v(x,t) \le 1 - s_0 + \epsilon \quad (7.16)$$

holds for $(x,t) \in \Omega \times \{t_k\}$ for some t_k. By comparison, we deduce that (7.16) holds in $\Omega \times [t_k, \infty)$. Hence,

$$\begin{cases} s_0 - \epsilon \le \liminf_{t\to\infty} u(x,t) \le \limsup_{t\to\infty} u(x,t) \le s_0 + \epsilon, \\ 1 - s_0 - \epsilon \le \liminf_{t\to\infty} v(x,t) \le \limsup_{t\to\infty} v(x,t) \le 1 - s_0 + \epsilon. \end{cases}$$

Since $\epsilon > 0$ is arbitrary, it holds that $(u,v) \to (s_0, 1-s_0)$. □

Theorem 7.2.5 *Suppose a_i, b_i, c_i are positive constants such that*

$$\frac{c_1}{c_2} < \frac{a_1}{a_2}, \quad \text{and} \quad \frac{b_1}{b_2} \le \frac{a_1}{a_2}.$$

Then for any solution (u,v) of (7.13) with nonnegative, nontrivial initial value, we have

$$(u,v) \to \left(\frac{a_1}{c_1}, 0\right) \quad \text{uniformly as } t \to \infty.$$

Proof As in the beginning of the proof of Theorem 7.2.1, we may assume without loss of generality that

$$a_1 = b_1, \quad a_2 = c_2, \quad \frac{c_1}{a_1} < 1 < \frac{b_2}{a_2}, \quad (7.17)$$

so that the homogeneous equilibria are exactly $E_0 = (0,0)$, $E_1 = (1,0)$ and $E_2 = (0,1)$ and there are no positive homogeneous equilibria.

In such a case, we can consider the same Lyapunov function

$$W(u,v) = \int_\Omega \left\{ \frac{1}{a_1}\,[u - 1 - \log u] + \frac{v}{a_2} \right\} \mathrm{d}x,$$

which is obtained by setting $(u^*, v^*) = (1, 0)$ in the previous Lyapunov function. By an analogous computation, we obtain

$$\begin{aligned}
\frac{\mathrm{d}}{\mathrm{d}t} W(u,v) &= \int_\Omega \left[(u-1)\left(1 - u - \frac{c_1}{a_1} v\right) + v\left(1 - v - \frac{b_2}{a_2} u\right)\right] - d_1 \int_\Omega \frac{|\nabla u|^2}{u^2} \\
&\le \int_\Omega \left[(u-1)\left(1 - u - \frac{c_1}{a_1} v\right) + v\,(1 - v - u)\right] \\
&= -\int_\Omega \left[|u - u^*|^2 + \left(\frac{c_1}{a_1} + 1\right)(u - u^*)v + v^2 \right] \\
&\le -\delta \int_\Omega [|u-1|^2 + v^2].
\end{aligned}$$

It then follows from LaSalle's invariance principle that all positive solutions (u, v) of (7.13) converge to $E_1 = (1, 0)$ as $t \to \infty$. This proves the theorem. □

Next, we give a result for general two-species models.

Theorem 7.2.6 *Consider the reaction-diffusion system with homogeneous Neumann conditions:*

$$\begin{cases}
u_t = d_1 \Delta u + f(u,v) & \text{in } \Omega \times (0,\infty), \\
v_t = d_2 \Delta v + g(u,v) & \text{in } \Omega \times (0,\infty), \\
\mathbf{n} \cdot \nabla u = 0 = \mathbf{n} \cdot \nabla v & \text{on } \partial\Omega \times (0,\infty), \\
u(x,0) = u_0(x), \quad v(x,0) = v_0(x) & \text{in } \Omega.
\end{cases} \tag{7.18}$$

Suppose the kinetic system admits a function $E(u, v)$ such that the following statements hold:

1. $E : \mathbb{R}^2 \to \mathbb{R}$ *is convex and there exists $(u^*, v^*) \in \mathbb{R}^2$ such that $E(u, v) \ge E(u^*, v^*)$ for all $u, v > 0$.*
2. $E(u, v) = E_1(u) + E_2(v)$ *for some functions E_i.*
3. E *is a Lyapunov function for the kinetic system, i.e.*

$$E_1'(\eta) f(\eta,\xi) + E_2'(\xi) g(\eta,\xi) < 0 \quad \textit{for all } (\eta,\xi) \ne (u^*, v^*).$$

Then

$$W(u,v) := \int_\Omega (E_1(u) + E_2(v))\,\mathrm{d}x$$

is a Lyapunov function for the PDE system (7.18). *Moreover, let (u, v) be a positive solution of* (7.18). *If*

$$\limsup_{t\to\infty} \|(u(\cdot,t), v(\cdot,t))\|_{C(\overline{\Omega})} < +\infty,$$

then $(u(\cdot,t), v(\cdot,t))$ *converges uniformly to the homogeneous equilibrium* (u^*, v^*) *as* $t \longrightarrow \infty$.

Proof Since $\sup_{t>0} \|(u(\cdot,t), v(\cdot,t))\|_{C(\overline{\Omega})} \le C_0$, standard parabolic regularity yields that $\{(u(\cdot,t), v(\cdot,t) : t \ge 0\}$ is precompact in $C(\overline{\Omega})$.

Next, we show $W(u, v)$ is a Lyapunov function.

$$\begin{aligned}
&\frac{\mathrm{d}}{\mathrm{d}t}W(u(\cdot,t), v(\cdot,t)) \\
&= \int_\Omega E_1'(u)[d_1\Delta u + f(u,v)]\,\mathrm{d}x + \int_\Omega E_2'(u)[d_2\Delta v + g(u,v)]\,\mathrm{d}x \\
&= -d_1\int_\Omega E_1''(u)|\nabla u|^2\,\mathrm{d}x - d_2\int_\Omega E_2''(v)|\nabla v|^2\,\mathrm{d}x \\
&\quad + \int_\Omega \left[E_1'(u)f(u,v) + E_2'(v)g(u,v)\right]\mathrm{d}x \\
&\le \int_\Omega \left[E_1'(u)f(u,v) + E_2'(v)g(u,v)\right]\mathrm{d}x \le 0.
\end{aligned}$$

Next, observe that

$$\dot{W}(u_0, v_0) = 0 \quad \text{if and only if} \quad (u_0, v_0) \equiv (u^*, v^*) \quad \text{in } \Omega.$$

So that $\mathfrak{M} = \{(u^*, v^*)\}$ and one may conclude by Theorem 7.2.3. □

Remark 7.2.7 The above results used the fact that the coefficients of the problem as well as the equilibrium solution are constant in space. See [70] for a recent generalization to nonconstant equilibria under certain assumptions.

7.3 Lotka–Volterra Systems with Heterogeneous Coefficients

How does the spatial heterogeneity of an environment affect the invasion of exotic species and the competition outcome of native and exotic species? By the term heterogeneity, we mean the spatial or temporal variability of various environmental conditions. An example is the spatial heterogeneity of light intensity in the ocean, an essential resource for the growth of a phytoplankton population which is a decreasing function with respect to the depth due to background attenuation [49].

Consider the following competition-diffusion model with spatially heterogeneous coefficients, which reflects the heterogeneity in an environment.

$$\begin{cases} u_t - d_1\Delta u = u(a_1(x) - b_1(x)u - c_1(x)v) & \text{in } \Omega \times (0,\infty), \\ v_t - d_2\Delta v = v(a_2(x) - b_2(x)u - c_2(x)v) & \text{in } \Omega \times (0,\infty), \\ \mathbf{n}\cdot\nabla u = \mathbf{n}\cdot\nabla v = 0 & \text{on } \partial\Omega \times (0,\infty), \end{cases} \tag{7.19}$$

with nonnegative initial data, where $a_i, b_i, c_i \in C^{\alpha}(\overline{\Omega})$, and are strictly positive in $\overline{\Omega}$. The following result concerns the global dynamics of (7.19) for small diffusion coefficients, as considered in [41].

Theorem 7.3.1 *Let (u, v) be a solution to (7.19) with nonnegative and nontrivial initial data. Suppose that*

$$\frac{b_1(x)}{b_2(x)} > \frac{a_1(x)}{a_2(x)} > \frac{c_1(x)}{c_2(x)}, \qquad x \in \overline{\Omega}. \tag{7.20}$$

Then there exists a $\delta > 0$ such that if $\max\{d_1, d_2\} < \delta$, then (7.19) has a unique positive equilibrium, denoted by $(\tilde{u}, \tilde{v})$, such that $(u, v) \to (\tilde{u}, \tilde{v})$ as $t \to \infty$. This equilibrium $(\tilde{u}, \tilde{v})$ is globally asymptotically stable among all positive solutions of (7.19). Furthermore,

$$(\tilde{u}, \tilde{v}) \to (u^*, v^*) := \left(\frac{a_2c_1 - a_1c_2}{b_2c_1 - b_1c_2}, \frac{a_2b_1 - a_1b_2}{b_2c_1 - b_1c_2}\right) \quad \textit{as } \max\{d_1, d_2\} \to 0.$$

Remark 7.3.2 It will be of interest to consider the case when one of the diffusion rates is small. We conjecture that under the assumption (7.20), there exists a $\delta > 0$ such that if $\min\{d_1, d_2\} < \delta$, then (7.19) has a unique positive equilibrium which is globally asymptotically stable among all positive solutions of (7.19).

Lemma 7.3.3 *Let $(\tilde{u}, \tilde{v})$ denote any positive equilibrium to (7.19). Suppose that (7.20) holds. Then $(\tilde{u}, \tilde{v}) \to (u^*, v^*)$ uniformly in $\bar{\Omega}$ as $d_1, d_2 \to 0$.*

Proof For the sake of clarity, set

$$f(x, u, v) = a_1(x) - b_1(x)u - c_1(x)v, \quad g(x, u, v) = a_2(x) - b_2(x)u - c_2(x)v.$$

Let $\overline{u}_0$ denote the unique positive solution of

$$d_1\Delta\overline{u}_0 + \overline{u}_0 f(x, \overline{u}_0, 0) = 0 \quad x \in \Omega, \qquad \frac{\partial \overline{u}_0}{\partial n} = 0 \quad x \in \partial\Omega.$$

By the comparison principle, $\tilde{u} \le \overline{u}_0$ in $\bar{\Omega}$. Hence, $\tilde{v}$ satisfies

$$d_2\Delta\tilde{v} + \tilde{v}g(x, \overline{u}_0, \tilde{v}) \le 0 \quad x \in \Omega, \qquad \frac{\partial \tilde{v}}{\partial n} = 0 \quad x \in \partial\Omega.$$

As $\overline{u}_0 \to a_1/b_1$ uniformly in $\bar{\Omega}$ when $d_1 \to 0$, we see that $g(x, \overline{u}_0, 0) \to a_2 - a_1b_2/b_1 > 0$ uniformly in $\bar{\Omega}$. Thus for sufficiently small d_1, the following problem has a unique positive solution, denoted by $\underline{v}_1$:

$$d_2\Delta\underline{v}_1 + \underline{v}_1 g(x, \overline{u}_0, \underline{v}_1) = 0 \quad x \in \Omega, \qquad \frac{\partial \underline{v}_1}{\partial n} = 0 \quad x \in \partial\Omega.$$

By the comparison principle, $\tilde{v} \ge \underline{v}_1$ in $\bar{\Omega}$. By the equation of $\tilde{u}$, we have

$$d_1\Delta\tilde{u} + \tilde{u}f(x,\tilde{u},\underline{v}_1) \geq 0 \quad x \in \Omega, \quad \frac{\partial \tilde{u}}{\partial n} = 0 \quad x \in \partial\Omega.$$

Therefore, $\tilde{u} \leq \overline{u}_1$ in $\bar{\Omega}$, where $\overline{u}_1$ is the unique positive solution of the equation

$$d_1\Delta\overline{u}_1 + \overline{u}_1 f(x,\overline{u}_1,\underline{v}_1) = 0 \quad x \in \Omega, \quad \frac{\partial \overline{u}_1}{\partial n} = 0 \quad x \in \partial\Omega.$$

Now, for each $k \geq 1$, define $\underline{v}_k$ and $\overline{u}_k$ successively as the unique positive solution of the equations

$$d_2\Delta\underline{v}_k + \underline{v}_k g(x,\overline{u}_{k-1},\underline{v}_k) = 0 \quad x \in \Omega, \quad \frac{\partial \underline{v}_k}{\partial n} = 0 \quad x \in \partial\Omega$$

and

$$d_1\Delta\overline{u}_k + \overline{u}_k f(x,\overline{u}_k,\underline{v}_k) = 0 \quad x \in \Omega, \quad \frac{\partial \overline{u}_k}{\partial n} = 0 \quad x \in \partial\Omega.$$

Furthermore, the following inequalities hold:

$$\tilde{u} \leq \cdots \leq \overline{u}_k \leq \cdots \leq \overline{u}_1 \leq \overline{u}_0, \quad \tilde{v} \geq \cdots \geq \underline{v}_k \geq \cdots \geq \underline{v}_1 \quad \text{in } \Omega.$$

Similarly, we can construct $\{\underline{u}_k\}_{k=1}^{\infty}$ and $\{\overline{v}_k\}_{k=0}^{\infty}$ as follows: $\overline{v}_0 := \tilde{v}$, and $\underline{u}_k$ and $\underline{v}_k$ successively as the unique positive solutions of the equations

$$d_1\Delta\underline{u}_k + \underline{u}_k f(x,\underline{u}_k,\overline{v}_{k-1}) = 0 \quad x \in \Omega, \quad \frac{\partial \underline{u}_k}{\partial n} = 0 \quad x \in \partial\Omega$$

and

$$d_2\Delta\overline{v}_k + \overline{v}_k g(x,\underline{u}_k,\overline{v}_k) = 0 \quad x \in \Omega, \quad \frac{\partial \overline{v}_k}{\partial n} = 0 \quad x \in \partial\Omega.$$

Furthermore, the following inequalities hold:

$$\tilde{u} \geq \cdots \geq \underline{u}_k \geq \cdots \geq \underline{u}_1, \quad \tilde{v} \leq \cdots \leq \overline{v}_k \leq \cdots \leq \overline{v}_1 \leq \overline{v}_0 \quad \text{in } \Omega.$$

Next, we define $\{\overline{U}_k\}_{k=0}^{\infty}$ and $\{\underline{V}_k\}_{k=1}^{\infty}$ successively as follows:

$$\overline{U}_0 = \frac{a_1}{b_1}, \quad g(x,\overline{U}_{k-1},\underline{V}_k) = f(x,\overline{U}_k,\underline{V}_k) = 0 \quad \text{in } \Omega, k \geq 1. \tag{7.21}$$

Similarly, $\{\underline{U}_k\}_{k=1}^{\infty}$ and $\{\overline{V}_k\}_{k=0}^{\infty}$ can be defined by

$$\overline{V}_0 = \frac{a_2}{b_2}, \quad f(x,\underline{U}_k,\overline{V}_k) = g(x,\underline{U}_{k-1},\overline{V}_k) = 0 \quad \text{in } \Omega, k \geq 1. \tag{7.22}$$

It is easy to show that as $d_1, d_2 \to 0$, $(\overline{u}_k,\underline{v}_k) \to (\overline{U}_k,\underline{V}_k)$ and $(\underline{u}_k,\overline{v}_k) \to (\underline{U}_k,\overline{V}_k)$ uniformly in Ω. Hence,

$$\underline{U}_1 \leq \cdots \leq \underline{U}_k \leq \cdots \leq \liminf_{d_1,d_2\to 0} \tilde{u} \leq \limsup_{d_1,d_2\to 0} \tilde{u} \leq \cdots \leq \overline{U}_k \leq \cdots \leq \overline{U}_1 \quad \text{in } \Omega,$$

and

$$\underline{V}_1 \le \cdots \le \underline{V}_k \le \cdots \le \liminf_{d_1,d_2\to 0} \tilde{v} \le \limsup_{d_1,d_2\to 0} \tilde{v} \le \cdots \le \overline{V}_k \le \cdots \le \overline{V}_1 \quad \text{in } \Omega.$$

Set

$$\overline{U} := \lim_{k\to\infty} \overline{U}_k, \quad \underline{U} := \lim_{k\to\infty} \underline{U}_k, \quad \overline{V} := \lim_{k\to\infty} \overline{V}_k, \quad \underline{V} := \lim_{k\to\infty} \underline{V}_k.$$

Passing to the limit in (7.21) and (7.22) we have

$$f(x,\overline{U},\underline{V}) = g(x,\overline{U},\underline{V}) = f(x,\underline{U},\overline{V}) = g(x,\underline{U},\overline{V}) = 0.$$

Hence, $\overline{U} = \underline{U} = u^*$ and $\overline{V} = \underline{V} = v^*$. In particular, this yields that $\overline{U}_k, \underline{U}_k \to u^*$ and $\overline{V}_k, \underline{V}_k \to v^*$ uniformly as $k \to \infty$. This in turn implies that $\tilde{u} \to u^*$ and $\tilde{v} \to v^*$ uniformly in $\bar{\Omega}$ as $d_1, d_2 \to 0$. □

Lemma 7.3.4 *There exists some positive constant δ such that if $0 < d_1, d_2 \le \delta$, then every positive equilibrium of* (7.19) *is linearly stable.*

Proof Let $(\tilde{u}, \tilde{v})$ denote any positive equilibrium of (7.19). The stability of $(\tilde{u}, \tilde{v})$ is determined by the sign of the principal eigenvalue, denoted by λ_1, of the linear problem

$$\begin{cases} d_1\Delta\varphi + a_{11}\varphi + a_{12}\psi + \lambda_1\varphi = 0 & \text{in } \Omega, \\ d_2\Delta\psi + a_{21}\varphi + a_{22}\psi + \lambda_1\psi = 0 & \text{in } \Omega, \\ \frac{\partial\varphi}{\partial n} = \frac{\partial\psi}{\partial n} = 0 & \text{on } \partial\Omega, \end{cases}$$

where $a_{ij} (i = 1, 2)$ are given by, respectively,

$$\begin{aligned} a_{11} &= f(x,\tilde{u},\tilde{v}) + \tilde{u} f_u(x,\tilde{u},\tilde{v}), \\ a_{12} &= \tilde{u} f_v(x,\tilde{u},\tilde{v}), \\ a_{21} &= \tilde{v} g_u(x,\tilde{u},\tilde{v}), \\ a_{22} &= g(x,\tilde{u},\tilde{v}) + \tilde{v} g_v(x,\tilde{u},\tilde{v}). \end{aligned}$$

The existence of λ_1 follows from Theorem 3.3.3. It suffices to show that if d_1, d_2 are sufficiently small, then $\lambda_1 < 0$. Since $\tilde{u} \to u^*$ and $\tilde{v} \to v^*$ uniformly in $\bar{\Omega}$ as $d_1, d_2 \to 0$, we have

$$a_{11} \to -b_1 u^*, \quad a_{12} \to -c_1 v^*, \quad a_{21} \to -b_2 u^*, \quad a_{22} \to -c_2 v^*$$

uniformly in $\bar{\Omega}$. Next, consider the transformation $(\tilde{\varphi}, \tilde{\psi}) = (\varphi, -\psi)$, then λ_1 is the principal eigenvalue of the cooperative system

$$\begin{cases} d_1\Delta\tilde{\varphi} + a_{11}\tilde{\varphi} - a_{12}\tilde{\psi} + \lambda_1\tilde{\varphi} = 0 & \text{in } \Omega, \\ d_2\Delta\tilde{\psi} - a_{21}\tilde{\varphi} + a_{22}\tilde{\psi} + \lambda_1\tilde{\psi} = 0 & \text{in } \Omega, \\ \frac{\partial\tilde{\varphi}}{\partial n} = \frac{\partial\tilde{\psi}}{\partial n} = 0 & \text{on } \partial\Omega. \end{cases}$$

By Corollary 3.2.11, it follows that

$$\lim_{\max\{d_i\}\to 0} \lambda_1 = -\max_{\overline{\Omega}} \Lambda_1(M^*(x)) \tag{7.23}$$

where the matrix $M^*(x)$ is given by

$$M^*(x) = \lim_{\max\{d_i\}\to 0} \begin{pmatrix} a_{11} & -a_{12} \\ -a_{21} & a_{22} \end{pmatrix} = \begin{pmatrix} -b_1 u^*(x) & c_1 v^*(x) \\ b_2 u^*(x) & -c_2 v^*(x) \end{pmatrix} \quad \text{for } x \in \overline{\Omega}.$$

Since

$$\mathrm{Trace}(M^*(x)) = -b_1 u^* - c_2 v^* < 0, \quad \mathrm{Det}(M^*(x)) = u^* v^* (b_1 c_2 - b_2 c_1) > 0,$$

it follows that, for each $x \in \overline{\Omega}$, both eigenvalues of $M^*(x)$ are negative. Since $\Lambda_1(M^*(x))$ is a simple eigenvalue, it depends continuously (in fact smoothly) on the coefficients. Hence,

$$\max_{\overline{\Omega}} \Lambda_1(M^*(x)) < 0.$$

By (7.23), this shows the linear stability of the positive equilibrium $(\tilde{u}, \tilde{v})$ when d_i is sufficiently small. □

Proof (of Theorem 7.3.1) It is a consequence of Lemmas 7.3.3 and 7.3.4: First of all, when both d_1 and d_2 are sufficiently small, both semi-trivial equilibria are linearly unstable under the assumption $\frac{b_1}{b_2} > \frac{a_1}{a_2} > \frac{c_1}{c_2}$; see Problem 7.5.2. Hence, by the theory of monotone dynamical systems, (7.19) has at least one positive equilibrium. By Lemma 7.3.4 and the theory of monotone dynamical systems (specifically Theorem 7.1.7), the positive equilibrium of (7.19) is unique and globally asymptotically stable among all nonnegative and nontrivial initial data. □

Theorem 7.3.5 *Let (u, v) be a solution to* (7.19). *Suppose that $\frac{\bar{b}_1}{\bar{b}_2} > \frac{\bar{a}_1}{\bar{a}_2} > \frac{\bar{c}_1}{\bar{c}_2}$, where $\bar{f} := \int_\Omega f/|\Omega|$. Then there exists a $\delta > 0$ such that if $\min\{d_1, d_2\} > \frac{1}{\delta}$,* (7.19) *has a unique positive equilibrium (u^*, v^*), which is linearly stable and globally asymptotically stable. Moreover,*

$$(u^*, v^*) \to \left(\frac{\bar{a}_2\bar{c}_1 - \bar{a}_1\bar{c}_2}{\bar{b}_2\bar{c}_1 - \bar{b}_1\bar{c}_2}, \frac{\bar{a}_2\bar{b}_1 - \bar{a}_1\bar{b}_2}{\bar{b}_2\bar{c}_1 - \bar{b}_1\bar{c}_2} \right) \quad \text{as } \min\{d_1, d_2\} \to \infty. \tag{7.24}$$

Proof By Problem E.3.2, (7.19) has at least one positive equilibrium (u^*, v^*) for $\max\{d_i\} \gg 1$. By Problem 7.5.5, every positive equilibrium (u^*, v^*) is linearly stable. By Theorem 7.1.7, there exists an $M > 0$ such that for $\min\{d_i\} \geq M$, (7.19) has a unique positive equilibrium (u^*, v^*). Moreover, (u^*, v^*) attracts all nonnegative, nontrivial solutions of (7.19). The formula (7.24) is proved in Problem 7.5.4. □

7.3.1 Slow vs fast diffusing populations

It was shown in previous subsections that the successful invasion of an exotic species is often attributed to its competitive advantage over the resident species. However, in a heterogeneous environment, with the help of (having less) diffusion, invasion is still possible even though the exotic species does not possess any apparent competitive advantage. In this section, we consider some classical competition models to illustrate this phenomenon.

The following semilinear system

$$\begin{cases} u_t = d_1\Delta u + u[m(x) - u - bv] & \text{in } \Omega \times (0,\infty), \\ v_t = d_2\Delta v + v[m(x) - cu - v] & \text{in } \Omega \times (0,\infty), \\ \mathbf{n}\cdot\nabla u = \mathbf{n}\cdot\nabla v = 0 & \text{on } \partial\Omega \times (0,\infty) \end{cases} \tag{7.25}$$

models two species that are competing for the same resources, where $u(x,t)$ and $v(x,t)$ represent the population densities of two competing species with respective dispersal rates d_1 and d_2, the function $m(x)$ represents the common intrinsic growth rate, and b and c are inter-specific competition coefficients. Here we assume that the two species has the same intrinsic growth rate $m(x)$. This is possible if, for example, the two species depend on the same spatially distributed resource. We shall assume that d_1, d_2, b and c are positive constants, and $u(x,0)$ and $v(x,0)$ are nonnegative functions that are not identically equal to zero.

We say that an equilibrium of (7.25) is a *coexistence equilibrium* if both components are positive, and it is a *semi-trivial equilibrium* if one component is positive and the other is zero. If m is nonnegative and nontrivial, (7.25) has two semi-trivial equilibria, denoted by $E_1 = (\theta_{d_1}, 0)$ and $E_2 = (0, \theta_{d_2})$, where $\theta_d = \theta(\cdot, d)$ is the unique positive solution of

$$d\Delta\theta + \theta(m-\theta) = 0 \quad \text{in } \Omega, \qquad \text{and} \qquad \mathbf{n}\cdot\nabla\theta = 0 \quad \text{on } \partial\Omega.$$

We first consider the case $b = c = 1$. If $m(x) \equiv 1$, then it was proved in Theorem 7.2.1 that $(u,v) \to (s, 1-s)$ for some $s \in [0,1]$ that depends on initial conditions. This means that diffusion plays no role in the outcome of the competition in homogeneous environments. When $m(x)$ is nonconstant, it was proved by Hastings [26] and Dockery et al. [19] that the slower diffuser always wins, as the following theorem illustrates.

Theorem 7.3.6 *Let* $0 < d_1 < d_2$, $b = c = 1$ *and* $m(x)$ *be nonnegative and nonconstant. Let* (u,v) *be a solution to (7.25), then* $(u,v) \to E_1 = (\theta_{d_1}, 0)$ *regardless of initial conditions.*

Proof The proof is divided into two steps.

Step 1. We claim that (7.25) has no positive equilibria. Suppose (7.25) has a positive equilibrium (u^*, v^*). We claim that $m - u^* - v^*$ is nonconstant. Suppose not, then $m - u^* - v^* = C$, and so

$$d_1\Delta u^* + Cu^* = 0 \text{ in } \Omega, \quad \mathbf{n}\cdot\nabla u^* = 0 \text{ on } \partial\Omega.$$

Hence u^* is a positive eigenfunction of the Neumann Laplacian in Ω. This implies that u^* is constant. Similarly, v^* is also a constant. This leads to the conclusion that $m = C + u^* + v^*$ is a constant, which is impossible. Hence $m - u^* - v^*$ is nonconstant in Ω.

For $d > 0$ and $h \in C(\overline{\Omega})$, denote by $\mu_1(d, h)$ the principal eigenvalue of

$$-d\Delta\phi - h\phi = \lambda\phi \quad \text{in } \Omega, \quad \text{and} \quad \mathbf{n} \cdot \nabla\phi = 0 \quad \text{on } \partial\Omega.$$

By regarding u^* and v^* as positive eigenfunctions, we deduce from Remark 1.3.7 that

$$\mu_1(d_i, m - u^* - v^*) = 0 \quad \text{for } i = 1, 2. \tag{7.26}$$

Since $m - u^* - v^*$ is nonconstant, it follows from Proposition 1.3.15 that $d \mapsto \mu_1(d, m - u^* - v^*)$ is strictly increasing. This is a contradiction with (7.26). Hence, (7.25) has no positive equilibrium when $d_1 < d_2$ and $b = c = 1$.

Step 2. We claim that $E_2 = (0, \theta_{d_2})$ is linearly unstable, i.e. u can invade when rare. First, one can argue similarly as above that $m - \theta_{d_2}$ is nonconstant. Second, it follows again from Remark 1.3.7 that $\mu_1(d_2, m - \theta_{d_2}) = 0$, since θ_{d_2} supplies a positive eigenfunction. Thirdly, we apply Proposition 1.3.15 to deduce that $d \mapsto \mu_1(d, m - \theta_{d_2})$ is strictly increasing. Since $d_1 < d_2$, we get

$$\mu_{E_2} = \mu_1(d_1, m - \theta_{d_2}) < \mu_1(d_2, m - \theta_{d_2}) = 0.$$

Thanks to Lemma 7.1.1, this implies that $(0, \tilde{v})$ is linearly unstable.

Applying one of Theorems 7.1.2 or 7.1.6, we deduce that $(u, v) \to E_1 = (\theta_{d_1}, 0)$ for every solution (u, v) of (7.25) with nonnegative, nontrivial initial conditions. □

Remark 7.3.7 For two-species competition models, if one species adopts random movement and the other species does not move, the global dynamics of such systems has been studied recently in [20, 61].

Remark 7.3.8 A challenging open problem, proposed by Dockery et al. [19], is to show that for any $N \geq 3$ and any set of diffusion rates $0 < d_1 < d_2 < \cdots < d_N$, all positive solutions of the N-species competition model

$$\begin{cases} \partial_t u_i = d_i \Delta u_i + u_i(m(x) - \sum_{j=1}^N u_j) & \text{in } \Omega \times (0, \infty), \ 1 \leq i \leq N, \\ \mathbf{n} \cdot \nabla u_i = 0 & \text{on } \partial\Omega \times (0, \infty), \ 1 \leq i \leq N, \\ u_i(x, 0) = u_{i,0}(x) & \text{in } \Omega, \ 1 \leq i \leq N, \end{cases}$$

converge to the equilibrium $E_1 = (\theta_{d_1}, 0, \ldots, 0)$.

7.3.2 Weak competition in a heterogeneous environment

Next, we discuss the weak competition case of (7.25), namely, we set $b, c \in (0, 1)$. If $0 < b, c < 1$ and $m(x) \equiv \overline{m}$ for some positive constant $\overline{m}$, by Theorem 7.2.1, every solution (u, v) of (7.25) converges to $(\frac{1-b}{1-bc}\overline{m}, \frac{1-c}{1-bc}\overline{m})$ for all diffusion rates d_1, d_2 and any initial data. However, the dynamics of (7.25) is less transparent when m is nonconstant. To this end, we start by studying the stability of the semi-trivial equilibrium $(\theta_{d_1}, 0)$ of (7.25). For the rest of this subsection, we focus on the case $0 < c < 1$.

Lemma 7.3.9 *If $m(x)$ is nonnegative and nonconstant, then there exists some constant $c_* = c_*(m, \Omega) \in (0, 1)$ such that the following hold:*

(a) *For $c \in (0, c_*]$, $(\theta_{d_1}, 0)$ is unstable for any $d_1, d_2 > 0$.*
(b) *For $c \in (c_*, 1)$, there exists a $d_* = d_*(c, m, \Omega) > 0$ such that* (i) *for $d_2 \in (0, d_*)$, $(\theta_{d_1}, 0)$ is unstable for any $d_1 > 0$; and* (ii) *for $d_2 > d_*$, $(\theta_{d_1}, 0)$ changes stability at least twice as d_1 increases from 0 to d_2.*

Lemma 7.3.9 applies to any $b \geq 0$. The most interesting case is when $c_* < c < 1$ and $d_2 > d_*$, where the followings hold:

(i) If $b > 1$, it is well known that without dispersal, species v always drives species u to extinction. However, with dispersal, for some ranges of dispersal rates, species v may fail to invade when rare.
(ii) If $b < 1$, it is well known that, without dispersal, species u and v always coexist. Surprisingly, for certain dispersal rates, species u is able to drive species v to extinction for arbitrary initial conditions.

Proof By Lemma 7.1.1, the stability of $(\theta_{d_1}, 0)$ is determined by the sign of the smallest eigenvalue, denoted by λ_1, of the problem

$$\begin{cases} d_2\Delta\varphi + (m - c\theta_{d_1})\varphi + \lambda\varphi = 0 & \text{in } \Omega, \\ \mathbf{n} \cdot \nabla\varphi = 0 & \text{on } \partial\Omega. \end{cases} \tag{7.27}$$

To determine the sign of λ_1, note that λ_1 is strictly increasing in d_2, and

$$\lim_{d_2 \to 0} \lambda_1 = \min_{\bar{\Omega}}(c\theta_{d_1} - m) \leq \min_{\bar{\Omega}}(\theta_{d_1} - m) < 0; \tag{7.28}$$

$$\lim_{d_2 \to +\infty} \lambda_1 = \frac{\int_\Omega \theta_{d_1}}{|\Omega|}\left(c - \frac{\int_\Omega m}{\int_\Omega \theta_{d_1}}\right). \tag{7.29}$$

Set

$$c_* = \inf_{d_1 > 0} \frac{\int_\Omega m}{\int_\Omega \theta_{d_1}}.$$

By Corollary 6.2, $c_* \in (0, 1)$.

For every $c \in (0, c_*]$ and any $d_1 > 0$, $\lim_{d_2 \to +\infty} \lambda_1 \leq 0$. Since λ_1 is strictly increasing in d_2, we see that $\lambda_1 < 0$ for any $d_1, d_2 > 0$. This proves part (a).

For every $c \in (c_*, 1)$, for simplicity assume that there exist two positive constants $\underline{d} < \overline{d}$ such that $c - \int_\Omega m / \int_\Omega \theta_{d_1} > 0$ for $d_1 \in (\underline{d}, \overline{d})$, and $c - \int_\Omega m / \int_\Omega \theta_{d_1} < 0$ for $d_1 \in (0, \underline{d}) \cup (\overline{d}, +\infty)$. Define $d^* = d^*(d_1) := 1/\lambda_1(m - c\theta_{d_1})$, i.e.

$$d^* = \sup_{\varphi \in \mathcal{S}} \frac{\int_\Omega (m - c\theta_{d_1})\varphi^2}{\int_\Omega |\nabla \varphi|^2},$$

where

$$\mathcal{S} = \{\varphi \in H^1(\Omega) : \int_\Omega (m - c\theta_{d_1})\varphi^2 > 0\}.$$

Now, $d^* = +\infty$ if and only if the set $\mathcal{S}$ contains the constant function $\varphi \equiv 1$, and the latter holds if and only if $c - \int_\Omega m / \int_\Omega \theta_{d_1} < 0$, i.e. $d \in (0, \underline{d}] \cup [\overline{d}, +\infty)$. In particular, $d^*(d_1)$ is finite when $d_1 \in (\underline{d}, \overline{d})$, and that $d^*(d_1) \to +\infty$ as $d_1 \to \underline{d}-$ or $d_1 \to \overline{d}+$. This allows us to define $d_* = \inf_{d_1 > 0} d^*(d_1)$. For $d_2 < d_*$, we have $d_2 < d^*(d_1)$ for all $d_1 > 0$. In this case, $\lambda_1 < 0$ for all $d_1 > 0$, which implies that $(\theta_{d_1}, 0)$ is unstable for any $d_1 > 0$ and $d_2 < d_*$. For $d_2 > d_*$, we likewise have $\lambda_1 < 0$ for $d_1 \in (0, \underline{d}) \cup (\overline{d}, +\infty)$; and $\lambda_1 > 0$ in some subinterval of $(\underline{d}, \overline{d})$. Therefore λ_1 changes sign twice as d_1 increases from 0 to d_2, i.e. part (b) is proved. □

Remark 7.3.10 It holds that $\lambda_1 < 0$ whenever $c < 1$ and $d_1 \geq d_2$; see Problem 7.5.6.

For every $c > 0$, define

$$\Sigma_c = \left\{(d_1, d_2) \in \mathbb{R}_+^2 : \ E_1 = (\theta_{d_1}, 0) \text{ is linearly stable}\right\}. \tag{7.30}$$

Note that $\Sigma_c \subset \{(d_1, d_2) \in \mathbb{R}_+^2 : d_1 < d_2\}$ since, by the comparison principle for principal eigenvalues, $\lambda_1 < 0$ for $d_1 \geq d_2$. Clearly, Σ_c is nonempty if and only if $c > c_*$.

In a series of works He and Ni classified the dynamics of a class of Lotka–Volterra competition-diffusion models which include system (7.25) as a special case [27, 28, 29, 30]. One of their main results can be stated as follows:

Theorem 7.3.11 *If $m(x)$ is nonnegative and nonconstant, $c \in (c_*, 1)$ and $0 < b \leq 1$, then $(\theta_{d_1}, 0)$ is globally asymptotically stable for any $(d_1, d_2) \in \overline{\Sigma}_c$; if $d_2 > d_1$ and $(d_1, d_2) \notin \overline{\Sigma}_c$, then system (7.25) has a unique positive equilibrium which is globally asymptotically stable.*

The most interesting case of Theorem 7.3.11 is when $c_* < c < 1$ and $d_2 > d_*$: Without dispersal, species u and v always coexist; with dispersal, for $(d_1, d_2) \in \overline{\Sigma}_c$, species u will eliminate species v for arbitrary initial data. By a previous result, $\Sigma_c \subset \{d_1 \leq d_2\}$, i.e. species u may exclude species v only if u is the slower diffuser. Furthermore, the set Σ_c is nonempty for every $c \in (c_*, 1)$. It is not difficult to see that $\Sigma_{c_1} \subset \Sigma_{c_2}$ for any $c_1 < c_2$ with $c_1, c_2 \in (c_*, 1)$. In fact, the set Σ_c converges to the set $\{(d_1, d_2) : 0 < d_1 < d_2\}$ as $c \to 1-$, and this gives another perspective upon why the slower diffuser wins the competition for the case when $b = c = 1$; see Theorem 7.3.6.

The key ingredient in the proof of Theorem 7.3.11 is the following lemma:

Lemma 7.3.12 *If $bc < 1$, then any positive equilibrium of (7.25), if it exists, is linearly stable.*

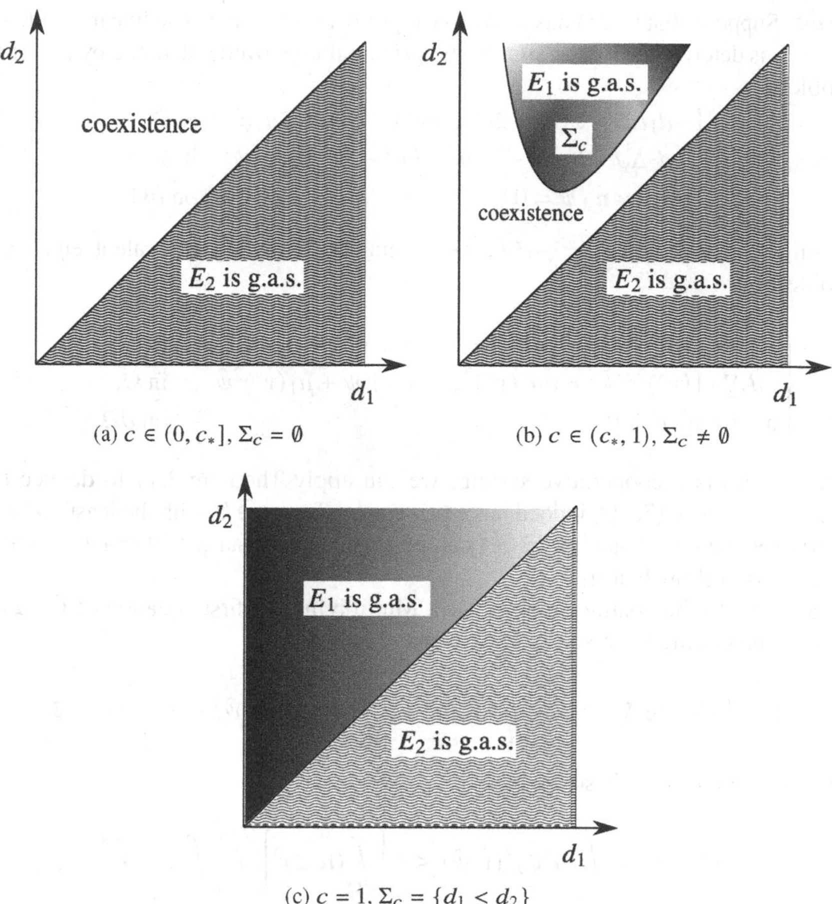

(a) $c \in (0, c_*], \Sigma_c = \emptyset$

(b) $c \in (c_*, 1), \Sigma_c \neq \emptyset$

(c) $c = 1, \Sigma_c = \{d_1 < d_2\}$

Fig. 7.2: Three diagrams illustrating the long-time dynamics of (7.25), with $b = 1$ and $c \in (0, 1)$. For $d_1 > d_2$, the equilibrium E_2 is always globally asymptotically stable (g.a.s.). The long-time dynamics when $d_1 < d_2$ depends on the value of c. When $c \in (0, c_*)$, the set Σ_c is empty, and both E_1 and E_2 are unstable for $d_1 < d_2$ and there exists a unique positive equilibrium that is g.a.s. When $c \in (c_*, 1)$, there exists a nonempty subset Σ_c of $\{d_1 < d_2\}$ such that for $(d_1, d_2) \in \Sigma_c$, E_1 is g.a.s., whereas for $(d_1, d_2) \notin \Sigma_c$ such that $d_1 < d_2$, there is a unique coexistence equilibrium that is g.a.s. Here $E_1 = (\theta_{d_1}, 0)$ and $E_2 = (0, \theta_{d_2})$ are respectively the semitrivial equilibria of (7.25). Finally, one can show that Σ_c is monotonically increasing in c, and converges to the set $\{d_1 < d_2\}$ as $c \nearrow 1$.

Proof Suppose that (7.25) has a positive equilibrium (u^*, v^*). The linear stability of (u^*, v^*) is determined by the sign of the principal eigenvalue, denoted by μ_1, of the problem

$$\begin{cases} -d_1\Delta\varphi - \varphi(m - 2u^* - cv^*) + cu^*\psi = \mu_1\varphi & \text{in } \Omega, \\ -d_2\Delta\psi + bv^*\varphi - \psi(m - bu^* - 2v^*) = \mu_1\psi & \text{in } \Omega, \\ \mathbf{n}\cdot\varphi = \mathbf{n}\cdot\psi = 0 & \text{on } \partial\Omega. \end{cases} \tag{7.31}$$

Setting $\varphi = u^*\tilde{\varphi}$ and $\psi = -v^*\tilde{\psi}$, we obtain the following equivalent eigenvalue problem:

$$\begin{cases} -d_1\nabla\cdot[(u^*)^2\nabla\tilde{\varphi}] = -(u^*)^3\tilde{\varphi} + c(u^*)^2v^*\tilde{\psi} + \mu_1(u^*)^2\tilde{\varphi} & \text{in } \Omega, \\ -d_2\nabla\cdot[(v^*)^2\nabla\tilde{\psi}] = bu^*(v^*)^2\tilde{\varphi} - (v^*)^3\tilde{\psi} + \mu_1(v^*)^2\tilde{\psi} & \text{in } \Omega, \\ \mathbf{n}\cdot\tilde{\varphi} = \mathbf{n}\cdot\tilde{\psi} = 0 & \text{on } \partial\Omega. \end{cases} \tag{7.32}$$

Since (7.32) is a cooperative system, we can apply Theorem 3.11 to deduce that (7.32) (and hence (7.31)) indeed has a principal value $\mu_1 \in \mathbb{R}$ with the least real part. Moreover, the eigenfunction $(\tilde{\varphi}, \tilde{\psi})$ can be chosen such that $\tilde{\varphi} > 0$ and $\tilde{\psi} > 0$ in Ω. It remains to show that $\mu_1 > 0$.

Suppose to the contrary that $\mu_1 \le 0$. Multiplying the first equation of (7.32) by $\tilde{\varphi}^2$ and integrating the result in Ω, we have

$$2d_1\int_\Omega (u^*)^2\tilde{\varphi}|\nabla\tilde{\varphi}|^2 = -\int_\Omega (u^*\tilde{\varphi})^3 + c\int_\Omega (u^*\tilde{\varphi})^2(v^*\tilde{\psi}) + \mu_1\int_\Omega (u^*)^2\tilde{\varphi}^3.$$

By assumption, $\mu_1 \le 0$, so we have

$$\int_\Omega (u^*\tilde{\varphi})^3 \le c\int_\Omega (u^*\tilde{\varphi})^2(v^*\tilde{\psi}) \le c\left[\int_\Omega (u^*\tilde{\varphi})^3\right]^{2/3}\left[\int_\Omega (v^*\tilde{\psi})^3\right]^{1/3},$$

where we used Hölder's inequality. Hence, we obtain

$$0 < \int_\Omega (u^*\tilde{\varphi})^3 \le c^3\int_\Omega (v^*\tilde{\psi})^3. \tag{7.33}$$

By arguing analogously, we obtain

$$0 < \int_\Omega (v^*\tilde{\psi})^3 \le b^3\int_\Omega (u^*\tilde{\varphi})^3. \tag{7.34}$$

However (7.33) and (7.34) are in contradiction with $bc < 1$. This proves that $\mu_1 > 0$, which means that (u^*, v^*) is linearly stable. □

Now, we prove Theorem 7.3.11.

Proof (of Theorem 7.3.11) We only need to consider $0 < d_1 < d_2$. We claim that $E_2 = (0, \theta_{d_2})$ is linearly unstable. Indeed, this follows from Lemma 7.1.1 and

$$\mu_{E_2} = \mu_1(d_1, m - b\theta_{d_2}) < \mu_1(d_1, m - \theta_{d_2}) \le \mu_1(d_2, m - \theta_{d_2}) = 0.$$

Let $d_2 > d_1$ and $(d_1, d_2) \notin \bar{\Sigma}_c$. Then both $E_1 = (\theta_{d_1}, 0)$ and $E_2 = (0, \theta_{d_2})$ are linearly unstable, and there exists at least one positive equilibrium (u^*, v^*). By Lemma 7.3.12, every positive equilibrium is linearly stable, and hence locally asymptotically stable. It follows from Theorem 7.1.7 that (u^*, v^*) is unique and is globally asymptotically stable.

Suppose next, that $(d_1, d_2) \in \Sigma_c$, then E_1 is linearly stable and E_2 is linearly unstable. We need to show that E_1 is globally asymptotically stable. In view of Theorem 7.1.6 and the instability of E_2, it remains to show the nonexistence of positive equilibria. Suppose to the contrary that there is one positive equilibrium (u^*, v^*), then it follows from Theorem 7.1.7 that (u^*, v^*) is unique and is globally asymptotically stable, which is impossible since E_1 is linearly stable and hence locally asymptotically stable. We conclude that E_1 is globally asymptotically stable when $(d_1, d_2) \in \Sigma_c$.

Finally, consider the case $(d_1, d_2) \in \partial\Sigma_c$. It is again enough to show that there is no positive equilibrium, so that the instability of E_2 implies E_1 is globally asymptotically stable, thanks to Theorem 7.1.6. Suppose to the contrary that there is one positive equilibrium (u^*, v^*), then Lemma 7.3.12 asserts that (u^*, v^*) is linearly stable and hence is nondegenerate. It then follows from the implicit function theorem that there is a neighborhood $\mathcal{N}$ of (d_1, d_2) in $\mathbb{R}^2_+$ such that the competition system has at least one positive equilibrium whenever the diffusion rate lies in $\mathcal{N}$. But this is in contradiction with the fact that $\mathcal{N} \cap \Sigma_c$ is nonempty, and that $(d_1, d_2) \in \Sigma_c$ implies E_1 is globally asymptotically stable. □

7.4 Competition in an Advective Environment

In advective environments, higher dispersal rates can evolve. In fact, populations with a higher dispersal rate can not only prevent the invasion of populations with a slower dispersal rate but can also invade and replace such populations. In this connection, consider the system

$$\begin{cases} u_t = d_1 u_{xx} - q u_x + u(r - u - v), & 0 < x < L,\ t > 0, \\ v_t = d_2 v_{xx} - q v_x + v(r - u - v), & 0 < x < L,\ t > 0, \\ d_1 u_x(0,t) - q u(0,t) = d_2 v_x(0,t) - q v(0,t) = 0, \\ u_x(L,t) = v_x(L,t) = 0, \end{cases} \tag{7.35}$$

with nonnegative, nontrivial initial data. The population is subject to the no-flux boundary condition and a free-flow boundary condition; see [59]. Throughout this section, we assume that q and r are positive constants.

Remark 7.4.1 By the transformation

$$\tilde{u}(x,t) = \mathrm{e}^{-qx/d_1} u(x,t) \text{ and } \tilde{v}(x,t) = \mathrm{e}^{-qx/d_2} v(x,t),$$

we obtain

$$\begin{cases} \tilde{u}_t = d_1\tilde{u}_{xx} + q\tilde{u}_x + \tilde{u}(r - \mathrm{e}^{qx/d_1}\tilde{u} - \mathrm{e}^{qx/d_2}\tilde{v}), & 0 < x < L,\ t > 0, \\ \tilde{v}_t = d_2\tilde{v}_{xx} + q\tilde{v}_x + \tilde{v}(r - \mathrm{e}^{qx/d_1}\tilde{u} - \mathrm{e}^{qx/d_2}\tilde{v}), & 0 < x < L,\ t > 0, \\ \tilde{u}_x(0,t) = \tilde{v}_x(0,t) = 0, \quad t > 0, \\ d_1\tilde{u}_x(L,t) + q\tilde{u}(L,t) = d_2\tilde{v}_x(L,t) + q\tilde{v}(L,t) = 0, \quad t > 0, \end{cases}$$

with nonnegative initial data. This latter system is a special case of (7.1) and satisfies all the hypotheses in Section 7.1. In particular, Theorems 7.1.2, 7.1.5, 7.1.6 and 7.1.7 are applicable to the semiflow governing $(\tilde{u}, \tilde{v})$, and also the semiflow governing (u, v) by extension.

Theorem 7.4.2 *Suppose that q, r are positive constants. If $d_1 > d_2$, then $(u^*, 0)$, whenever it exists, is globally asymptotically stable among all nonnegative and nontrivial initial data.*

The proof of this theorem is preceded by a number of lemmas. We begin by considering the eigenvalue problem

$$\begin{cases} \tau\phi_{xx} - q\phi_x + h(x)\phi + \lambda\phi = 0, & 0 < x < L, \\ \tau\phi_x(0) = q\phi(0), & \phi_x(L) = 0. \end{cases} \tag{7.36}$$

For any $\tau > 0$, let $\lambda(\tau)$ denote the principal eigenvalue and ϕ the corresponding eigenfunction, uniquely determined by $\max\limits_{0\le x\le L} \phi(x) = 1$.

Lemma 7.4.3 *Suppose $h(x) > 0$ in $(0, L)$. Then $\lambda(\tau)$ has at most one positive root. Moreover, if $\lambda(\tau^*) = 0$ for some $\tau^* > 0$, then $\frac{\partial\lambda}{\partial\tau}(\tau^*) < 0$.*

Proof By direct calculations (see Problem 7.5.7),

$$\frac{\partial\lambda}{\partial\tau} = \frac{\int(\mathrm{e}^{-\alpha x}\phi)_x\phi_x\mathrm{d}x}{\int \mathrm{e}^{-\alpha x}\phi^2\mathrm{d}x}, \tag{7.37}$$

where $\alpha = q/\tau$. It is clear that our lemma is a consequence of the following assertion:

Claim: If $\lambda(\tau^*) = 0$ for some $\tau^* > 0$, then

$$\left.\frac{\partial\lambda}{\partial\tau}\right|_{\tau=\tau^*} < 0. \tag{7.38}$$

To establish (7.38), let ϕ^* denote the eigenfunction corresponding to $\lambda(\tau^*)$, uniquely determined by $\max \phi^* = 1$. As $\lambda(\tau^*) = 0$, ϕ^* satisfies

$$\begin{cases} \tau^*\phi^*_{xx} - q\phi^*_x + h(x)\phi^* = 0, & 0 < x < L, \\ \tau^*\phi^*_x(0) - q\phi^*(0) = \phi^*_x(L) = 0. \end{cases} \tag{7.39}$$

Integrating equation (7.39) from zero to x, we find

$$\tau^* \phi_x^*(x) - q\phi^*(x) + \int_0^x \phi^*(y)h(y)\mathrm{d}y = 0. \tag{7.40}$$

As $h > 0$ in $(0, L)$, we conclude $0 > \tau^* \phi_x^* - q\phi^* = \tau^* \mathrm{e}^{\alpha^* x}(\mathrm{e}^{-\alpha^* x}\phi^*)_x$, where $\alpha^* = q/\tau^*$.

Next we show that $\phi_x^* > 0$ in $[0, L)$. If $\phi^*(0) = 0$, then by the boundary condition of ϕ^* at $x = 0$, we also have $\phi_x^*(0) = 0$. By the equation of ϕ^* and the uniqueness of ODEs, we have $\phi^* \equiv 0$ in $(0, L)$, which is a contradiction. Hence, $\phi^*(0) > 0$. This implies that $\phi_x^*(0) > 0$. If $\phi_x^* > 0$ in $[0, L)$ does not hold, there exists some $x_0 \in (0, L)$ such that $\phi_x^* > 0$ in $[0, x_0)$ and $\phi_x^*(x_0) = 0$. Since $\phi_x^*(L) = 0$, there must exist some $x_1 \geq x_0$ such that $\phi_x^*(x_1) = 0$ and $\phi_{xx}^*(x_1) \geq 0$. By the equation of ϕ^*, $h(x_1) \leq 0$, which is a contradiction. Hence, $\phi_x^* > 0$ in $[0, L)$.

Finally, using the facts that $(\mathrm{e}^{-\alpha^* x}\phi^*)_x < 0$ and $\phi_x^* > 0$ in $(0, L)$, we see that (7.37) implies (7.38). □

Lemma 7.4.4 *If $d_1 > d_2$, then $(u^*, 0)$, whenever it exists, is linearly stable.*

Proof Let u^* be the unique positive solution of

$$\begin{cases} d_1 u_{xx}^* - q u_x^* + u^*(r - u^*) = 0, & 0 < x < L, \\ d_1 u_x^*(0) - q u^*(0) = 0 = u_x^*(L). \end{cases} \tag{7.41}$$

It remains to show that $\lambda(d_2) > 0$, where $\lambda(\tau)$ is the principal eigenvalue of (7.36) with $h(x) = r - u^*$.

Note that $\lambda(d_1) = 0$, since u^* supplies a positive eigenfunction.

Since $\overline{u} \equiv r$ is a superequilibrium, it follows from Corollary 5.1.9 that $u^* < r$ in $[0, L]$ (see also Problem 7.5.8), i.e. $h > 0$. It follows from Lemma 7.4.3 that d_1 is the unique positive root of $\lambda(\tau)$, and that $\lambda(d_2) > 0$ since $d_2 \in (0, d_1)$. □

Next, we will show that (7.35) has no coexistence (positive) equilibrium. First, we establish the following *a priori* estimate on any positive equilibrium of (7.35).

Lemma 7.4.5 *Let (u, v) be any positive equilibrium of* (7.35). *Then $u_x, v_x > 0$ in $[0, L)$ and $u + v < r$ in $[0, L]$.*

Proof Differentiating the equations of u and v we obtain

$$\begin{cases} d_1 u_{xxx} - q u_{xx} + u_x(r - 2u - v) - u v_x = 0, & 0 < x < L, \\ d_2 v_{xxx} - q v_{xx} + v_x(r - u - 2v) - v u_x = 0, & 0 < x < L, \\ d_1 u_x - q u = d_2 v_x - q v = 0 \quad \text{at } x = 0, \\ u_x = v_x = 0 \quad \text{at } x = L. \end{cases} \tag{7.42}$$

Set $w = u_x/u$ and $z = v_x/v$. Then w and z satisfy

$$\begin{cases} d_1 w_{xx} + (2d_1 u_x/u - q)w_x - uw - vz = 0, & 0 < x < L, \\ d_2 z_{xx} + (2d_2 v_x/v - q)z_x - uw - vz = 0, & 0 < x < L, \\ w(0) = q/d_1 > 0,\ z(0) = q/d_2 > 0,\ w(L) = z(L) = 0. \end{cases} \tag{7.43}$$

We first show that $u(L) + v(L) \neq r$. To this end, we argue by contradiction: suppose that $u(L) + v(L) = r$. By the boundary condition $u_x(L) = v_x(L) = 0$ and equations of u and v, we obtain $u_{xx}(L) = v_{xx}(L) = 0$. That is, $w_x(L) = z_x(L) = 0$. As $w(L) = z(L) = 0$, by the uniqueness of ODEs we obtain $w = z = 0$ in $[0, L]$, i.e., both u and v are positive constants. However, this contradicts the boundary conditions at $x = 0$. Hence, it suffices to consider two cases: (i) $u(L)+v(L) > r$; (ii) $u(L) + v(L) < r$.

Case (i). By the equations of u, v and the boundary conditions $u_x(L) = v_x(L) = 0$, there exists some $\delta > 0$ such that $u_{xx} > 0$ and $v_{xx} > 0$ for $L - \delta \le x \le L$. Hence, $u_x < 0$ and $v_x < 0$ in $[L - \delta, L)$. Since $u_x(0) > 0$ and $v_x(0) > 0$, u_x and v_x must have some root in $(0, L - \delta)$. Without loss of generality, we may assume that there exists some $x_1 \in (0, L - \delta)$ such that $u_x(x_1) = 0$, $v_x(x_1) \le 0$, $u_x < 0$ and $v_x < 0$ in (x_1, L). Recall that $w = u_x/u$ in (x_1, L). Choose $x_2 \in (x_1, L)$ such that $w(x_2) = \min_{x_1 \le x \le L} w(x)$. Hence, $w(x_2) < 0$, $w_x(x_2) = 0$, and $w_{xx}(x_2) \ge 0$. By the equation of w we obtain $z(x_2) > 0$, which is a contradiction. This rules out Case (i).

Case (ii). By the equations of u, v and the boundary conditions $u_x(L) = v_x(L) = 0$, there exists some $\delta > 0$ such that $u_{xx} < 0$ and $v_{xx} < 0$ for $L - \delta \le x \le L$. Hence, $u_x > 0$ and $v_x > 0$ in $[L - \delta, L)$. We claim that $u_x > 0$ in $[0, L)$. Suppose to the contrary that there exists some $x_3 \in (0, L - \delta)$ such that $u_x(x_3) = 0$ and $u_x > 0$ in (x_3, L). Choose $x_4 \in (x_3, L)$ such that $w(x_4) = \max_{x_3 \le x \le L} w(x)$. Hence, $w(x_4) > 0$, $w_x(x_4) = 0$, and $w_{xx}(x_4) \le 0$. By the equation of w we obtain $z(x_4) < 0$. Hence, as $v_x > 0$ in $[L - \delta, L]$, there exists some $x_5 \in (x_4, L - \delta)$ such that $z(x_5) = 0$, $z > 0$ in (x_5, L). Choose $x_6 \in (x_5, L)$ such that $z(x_6) = \max_{x_5 \le x \le L} z(x)$. Hence, $z(x_6) > 0$, $z_x(x_6) = 0$, and $z_{xx}(x_6) \le 0$. By the equation of z we obtain $w(x_6) < 0$, which is a contradiction as $u_x > 0$ in (x_3, L). Hence, $u_x > 0$ in $[0, L)$. Similarly, $v_x > 0$ in $[0, L)$. Furthermore, $u(x) + v(x) \le u(L) + v(L) < r$ for any $x \in [0, L]$. This completes the proof. □

Let (u^*, v^*) denote any positive equilibrium of system (7.35). Consider the eigenvalue problem

$$\begin{aligned} &\tau\phi_{xx} - q\phi_x + \phi(r - u^* - v^*) + \lambda\phi = 0, \quad 0 < x < L, \\ &\tau\phi_x(0) = q\phi(0), \quad \phi_x(L) = 0. \end{aligned} \tag{7.44}$$

For any $\tau > 0$, let $\lambda_1(\tau)$ denote the largest eigenvalue of problem (7.44).

Lemma 7.4.6 *The dominant eigenvalue $\lambda_1(\tau)$ has at most one positive root.*

Proof This follows from Lemmas 7.4.3 and 7.4.5. □

Lemma 7.4.7 *Suppose that $d_1 \neq d_2$. Then system* (7.35) *has no positive equilibrium.*

Proof Let (u^*, v^*) denote any positive equilibrium of system (7.35). Then u^*, v^* satisfy

$$\begin{cases} d_1 u^*_{xx} - q u^*_x + u^*(r - u^* - v^*) = 0, \quad 0 < x < L, \\ d_2 v^*_{xx} - q v^*_x + v^*(r - u^* - v^*) = 0, \quad 0 < x < L, \\ d_1 u^*_x(0) - q u^*(0) = d_2 v^*_x(0) - q v^*(0) = 0, \\ u^*_x(L) = v^*_x(L) = 0. \end{cases} \tag{7.45}$$

Hence, $\lambda_1(d_1) = \lambda_1(d_2) = 0$. By Lemma 7.4.6, $d_1 = d_2$, which is a contradiction. □

Proof **(of Theorem 7.4.2)** By the maximum principle for cooperative systems and standard theory for parabolic equations, if the initial conditions of (7.35) are non-negative and not identically zero, system (7.35) has a unique positive smooth solution, which exists for all time, and it defines a smooth dynamical system on $C([0,L]) \times C([0,L])$. The stability of equilibria of (7.35) is understood with respect to the topology of $C([0,L]) \times C([0,L])$. Furthermore, system (7.35) is a strongly monotone dynamical system.

Assume that $d_1 > d_2$ and that the semi-trivial state $(u^*, 0)$ exists. If the semi-trivial equilibrium $(0, v^*)$ also exists, Theorem 7.4.2 follows from Theorem 7.1.6, Lemmas 7.4.4 and 7.4.7. If $(0, v^*)$ does not exist, the solution of the single species equation

$$\begin{cases} V_t = d_2 V_{xx} - qV_x + V(r - V), & 0 < x < L,\ t > 0, \\ d_2 V_x(0,t) - qV(0,t) = 0, & V_x(L,t) = 0 \end{cases} \tag{7.46}$$

satisfies $V(x,t) \longrightarrow 0$ uniformly in x as $t \longrightarrow \infty$. Let $(u(x,t), v(x,t))$ denote the solution of (7.35) with $v(x,0) = V(x,0)$. Since $u(x,t) > 0$, $v(x,t)$ satisfies

$$\begin{cases} v_t \le d_2 v_{xx} - qv_x + v(r - v), & 0 < x < L,\ t > 0, \\ d_2 v_x(0,t) - qv(0,t) = 0, & v_x(L,t) = 0. \end{cases} \tag{7.47}$$

By the comparison principle for parabolic equations, $v(x,t) \le V(x,t)$ for any $x \in [0,L]$ and $t \ge 0$. Hence, $v(x,t) \longrightarrow 0$ uniformly in x as $t \longrightarrow \infty$, which in turn implies that $u(x,t) \longrightarrow u^*$ uniformly in x as $t \longrightarrow \infty$. □

In the previous model (7.35), the no-flux condition is imposed at the upper upstream and the free-flow condition is imposed at the downstream end. In the general situation, it might be more realistic to impose a certain rate at which the population is lost at the downstream boundary as individuals are being washed away. The following model was introduced in [59].

$$\begin{cases} \partial_t u - \alpha \partial_{xx} u + q\partial_x u = u(r - u - v) & \text{in } (0,L) \times (0,\infty), \\ \partial_t v - \beta \partial_{xx} v + q\partial_x v = v(r - u - v) & \text{in } (0,L) \times (0,\infty), \\ \alpha \partial_x u - qu = 0 = \beta \partial_x v - qv = 0 & \text{on } \{0\} \times (0,\infty), \\ \alpha \partial_x u + (b-1)qu = 0 = \beta \partial_x v - (b-1)qv = 0 & \text{on } \{L\} \times (0,\infty), \\ u(x,0) = u_0(x), \quad v(x,0) = v_0(x) & \text{in } \Omega, \end{cases} \tag{7.48}$$

where $\alpha, \beta > 0$ are the unconditional dispersal rates and $r > 0$ is the intrinsic growth rate of the population. Both species are impacted by an advection with a common rate $q > 0$ in the direction from $x = 0$ towards $x = L$. The nondimensional parameter $b \ge 0$ informs the rate of population loss at the end of the river $x = L$. If $b = 0$, we recover no-flux boundary condition at $x = 0, L$; the case $b = 1$ is called the free-flow condition, and the case $b = \infty$ corresponds to the lethal boundary condition.

Finally, we state a more general result where the faster diffusion is selected for.

Theorem 7.4.8 *Suppose $q > 0$ and r is a positive constant, and that $0 \leq b \leq 1$, then the faster diffusion is selected; i.e. if $0 < \beta < \alpha$, then the equilibrium $E_1 = (\tilde{u}, 0)$, whenever it exists, is globally asymptotically stable.*

Theorem 7.4.8 was first proved in the case $b = 1$; see [59, Theorem 6.1]. Subsequently, it was generalized in [64, Theorem 3.1] to the case $0 \leq b < 1$.

7.5 Further Reading

In Section 7.1, we applied the theory of monotone dynamical systems to deduce some useful abstract theorems for the dynamics of diffusive Lotka–Volterra systems. The theory of monotone dynamical systems grew out of the works of Hirsch [33, 34, 35, 36] and Matano [67, 68, 69]. The abstract treatment of competitive systems was initiated by Hess and Lazer [32], who first observed that the dynamics of two-species competition has common features regardless of the fine details of the model, or its dimension. Hsu, Waltman and Ellermeyer [38] considered continuous time competitive systems using monotone dynamical systems theory, which was later extended in [37]. Further progress, particularly on the bistable dynamics and saddle point behaviors was obtained in [44, 57, 74]. Building on the framework of [37], we show that diffusive Lotka–Volterra systems satisfy an additional hypothesis and deduce an improved trichotomy result for long-time dynamics (Theorem 7.1.6). This improved an earlier result in [53]. For further references, see the monograph [72] and the survey paper [73] for more recent developments in monotone dynamical systems, and the monograph [75] concerning the dynamical systems approach to population persistence.

For the case of constant coefficients, Theorems 7.2.1 and 7.2.5 in Section 7.2 addressed the questions of coexistence and competitive exclusion. In particular, there exist no nonconstant positive equilibria. There is some important progress on the bistable case where both semi-trivial equilibria are locally stable: Kishimoto and Weinberger proved in [46] that if the underlying physical domain is convex, then there is no stable nonconstant positive equilibria. Unstable nonconstant positive equilibria, however, could exist in the bistable case; see [22, 45]. When the interspecific competition coefficients are sufficiently large, the spatial segregation of two competing species could occur; see [17, 18]. A question concerning the global dynamics in the bistable case is to determine the attraction domains of two locally stable semi-trivial equilibria and characterize the dynamics on the corresponding separatrix which serves as the boundary of the attraction domains. We refer to [43, 44, 71] for the related works.

For Lotka–Volterra competition-diffusion models in spatially heterogeneous environments, a large part of the study originated in the book of Cantrell and Cosner [7] on ecological dynamics and the applications to the evolution of dispersal in spatially homogeneous but temporally constant environments [19, 26]. Most of the earlier works are more focused on the ecological dynamics side; see [5, 6, 8, 9, 40, 42, 60]. An attempt to connect ecological and evolutionary dynamics was initiated in [58] in the context of the two-species Lotka–Volterra competition model, which was significantly extended by the works [27, 28, 30, 55]; for instance, a complete understanding of the full dynamics of two-species competition-diffusion systems in the weak competition case was established in the work [29] for a class of Lotka–Volterra models.

In [3] Belgacem and Cosner considered the combined movement of unbiased diffusion with the directed movement upward along the resource gradient for a single species. They showed that such combined movement could be beneficial to the persistence of a single species; see also [16] for further developments. Cantrell et al. considered a two-species competition model in which one species adopts unconditional diffusion while the other species adopts a combination of unconditional diffusion and directed movement upward along the resource gradient in [10]. They asked whether the directed movement upward along the resource gradient could convey competitive advantage to the species. The investigations in [11] suggested that the strong directed movement upward along the resource gradient could force the species with directed movement to concentrate at the locations with the most available resources, so that the other species with unconditional dispersal could use the resources elsewhere to coexist; see the works [13, 47, 48, 54, 56] for further developments. The study in [1] provided the global bifurcation diagrams of the positive equilibria in multiple parameters, revealing the complexity of the dynamics of the two-species competition-diffusion models with advection. For two-species competition-diffusion models in which both species adopt combinations of unconditional diffusion and directed movement upward along the resource gradient, we refer to [12, 24, 50, 51], in which it is shown that intermediate diffusion or advection rates could be evolutionarily stable; a general philosophy is that the evolutionarily stable dispersal rates are proportional to the advection rates, suggesting that an optimal dispersal strategy is to bet and hedge, i.e. to apply the directed movement to locate the most favorable resources and to use the unconditional dispersal to find resources elsewhere. We refer to Chapter 9 for a discussion of evolutionary stable strategy and other notions of adaptive dynamics, and to the surveys [14, 15] for further discussions.

In their seminal work [76] Speirs and Gurney used reaction-diffusion-advection models to address the drift paradox, i.e. how can species persist in the water columns without being washed out. They found that when the lethal boundary condition is imposed at the downstream end, a necessary and sufficient condition for a single species to persist is that the river is sufficiently long and the diffusion rate falls into a proper range: a population with a small diffusion rate will not be able to resist the drift to the lethal downstream end, while one that has a large diffusion rate will also be losing too many individuals at the downstream end. When the free flow boundary is imposed at the downstream end, the persistence criteria was investigated in [77]. Recently, more general boundary conditions have been considered in [25], where a complete classification of the persistence criteria is found; see Proposition 5.3.2 (with an illustration in Fig. 5.1) and also the paper [82]. We refer to [65, 66] for further discussions of the modeling of single species population dynamics in rivers. The Allee effect on the persistence of a single species in rivers was considered in

[80, 81]. For two-species competition models with constant coefficients, it is shown in [59] that a faster dispersal rate is always selected when the free-flow boundary condition is imposed at the downstream end. This was subsequently extended to more general boundary conditions in [64], including no-flux boundary conditions at the downstream end. For two-species competition models with spatially heterogeneous coefficients, we refer to the works [52, 62, 63, 83, 84] for both ecological and evolutionary dynamics. These works are primarily focused on the situation when the resource distributions are static. In contrast, there are relatively fewer works on two-species competition models with nutrient dynamics; see [2, 78, 79].

Problems

7.5.1 Consider the following system of ordinary differential equations.

$$x_1' = x_1(1 - x_1 - x_2), \quad x_2' = x_2(1 - \mu x_1 - x_2)^3. \tag{7.49}$$

Suppose $\mu > 1$. Show that the semiflow generated by (7.49) satisfies (H1')–(H4') and has no positive equilibrium; $E_2 = (0, 1)$ is locally asymptotically stable. Moreover, $E_1 = (1, 0)$ attracts some internal trajectories. [Hint: E_1 attracts all trajectories initiating in $\{(x_1, x_2) \in \mathbb{R}^2 : x_1 > 1, 0 < x_2 < (x_1 - 1)^2\}$.]

7.5.2 Assume the coefficients $a_i, b_i, c_i \in C(\overline{\Omega})$ satisfy

$$\frac{b_1}{b_2} > \frac{a_1}{a_2} > \frac{c_1}{c_2} \quad \text{in } \Omega.$$

Show that the semi-trivial equilibria of (7.19) are linearly unstable provided that both d_1 and d_2 are sufficiently small.

7.5.3 Consider the following predator-prey model in a bounded smooth domain Ω with homogeneous Neumann boundary condition

$$\begin{cases} \partial_t u - d_1 \Delta u = u\,(a_1 - b_1 u - c_1 v) & \text{in } \Omega \times (0, T), \\ \partial_t v - d_2 \Delta v = a_2 u - b_2 v & \text{in } \Omega \times (0, T), \\ \mathbf{n} \cdot \nabla u = \mathbf{n} \cdot \nabla v = 0 & \text{on } \partial\Omega \times (0, T), \\ u(x, 0) = u_0(x) > 0, \quad v(x, 0) = v_0(x) > 0 & \text{in } \overline{\Omega}, \end{cases} \tag{7.50}$$

where a_i, b_i, c_i, d_i are all positive constants. Show that, as $t \to \infty$,

$$(u, v) \to (u^*, v^*) := \left(\frac{a_1 b_2}{b_1 b_2 + c_1 a_1}, \frac{a_1 a_2}{b_1 b_2 + c_1 a_1} \right) \quad \text{uniformly in } \Omega.$$

[Hint: Prove that

$$E(u, v) = \int_{u^*}^{u} \frac{\eta - u^*}{\eta} \mathrm{d}\eta + \frac{c_1}{2a_2}(v - v^*)^2$$

is a Lyapunov function, and use Theorem 7.2.6.]

7.5.4 Suppose (u^*, v^*) is a positive equilibrium of (7.19), i.e. it is a positive solution of

$$\begin{cases} d_1\Delta u^* + u^*(a_1(x) - b_1(x)u^* - c_1(x)v^*) = 0 & \text{in } \Omega, \\ d_2\Delta v^* + v^*(a_2(x) - b_2(x)u^* - c_2(x)v^*) = 0 & \text{in } \Omega, \\ \mathbf{n}\cdot u^* = \mathbf{n}\cdot v^* = 0 & \text{on } \partial\Omega, \end{cases} \tag{7.51}$$

where $a_i, b_i, c_i \in C^\alpha(\overline{\Omega})$, and are strictly positive in $\overline{\Omega}$, and that the spatially averaged system satisfies the weak competition hypothesis, i.e.

$$\frac{\bar{b}_1}{\bar{b}_2} > \frac{\bar{a}_1}{\bar{a}_2} > \frac{\bar{c}_1}{\bar{c}_2}, \tag{7.52}$$

where $\bar{g} = \frac{1}{|\Omega|}\int_\Omega g\,\mathrm{d}x$ for any function $g \in C(\overline{\Omega})$.

1. Show that $E_1 = (\tilde{u}, 0)$ and $E_2 = (0, \tilde{v})$ are both linearly unstable for $d_1, d_2 \to \infty$. In particular, (7.51) has at least one positive solution for $\min\{d_i\} \gg 1$. [Hint: $\tilde{u} \to \bar{a}_1/\bar{b}_1$ as $d_1 \to \infty$ and $\tilde{v} \to \bar{a}_2/\bar{c}_2$ as $d_2 \to \infty$. Then consider the linearized eigenvalue problem.]
2. Prove the a priori L^∞ estimate, i.e. there is a C which is uniform in $\min\{d_1, d_2\} \geq 1$ such that $\|u^*\|_{L^\infty(\Omega)} + \|v^*\|_{L^\infty(\Omega)} \leq C$.
3. Prove that, passing to a subsequence if necessary, there exist constants $\bar{U}, \bar{V}$ such that $(u^*, v^*) \to (\bar{U}, \bar{V})$ in $C(\overline{\Omega})$.
4. Show that $\bar{U} = \bar{V} = 0$ is impossible. [Hint: By integrating (7.51) over Ω, we obtain
$$\int_\Omega u^*(a_1 - b_1u^* - c_1v^*)\,\mathrm{d}x = 0.$$
Then use the fact that $\inf_\Omega a_1 > 0$.]
5. Prove that
$$\bar{U}(\bar{a}_1 - \bar{b}_1\bar{U} - \bar{c}_1\bar{V}) = 0 = \bar{V}(\bar{a}_2 - \bar{b}_2\bar{U} - \bar{c}_2\bar{V}), \tag{7.53}$$
and deduce that that $\bar{U} > 0$ and $\bar{V} > 0$. [Hint: Suppose $\bar{U} = 0$, then it follows from the above step that $\bar{V} > 0$. By (7.53), we have $\bar{V} = \bar{a}_2/\bar{c}_2$. Setting $w = u^*/\|u^*\|_{L^\infty(\Omega)}$, we have $\|w\|_{L^\infty(\Omega)} = 1$ and
$$\Delta w = -\frac{1}{d_1}w(a_1 - b_1u^* - c_1v^*) \quad \text{in } \Omega, \quad \mathbf{n}\cdot\nabla w = 0 \quad \text{on } \partial\Omega.$$
Observe that $w \to 1$ in $C(\overline{\Omega})$ and deduce that
$$\int_\Omega (a_1 - c_1\bar{V})\,\mathrm{d}x = \lim_{\min\{d_i\}\to\infty}\int_\Omega w(a_1 - b_1u^* - c_1v^*)\,\mathrm{d}x = 0.$$
This implies that $\bar{a}_1 = \bar{c}_1\bar{V} = \frac{\bar{c}_1\bar{a}_2}{\bar{c}_2}$, which is impossible.]
6. Conclude that $(u^*, v^*) \to (\bar{U}, \bar{V})$ in $C(\overline{\Omega})$ as $\min\{d_i\} \to \infty$, where $(\bar{U}, \bar{V})$ is the unique positive root of (7.53). [Hint: The uniqueness of the subsequential limit implies the convergence of the full sequence as $\min\{d_i\} \to \infty$.]

7.5.5 (Every positive solution of (7.51) is linearly stable when $\min\{d_i\}$ is sufficiently large.) Assuming the same hypotheses as Problem E.3.2, show that there exists an $M > 0$ such that if $\min\{d_i\} \geq M$, then any positive solution (u^*, v^*) of (7.51) is linearly stable. Recall that the linear stability of (u^*, v^*) is determined by the eigenvalue problem

$$\begin{cases} d_1\Delta\phi + (a_1 - 2b_1u^* - c_1v^*)\phi - c_1u^*\psi + \lambda_1\phi = 0 & \text{in } \Omega, \\ d_2\Delta\psi - c_2v^*\phi + (a_2 - b_2u^* - 2c_2v^*)\psi + \lambda_1\psi = 0 & \text{in } \Omega, \\ \mathbf{n}\cdot\nabla\phi = \mathbf{n}\cdot\nabla\psi = 0 & \text{in } \partial\Omega. \end{cases} \tag{7.54}$$

Without loss of generality, we may assume $|\Omega| = 1$.

1. Show that (7.54) has a principal eigenvalue λ_1, (i.e. $\lambda_1 \in \mathbb{R}$ is the eigenvalue with the least real part, and the corresponding eigenfunction (ϕ, ψ) can be chosen such that $\phi > 0 > \psi$ in $\overline{\Omega}$) by verifying the hypotheses of Theorem 3.3.3. In addition, we may normalize the eigenfunctions so that $\frac{1}{|\Omega|}\int_\Omega(\phi^2 + \psi^2)\,\mathrm{d}x = 1$.
2. Show that, passing to a subsequence if necessary, $(\phi, \psi) \to (\bar{\Phi}, \bar{\Psi})$ as $\min\{d_i\} \to \infty$, for some constants $\bar{\Phi} \geq 0 \geq \bar{\Psi}$ such that $\bar{\Phi}^2 + \bar{\Psi}^2 = 1$.
3. Show that there exists an $M \geq 0$, which depends on the L^∞ bounds of (u^*, v^*) but is independent of $d_i > 0$, such that $\lambda_1 \geq -M$.
4. Show that
$$\liminf_{\min\{d_i\}\to\infty} \lambda_1 > 0.$$

   [Hint: Assume to the contrary that, passing to a subsequence if necessary, $\lambda_1 \to \Lambda_1 \leq 0$ as $\min\{d_i\} \to \infty$. By integrating (7.54), we have

$$\begin{cases} \int_\Omega [(a_1 - 2b_1u^* - c_1v^* + \lambda_1)\phi - c_1u^*\psi]\ \mathrm{d}x = 0, \\ \int_\Omega [(a_2 - b_2u^* - 2c_2v^* + \lambda_1)\psi - b_2v^*\phi]\ \mathrm{d}x = 0. \end{cases}$$

   Passing to the limit, so that

$$\lambda_1 \to \Lambda_1, \quad (u^*, v^*) \to (\bar{U}, \bar{V}), \quad (\phi, \psi) \to (\bar{\Phi}, \bar{\Psi}),$$

   and using (7.53), deduce that

$$\begin{pmatrix} \Lambda_1 - \bar{b}_1\bar{U} & -\bar{c}_1\bar{U} \\ -b_2\bar{V} & \Lambda_1 - \bar{c}_2\bar{V} \end{pmatrix}\begin{pmatrix} \bar{\Phi} \\ \bar{\Psi} \end{pmatrix} = \begin{pmatrix} 0 \\ 0 \end{pmatrix}$$

   and therefore

$$\det\begin{pmatrix} \Lambda_1 - \bar{b}_1\bar{U} & -\bar{c}_1\bar{U} \\ -b_2\bar{V} & \Lambda_1 - \bar{c}_2\bar{V} \end{pmatrix} = 0.$$

   By (7.52), we deduce that $\Lambda_1 > 0$. This is a contradiction.]

7.5.6 Let λ_1 be the principal eigenvalue of (7.27). Prove that $\lambda_1 < 0$ whenever $c < 1$ and $d_1 \geq d_2$. [Hint: Observe that λ_1 is monotone increasing in c as well as in d_2, and that $\lambda_1 = 0$ when $c = 1$ and when d_2 is increased to d_1.]

7.5.7 Let λ be the principal eigenvalue of (7.36). Show that

$$\frac{\partial \lambda}{\partial \tau} = \frac{\int (\mathrm{e}^{-\alpha x}\phi)_x \phi_x \mathrm{d}x}{\int \mathrm{e}^{-\alpha x}\phi^2 \mathrm{d}x}, \tag{7.55}$$

where $\alpha = q/\tau$.

7.5.8 Suppose (7.41) has a unique positive solution u^*, where d_1, q, r, L are positive constants. Show that $u^*(x) < r$ in $[0, L]$.

References

1. I. Averill, K.-Y. Lam, and Y. Lou, *The role of advection in a two-species competition model: a bifurcation approach*, Mem. Amer. Math. Soc, 245 (2017). pp. 117.
2. M. Ballyk, L. Dung, D. A. Jones, and H. L. Smith, *Effects of random motility on microbial growth and competition in a flow reactor*, SIAM J. Appl. Math, 59 (1998), pp. 573–596.
3. F. Belgacem and C. Cosner, *The effects of dispersal along environmental gradients on the dynamics of populations in heterogeneous environment*, Canadian Appl. Math. Quarterly, 3 (1995), pp. 379–397.
4. P. N. Brown, *Decay to uniform states in ecological interactions*, SIAM J. Appl. Math., 38 (1980), pp. 22–37.
5. R. S. Cantrell and C. Cosner, *The effects of spatial heterogeneity in population dynamics*, J. Math. Biol., 29 (1991), pp. 315–338.
6. ———, *On the effects of spatial heterogeneity on the persistence of interacting species*, J. Math. Biol., 37 (1998), pp. 103–145.
7. R. S. Cantrell and C. Cosner, *Spatial ecology via reaction-diffusion equations*, Wiley Series in Mathematical and Computational Biology, John Wiley & Sons, Ltd., Chichester, 2003.
8. R. S. Cantrell, C. Cosner, and V. Huston, *Permanence in ecological systems with spatial heterogeneity*, Proc. Roy. Soc. Edinburgh Sect. A, 123 (1993), pp. 533–559.
9. ———, *Ecological models, permanence and spatial heterogeneity*, Rocky Mountain J. Math., 26 (1996), pp. 1–35.
10. R. S. Cantrell, C. Cosner, and Y. Lou, *Movement towards better environments and the evolution of rapid diffusion*, Math. Biosci., 240 (2006), pp. 199–214.
11. ———, *Advection-mediated coexistence of competing species*, Proc. Roy. Soc. Edinburgh Sect. A, 137 (2007), pp. 497–518.
12. X. Chen, K.-Y. Lam, and Y. Lou, *Dynamics of a reaction-diffusion-advection model for two competing species*, Discrete Contin. Dyn. Syst., 32 (2012), pp. 3841–3859.
13. X. Chen and Y. Lou, *Principal eigenvalue and eigenfunctions of an elliptic operator with large advection and its application to a competition model*, Indiana Univ. Math. J., 57 (2008), pp. 627–657.
14. C. Cosner, *Beyond diffusion: Conditional dispersal in ecological models, infinite dimensional dynamical systems*, Fields Inst. Commun., 64 (2013), pp. 305–317.
15. ———, *Reaction-diffusion-advection models for the effects and evolution of dispersal*, Discrete Contin. Dyn. Syst., 34 (2014), pp. 1701–1745.
16. C. Cosner and Y. Lou, *Does movement toward better environments always benefit a population?*, J. Math. Anal. Appl., 277 (2003), pp. 489–503.
17. E. N. Dancer and Y. H. Du, *Competing species equations with diffusion, large interactions, and jumping nonlinearities*, J. Differential Equations, 114 (1994), pp. 434–475.
18. E. N. Dancer, D. Hilhorst, M. Mimura, and L. A. Peletier, *Spatial segregation limit of a competition-diffusion system*, European J. Appl. Math., 10 (1999), pp. 97–115.

19. J. Dockery, V. Hutson, K. Mischaikow, and M. Pernarowski, *The evolution of slow dispersal rates: a reaction diffusion model*, J. Math. Biol., 37 (1998), pp. 61–83.
20. J. Doumate and R. Salako, *Asymptotic behavior of solutions of an ODE-PDE hybrid competition system*, J. Differential Equations, 334 (2022), pp. 216–225.
21. D. Gilbarg and N. S. Trudinger, *Elliptic partial differential equations of second order*, Classics in Mathematics, Springer-Verlag, Berlin, 2001. Reprint of the 1998 edition.
22. C. Gui and Y. Lou, *Uniqueness and nonuniqueness of coexistence states in the Lotka–Volterra competition model*, Comm. Pure Appl. Math., 47 (1994), p. 1571–1594.
23. J. K. Hale, *Dynamical systems and stability*, J. Math. Anal. Appl., 26 (1969), pp. 39–59.
24. R. Hambrock and Y. Lou, *The evolution of conditional dispersal strategy in spatially heterogeneous habitats*, Bull. Math. Biol., 71 (2009), pp. 1793–1817.
25. W. Hao, K.-Y. Lam, and Y. Lou, *Ecological and evolutionary dynamics in advective environments: critical domain size and boundary conditions*, Discrete Contin. Dyn. Syst. Ser. B, 26 (2021), pp. 367–400.
26. A. Hastings, *Can spatial variation alone lead to selection for dispersal?*, Theor. Pop. Biol., 24 (1983), pp. 244–251.
27. X. He and W.-M. Ni, *The effects of diffusion and spatial variation in Lotka–Volterra competition-diffusion system I: Heterogeneity vs. homogeneity*, J. Differential Equations, 254 (2013), pp. 528–546.
28. ———, *The effects of diffusion and spatial variation in Lotka–Volterra competition-diffusion system II: The general case*, J. Differential Equations, 254 (2013), pp. 4088–4108.
29. ———, *Global dynamics of the Lotka–Volterra competition-diffusion system: diffusion and spatial heterogeneity I*, Comm. Pure Appl. Math., 69 (2016), pp. 981–1014.
30. ———, *Global dynamics of the Lotka–Volterra competition-diffusion system with equal amount of total resources II,*, Calc. Var. Partial Differential Equations, 55 (2016).
31. D. Henry, *Geometric theory of semilinear parabolic equations*, vol. 840 of Lecture Notes in Mathematics, Springer-Verlag, Berlin-New York, 1981.
32. P. Hess and A. C. Lazer, *On an abstract competition model and applications*, Nonlinear Anal., 16 (1991), pp. 917–940.
33. M. W. Hirsch, *Systems of differential equations which are competitive or cooperative. I. Limit sets*, SIAM J. Math. Anal., 13 (1982), pp. 167–179.
34. ———, *The dynamical systems approach to differential equations*, Bull. Amer. Math. Soc. (N.S.), 11 (1984), pp. 1–64.
35. ———, *Stability and convergence in strongly monotone dynamical systems*, J. Reine Angew. Math., 383 (1988), pp. 1–53.
36. ———, *Positive equilibria and convergence in subhomogeneous monotone dynamics*, in Comparison methods and stability theory (Waterloo, ON, 1993), vol. 162 of Lecture Notes in Pure and Appl. Math., Dekker, New York, 1994, pp. 169–188.
37. S. B. Hsu, H. L. Smith, and P. Waltman, *Competitive exclusion and coexistence for competitive systems on ordered Banach spaces*, Trans. Amer. Math. Soc., 348 (1996), pp. 4083–4094.
38. S.-B. Hsu, P. Waltman, and S. F. Ellermeyer, *A remark on the global asymptotic stability of a dynamical system modeling two species competition*, Hiroshima Math. J., 24 (1994), pp. 435–445.
39. V. Hutson, J. López-Gómez, K. Mischaikow, and G. Vickers, *Limit behaviour for a competing species problem with diffusion*, in Dynamical systems and applications, vol. 4 of World Sci. Ser. Appl. Anal., World Sci. Publ., River Edge, NJ, 1995, pp. 343–358.
40. V. Hutson, Y. Lou, and K. Mischaikow, *Spatial heterogeneity of resources versus Lotka–Volterra dynamics*, J. Differential Equations, 185 (2002), pp. 97–136.
41. ———, *Convergence in competition models with small diffusion coefficients*, J. Differential Equations, 211 (2005), pp. 135–161.
42. V. Hutson, Y. Lou, K. Mischaikow, and P. Polacik, *Competing species near the degenerate limit*, SIAM J. Math. Anal., 35 (2003), pp. 453–491.
43. M. Iida, T. Muramatsu, H. Ninomiya, and E. Yanagida, *Diffusion-induced extinction of a superior species in a competition system*, Japan J. Indust. Appl. Math. J., 15 (1998), pp. 233–252.

44. J. Jiang, X. Liang, and X.-Q. Zhao, *Saddle-point behavior for monotone semiflows and reaction-diffusion models*, J. Differential Equations, 203 (2004), pp. 313–330.
45. Y. Kan-on, *Bifurcation structure of stationary solutions of a Lotka–Volterra competition model with diffusion*, SIAM J. Math. Anal., 29 (1998), pp. 424–436.
46. K. Kishimoto and H. Weinberger, *The spatial homogeneity of stable equilibria of some reaction-diffusion systems on convex domains*, J. Differential Equations, 58 (1985), pp. 15–21.
47. K.-Y. Lam, *Concentration phenomena of a semilinear elliptic equation with large advection in an ecological model*, J. Differential Equations, 250 (2011), pp. 161–181.
48. ———, *Limiting profiles of semilinear elliptic equations with large advection in population dynamics ii*, SIAM J. Math. Anal., 44 (2012), pp. 1808–1830.
49. K.-Y. Lam, S. Liu, and Y. Lou, *Selected topics on reaction-diffusion-advection models from spatial ecology*, Math. Appl. Sci. Eng., 1 (2020), pp. 150–180.
50. K.-Y. Lam and Y. Lou, *Evolution of dispersal: evolutionarily stable strategies in spatial models*, J. Math. Biol., 68 (2014), pp. 851–877.
51. ———, *Evolutionarily stable and convergent stable strategies in reaction- diffusion models for conditional dispersal*, Bull. Math. Biol., 76 (2014), pp. 261–291.
52. K.-Y. Lam, Y. Lou, and F. Lutscher, *Evolution of dispersal in closed advective environments*, J. Biol. Dyn., 9 (2015), pp. 188–212.
53. K.-Y. Lam and D. Munther, *A remark on the global dynamics of competitive systems on ordered Banach spaces*, Proc. Amer. Math. Soc., 144 (2016), pp. 1153–1159.
54. K.-Y. Lam and W. M. Ni, *Limiting profiles of semilinear elliptic equations with large advection in population dynamics*, Discrete Contin. Dyn. Syst., 28 (2010), pp. 1051–1067.
55. ———, *Uniqueness and complete dynamics of the Lotka–Volterra competition diffusion system*, SIAM J. Appl. Math., 72 (2012), pp. 1695–1712.
56. ———, *Advection-mediated competition in general environments*, J. Differential Equations, 257 (2014), pp. 3466–3500.
57. X. Liang and J. Jiang, *On the finite-dimensional dynamical systems with limited competition*, Trans. Amer. Math. Soc., 354 (2002), pp. 3535–3554.
58. Y. Lou, *On the effects of migration and spatial heterogeneity on single and multiple species,*, J. Differential Equations, 223 (2006), pp. 400–426.
59. Y. Lou and F. Lutscher, *Evolution of dispersal in advective environments*, J. Math Biol., 69 (2014), pp. 1319–1342.
60. Y. Lou, S. Martinez, and P. Polacik, *Loops and branches of coexistence states in a Lotka–Volterra competition model*, J. Differential Equations, 230 (2006), pp. 720–742.
61. Y. Lou and R. Salako, *Dynamics of a parabolic-ode competition system in heterogeneous environments*, Proc. Amer. Math. Soc., 148 (2020), pp. 3025–3028.
62. Y. Lou, D. Xiao, and P. Zhou, *Qualitative analysis for a Lotka–Volterra competition system in advective homogeneous environment*, Discrete Contin. Dyn. Syst., 36 (2016), pp. 953–969.
63. Y. Lou, X. Zhao, and P. Zhou, *Global dynamics of a Lotka–Volterra competition-diffusion-advection system in heterogeneous environments*, J. Math. Pure. Appl., 121 (2019), pp. 47–82.
64. Y. Lou and P. Zhou, *Evolution of dispersal in advective homogeneous environments: The effect of boundary conditions*, J. Differential Equations, 259 (2015), pp. 141–171.
65. F. Lutscher, E. McCauley, and M. Lewis, *Effects of heterogeneity on spread and persistence in rivers*, Bull. Math. Biol., 68 (2006), pp. 2129–2160.
66. F. Lutscher, E. Pachepsky, and M. Lewis, *The effect of dispersal patterns on stream populations*, SIAM Rev., 47 (2005), pp. 749–772.
67. H. Matano, *Asymptotic behavior and stability of solutions of semilinear diffusion equations*, Publ. Res. Inst. Math. Sci., 15 (1979), pp. 401–454.
68. ———, *Existence of nontrivial unstable sets for equilibriums of strongly order-preserving systems*, J. Fac. Sci. Univ. Tokyo Sect. IA Math., 30 (1984), pp. 645–673.
69. H. Matano, *Strongly order-preserving local semidynamical systems—theory and applications*, in Semigroups, theory and applications, Vol. I (Trieste, 1984), vol. 141 of Pitman Res. Notes Math. Ser., Longman Sci. Tech., Harlow, 1986, pp. 178–185.

70. W. Ni, J. Shi, and M. Wang, *Global stability of nonhomogeneous equilibrium solution for the diffusive Lotka–Volterra competition model*, Calc. Var. Partial Differential Equations, 59 (2020), p. 28. Paper No. 132.
71. H. Ninomiya, *Separatrices of competition-diffusion equations*, J. Math. Kyoto Univ., 35 (1995), pp. 539–567.
72. H. L. Smith, *Monotone dynamical systems*, vol. 41 of Mathematical Surveys and Monographs, American Mathematical Society, Providence, RI, 1995. An introduction to the theory of competitive and cooperative systems.
73. ———, *Monotone dynamical systems: reflections on new advances & applications*, Discrete Contin. Dyn. Syst., 37 (2017), pp. 485–504.
74. H. L. Smith and H. R. Thieme, *Stable coexistence and bi-stability for competitive systems on ordered Banach spaces*, J. Differential Equations, 176 (2001), pp. 195–222.
75. ———, *Dynamical systems and population persistence*, vol. 118 of Graduate Studies in Mathematics, American Mathematical Society, Providence, RI, 2011.
76. D. C. Speirs and W. S. C. Gurney, *Population persistence in rivers and estuaries*, Ecology, 82 (2001), pp. 1219–1237.
77. O. Vasilyeva and F. Lutscher, *Population dynamics in rivers: analysis of steady states*, Can. Appl. Math. Quart., 18 (2011), pp. 439–469.
78. X. Wang, *Qualitative behavior of solutions of chemotactic diffusion systems: effects of motility and chemotaxis and dynamics*, SIAM J. Math. Anal., 31 (2000), pp. 535–560.
79. X. Wang and Y. Wu, *Qualitative analysis on a chemotactic diffusion model for two species competing for a limited resource*, Quart. Appl. Math, 60 (2002), pp. 505–531.
80. Y. Wang and J. Shi, *Persistence and extinction of population in reaction-diffusion- advection model with weak allee effect growth*, SIAM J. Appl. Math., 79 (2019), pp. 1293–1313.
81. Y. Wang, J. Shi, and J. Wang, *Persistence and extinction of population in reaction- diffusion-advection model with strong allee effect growth*, J. Math. Biol., 78 (2019), pp. 2093–2140.
82. F. Xu, W. Gan, and D. Tang, *Population dynamics and evolution in river ecosystems*, Nonlinear Anal. Real World App., 51 (2020), p. 102983.
83. X.-Q. Zhao and P. Zhou, *On a Lotka–Volterra competition model: the effects of advection and spatial variation*, Calc. Var. Partial Differential Equations, 55 (2016), p. 25. Art. 73.
84. P. Zhou, *On a Lotka–Volterra competition system: diffusion vs advection*, Calc. Var. Partial Differential Equations, 55 (2016), p. 29. Art. 137.

Chapter 8
Dynamics of Phytoplankton Populations

Abstract In this chapter we analyze reaction-diffusion-advection models for the growth of single or multiple phytoplankton species in a eutrophic, vertical water column. In such environments, the nutrients are in abundance and the phytoplankton species is limited by the light only. A strong comparison principle for the cumulative distribution function of the single phytoplankton species is established so that the monotone dynamical system theory is applicable to some specific positive cone which is unique to the model. This allows us to fully determine the global dynamics of the nonlinear scalar equation for the growth of a single phytoplankton species. Next we study two phytoplankton species which are limited by, and competing for, the light only. Again, a strong comparison principle for two phytoplankton species is proved. As an application, we show that for the model of two phytoplankton species which are identical except for their buoyant/sinking rates, the more buoyant (or less prone to sinking) phytoplankton species drives the other species to extinction. Finally, as an application of the principal Floquet bundle theory that is discussed in Chapter 4, we consider some N-species phytoplankton models and prove that in general the competition exclusion takes place when the species are similar to each other.

8.1 Introduction

Phytoplankton are microscopic plant-like photosynthetic organisms that drift in the water columns of lakes and oceans. They grow abundantly around the globe and are the foundation of the marine food chain. Since they transport a significant amount of atmospheric carbon dioxide into the deep oceans, they play a crucial role in climate dynamics. Nutrients and light are the essential resources for the growth of phytoplankton. There are three possible ways for phytoplankton to compete for nutrients and light. At one extreme, in oligotrophic ecosystems with an ample supply of light, species compete for limiting nutrients [27]. At the other extreme, in eutrophic ecosystems with ample nutrient supply, species compete for light [8, 17, 18, 36]. In

K. -Y. Lam, Y. Lou, *Introduction to Reaction-Diffusion Equations*, Lecture Notes on Mathematical Modelling in the Life Sciences, https://doi.org/10.1007/978-3-031-20422-7_8

some ecosystems of intermediate conditions, they compete for both nutrients and light [26, 31, 38]. In the water column, phytoplankton diffuse by water turbulence, and also sink or buoy, depending on whether they are heavier than water or not [8].

We discuss a model proposed by Huisman et al. [18, 19, 20]. Consider a water column with unit cross-sectional area and with N phytoplankton species, for some $N \geq 1$. Let x denote the depth within the water column with length L, where x varies from 0 (the water surface) to L (the bottom), and let $u_i(x,t)$ denote the population density of the i-th species at the location x and time t.

$$\partial_t u_i = D_i \partial_{xx} u_i - \alpha_i \partial_x u_i + u_i \left[g_i(I(x,t)) - d_i \right] \quad \text{for } 0 < x < L,\ t > 0, \tag{8.1}$$

for $i = 1, \ldots, N$, and with no-flux boundary conditions

$$D_i \partial_x u_i - \alpha_i u_i = 0 \quad \text{for } x = 0,\ L,\ t > 0,\ i = 1, \ldots, N, \tag{8.2}$$

and initial data

$$u_i(x,0) = u_{i,0}(x) \quad \text{for } 0 \leq x \leq L,\ i = 1, \ldots, N. \tag{8.3}$$

Here $D_i > 0$ is the diffusion coefficient caused by turbulence, $\alpha_i \in \mathbb{R}$ is the sinking (if $\alpha_i < 0$) or buoyant (if $\alpha_i > 0$) velocity, $d_i > 0$ is the loss rate. The term $g_i(I)$ represents the specific growth rate of the i-th phytoplankton species, which depends on the light intensity $I(x,t)$ at that specific location. By the Lambert–Beer law, the light intensity $I(x,t)$ takes the form

$$I(x,t) = I_0 \exp\left(-k_0 x - \sum_{j=1}^{N} k_j \int_0^x u_j(y,t)\, \mathrm{d}y \right), \tag{8.4}$$

where $I_0 > 0$ is the incident light intensity, $k_0 > 0$ is the background turbidity, and k_j is the absorption coefficient of the j-th phytoplankton species.

The system (8.1)–(8.4) is intended to model a eutrophic water column, where nutrient is in abundance, and phytoplankton species compete for light via shading. The integral appearing in (8.4) is due to the fact that the light is able to reach to depth x only after being absorbed by water and the biomass population at depth between 0 and x. In other words, the competition for light is nonlocal in space. The functions g_i are smooth and satisfy

$$g_i(0) = 0, \quad g_i'(I) > 0 \quad \text{for } I > 0 \quad \text{and } g_i \in L^\infty([0,\infty)). \tag{8.5}$$

Typical examples of g_i include

$$g_i(I) = \frac{m_i I}{a_i + I} \quad \text{and} \quad g_i(I) = \frac{m_i}{a_i}(1 - \mathrm{e}^{-a_i I}),$$

where m_i, a_i are positive constants.

Proposition 8.1.1 *For continuous, nonnegative initial data* $(u_{i,0})_{i=1}^N$*, the problem* (8.1)–(8.4) *has a unique solution*

$$u = (u_i)_{i=1}^N \in C([0,\infty); C([0,L]; \mathbb{R}_+^2)) \cap C^1_{loc}((0,\infty); C^2([0,L]; \mathbb{R}_+^2)),$$

which depends continuously on initial data. Moreover, if $u_{i,0} \not\equiv 0$ *for some* i*, then* $u_i(x,t) > 0$ *for* $(x,t) \in [0,L] \times (0,\infty)$.

Proof By the maximum principle, the solutions are nonnegative and uniformly bounded in $C([0,L] \times [0,T])$ for each $T > 0$. The rest follows from [12, Chap. 3].□

By letting $\check{u}_i = k_i u_i$ and by $\check{g}(s) = g(I_0 s)$, we may assume without loss of generality that $I_0 = 1$ and $k_i = 1$ for $i = 1, \ldots, N$, which we enforce for the remainder of this chapter. For this particular model, the coefficients k_i do not affect the dynamics of the system after suitable transformation, but in practice they affect the relative population size of each phytoplankton species.

8.2 Single Species in a Eutrophic Water Column

In this section, we consider the case of a single species. In this case, (8.1)–(8.4) become

$$\begin{cases} \partial_t u = D\partial_{xx} u - \alpha \partial_x u + u\,[g(I(x,t)) - d] & \text{for } 0 < x < L,\ t > 0, \\ D\partial_x u - \alpha u = 0 & \text{for } x = 0, L,\ t > 0, \\ u(x,0) = u_0(x) & \text{for } 0 \le x \le L, \\ I(x,t) = \exp(-k_0 x - \int_0^x u(y,t)\,dy) & \text{for } 0 \le x \le L,\ t > 0. \end{cases} \tag{8.6}$$

Note that we are assuming without loss of generality that $I_0 = 1$ and $k_1 = 1$.

8.2.1 Monotonicity of the single species model

Define

$$\mathcal{K} := \left\{ u_0 \in C([0,L]) : \int_0^x u_0(y)\,dy \ge 0 \quad \text{for } 0 \le x \le L \right\}, \tag{8.7}$$

which has nonempty interior Int $\mathcal{K}$ in $C([0,L])$. We leave as an exercise (see Problem 8.5.1) for the readers to show that

$$\operatorname{Int} \mathcal{K} = \left\{ u_0 \in C([0,L]) : u_0(0) > 0 \text{ and } \int_0^x u_0(y)\,dy > 0 \quad \text{for } 0 < x \le L \right\}. \tag{8.8}$$

Definition 8.2.1 We say that $X = C([0, L])$ is a Banach space with order induced by the solid cone $\mathcal{K}$, such that

$$u_0 \le_{\mathcal{K}} \bar{u}_0 \qquad (\text{resp. } u_0 <_{\mathcal{K}} \bar{u}_0, \qquad u_0 \ll_{\mathcal{K}} \bar{u}_0)$$

if and only if

$$\bar{u}_0 - u_0 \in \mathcal{K} \qquad (\text{resp. } \bar{u}_0 - u_0 \in \mathcal{K} \setminus \{0\}, \qquad \bar{u}_0 - u_0 \in \operatorname{Int} \mathcal{K}).$$

The following result says that (8.6) generates a strongly monotone semiflow in the Banach space $X = C([0, L])$ with order induced by the cone $\mathcal{K}$.

Proposition 8.2.2 *Suppose u_i $(i = 1, 2)$ are two nonnegative solutions of* (8.6). *If*

$$u_1(\cdot, 0) <_{\mathcal{K}} u_2(\cdot, 0),$$

then

$$u_1(\cdot, t) \ll_{\mathcal{K}} u_2(\cdot, t) \quad \textit{for } t > 0.$$

To show Proposition 8.2.2, we consider the cumulative distribution function

$$U(x, t) := \int_0^x u(y, t)\, \mathrm{d}y,$$

so that $U(0, t) \equiv 0$ for $t \ge 0$, and $\partial_x U(x, t) = u(x, t)$. In this way, (8.6) is transformed into the following nonlocal system; see [32].

$$\begin{cases} \partial_t U = D\partial_{xx}U - \alpha\partial_x U + G[U, \partial_x U], & \text{for } 0 \le x \le L,\ t > 0, \\ U(0, t) = 0, \quad D\partial_{xx}U(L, t) - \alpha\partial_x U(L, t) = 0 & \text{for } t > 0, \\ U(x, 0) = \int_0^x u_0(y)\, \mathrm{d}y & \text{for } 0 \le x \le L, \end{cases} \tag{8.9}$$

where, letting $F(x, U) = \int_0^U g(\mathrm{e}^{-k_0 x - z})\, \mathrm{d}z - dU$,

$$\begin{aligned} G[U, \partial_x U](x, t) &= \int_0^x [g(I(y, t)) - d]u(y, t)\, \mathrm{d}y \\ &= \int_0^x [g(\exp(-k_0 y - U(y, t)) - d]\partial_x U(y, t)\, \mathrm{d}y \\ &= \int_0^x \left\{ \frac{\mathrm{d}}{\mathrm{d}y}[F(y, U(y, t))] - \partial_x F(y, U(y, t)) \right\} \mathrm{d}y \\ &= F(x, U(x, t)) - \int_0^x \partial_x F(y, U(y, t))\, \mathrm{d}y. \end{aligned} \tag{8.10}$$

For (8.9), we define the Banach space

$$Y = \{\phi \in C^1([0, L]) :\ \phi(0) = 0\},$$

with the usual C^1 norm and the cone P in Y

$$P = \{\phi \in Y : \phi(x) \geq 0 \quad \text{for } 0 \leq x \leq L\},$$

whose interior is nonempty and is given by

$$\text{Int}\, P = \{\phi \in Y : \phi'(0) > 0, \text{ and } \phi(x) > 0 \quad \text{for } 0 < x \leq L\}.$$

The cone P generates the partial order relations $\leq_P$, $<_P$ and $\ll_P$ in Y.

By construction, the solutions U of (8.9) live in the convex set

$$E = \{\phi \in C^1([0, L]) : \phi(0) = 0, \quad \text{and} \quad \phi'(x) \geq 0 \quad \text{for } 0 \leq x \leq L\}.$$

Lemma 8.2.3 *Let $u_i \in C([0, L] \times [0, \infty)) \cap C^{2,1}([0, L] \times (0, \infty))$ $(i = 1, 2)$ be given such that*

$$\begin{cases} \partial_t u_2 \geq D\partial_{xx} u_2 - \alpha \partial_x u_2 + [g(\exp(-k_0 - \int_0^x u_2)) - d]u_2 & \text{in } [0, L] \times (0, \infty), \\ D\partial_x u_2 - \alpha u_2 = 0 & \text{in } \{0, L\} \times (0, \infty), \end{cases} \tag{8.11}$$

and u_1 satisfies the reversed inequality and the same boundary conditions. If, for some $t^* > 0$,

$$u_1(\cdot, t) \leq_K u_2(\cdot, t) \quad \text{for } 0 \leq t \leq t^* \tag{8.12}$$

and $u_2(\cdot, t^*) - u_1(\cdot, t^*) \notin \text{Int}\, \mathcal{K}$, i.e.

$$u_2(0, t^*) - u_1(0, t^*) = 0 \;\; \text{or} \;\; \int_0^{x'} (u_2(y, t^*) - u_1(y, t^*))\, dy = 0 \;\; \text{for some } 0 < x' \leq L,$$

then

$$u_1(x, t) \equiv u_2(x, t) \quad \text{for } 0 \leq x \leq L, \ 0 \leq t \leq t^*. \tag{8.13}$$

Remark 8.2.4 By examining the proof, we can actually relax (8.11) by replacing the $\leq$ in the classical sense, by the generalized sense, or even with $\leq_{\mathcal{K}}$. This yields, in particular, a natural notion of super- (and sub-) solution for (8.6). (See Problem 8.5.3.) Since we do not need this fact here, we refer the readers to [25] for more details.

Proof Define

$$U_i(x, t) = \int_0^x u_i(y, t)\, dy \quad \text{for } i = 1, 2.$$

Then

$$U_1(x, t) \leq U_2(x, t) \quad \text{for } 0 \leq x \leq L, \ 0 \leq t \leq t^*, \tag{8.14}$$

and at least one of the following alternatives holds:

1. $U_1(x^*, t^*) = U_2(x^*, t^*)$ for some $0 < x^* \leq L$;
2. $(U_2 - U_1)_x(0, t^*) = 0$.

Suppose the first case holds. Define $W(x, t) = U_2(x, t) - U_1(x, t)$. Then

$$
\begin{aligned}
&\partial_t W - D\partial_{xx} W + \alpha \partial_x W \\
&\geq F(x, U_2(x,t)) - F(x, U_1(x,t)) + \int_0^x [\partial_x F(y, U_1(y,t)) - \partial_x F(y, U_2(y,t))] \, \mathrm{d}y \\
&= h(x,t)W + \int_0^x \left[\int_{U_2(y,t)}^{U_1(y,t)} -k_0 \mathrm{e}^{-k_0 y - z} g'(\mathrm{e}^{-k_0 y - z}) \, \mathrm{d}z \right] \mathrm{d}y \\
&\geq h(x,t)W, \qquad (8.15)
\end{aligned}
$$

where

$$
h(x,t) = \int_0^1 \{ g(\exp(-k_0 x - \xi U_2(x,t) - (1-\xi)U_1(x,t))) - d \} \, \mathrm{d}\xi,
$$

and we used (8.9) and (8.10) for the first equality, and $g' \geq 0$ for the last inequality. Thus $W = U_2 - U_1$ is a nonnegative solution to the following linear differential inequality:

$$
\partial_t W - D\partial_{xx} W + \alpha \partial_x W - h(x,t)W \geq 0, \quad \text{for } 0 < x \leq L, \ 0 < t \leq t^*. \qquad (8.16)
$$

We claim that $W \equiv 0$ in $[0, L] \times [0, t^*]$. If not, then the strong maximum principle for parabolic equations implies that $W(x, t^*) > 0$ for $0 < x < L$. Therefore, if there exists some $x^* \in (0, L]$ such that $W(x^*, t^*) = 0$, we must have $x^* = L$, i.e. $W(L, t^*) = 0$, and hence $\partial_t W(L, t^*) \leq 0$. Note also that $D\partial_{xx} W(L, t^*) = \alpha \partial_x W(L, t^*)$ by the boundary conditions in (8.9). Hence, we deduce that the last equality holds in (8.15) at $(x, t) = (L, t^*)$, which means

$$
\int_0^L \left[\int_{U_2(y,t^*)}^{U_1(y,t^*)} -k_0 \mathrm{e}^{-k_0 x - z} g'(\mathrm{e}^{-k_0 - z}) \, \mathrm{d}z \right] \mathrm{d}y = 0.
$$

Since the integrand is strictly negative, we deduce that $U_1(x, t^*) \equiv U_2(x, t^*)$ for all $0 \leq x \leq L$. This is a contradiction.

Next, consider the second case, i.e. $(U_2 - U_1)_x(0, t^*) = 0$. We claim that necessarily there is a sequence $t_j \nearrow t^*$ such that case 1 holds, so that we can deduce similarly that $U_1 \equiv U_2$ in $[0, L] \times [0, t_j]$ for all j, whence $U_1 \equiv U_2$ in $[0, L] \times [0, t^*]$ by continuity. To see the claim, assume for contradiction that $U_2 > U_1$ in $(0, L] \times [t^* - \delta, t^*]$ for some δ. Then, observe that the boundary condition ensures $W(0, t^*) = 0$. Since W satisfies the differential inequality (8.16), we may apply Hopf's boundary point lemma to deduce that $(U_2 - U_1)_x(0, t^*) > 0$. This is a contradiction. Thus the claim is established, which concludes the proof of the second case. □

We are in position to prove Proposition 8.2.2.

Proof (of Proposition 8.2.2) Let u_i $(i = 1, 2)$ be given, and define $U_i(x, t) = \int_0^x u_i(y, t) \, \mathrm{d}y$ as before.

Step 1. We claim that $U_1(x, t) \leq U_2(x, t)$ in $[0, L] \times (0, \infty)$.

We first prove the claim under the additional assumption that

$$\partial_x(U_2 - U_1)(0,0) > 0 \quad \text{and} \quad U_1(x,0) < U_2(x,0) \quad \text{for } 0 < x \le L. \tag{8.17}$$

Let t^* be the maximal time such that for $0 < x \le L$ and $0 \le t < t^*$,

$$\partial_x(U_2 - U_1)(0,t) > 0 \quad \text{and} \quad U_1(x,t) < U_2(x,t). \tag{8.18}$$

By (8.17) and Proposition 8.1.1, $t^* > 0$. It remains to show that $t^* = \infty$. Suppose to the contrary that $0 < t^* < \infty$, then we can apply Lemma 8.2.3 to deduce that $U_1 \equiv U_2$ in $[0,L] \times [0,t^*]$, which is in contradiction with (8.17). This proves Step 1 under the additional assumption (8.18).

In general, let $\delta > 0$ and consider the solution u_2^δ of (8.6) with initial data $u_2(x,0) + \delta$, then $U_2^\delta(x,0) := \int_0^x u_2(y,0)\,\mathrm{d}y + \delta x$ satisfies

$$\partial_x(U_2^\delta - U_1)(0,0) > 0 \quad \text{and } U_1(x,0) < U_2^\delta(x,0) \quad \text{for } 0 < x \le L.$$

So that the above arguments apply to yield

$$U_1(x,t) < U_2(x,t) + \delta x \quad \text{in } [0,L] \times (0,\infty).$$

Step 1 is completed as we take $\delta \to 0$.

Step 2. Having proved that $U_1 \le U_2$ in $[0,L] \times [0,\infty)$, i.e., $u_1(\cdot,t) \le_{\mathcal{K}} u_2(\cdot,t)$ for $t \ge 0$, and noting that $u_1(\cdot,0) \not\equiv u_2(\cdot,0)$, we can apply Lemma 8.2.3 once again to deduce that $u_1(\cdot,t) \ll_{\mathcal{K}} u_2(\cdot,t)$ for $t > 0$. □

8.2.2 Long-time dynamics of the single species model

Let μ_1 be the principal eigenvalue of

$$\begin{cases} D\varphi'' - \alpha\varphi' + [g(\mathrm{e}^{-k_0 x}) - d]\varphi + \mu\varphi = 0 & \text{in } [0,L], \\ D\varphi' - \alpha\varphi = 0 & \text{for } x = 0, L. \end{cases} \tag{8.19}$$

The long-time dynamics of (8.6) can be classified as follows:

Theorem 8.2.5 *Let μ_1 be the principal eigenvalue of* (8.19).

1. *If $\mu_1 \ge 0$, then every nonnegative solution u of* (8.6) *satisfies $\|u(\cdot,t)\|_{C([0,L])} \to 0$ as $t \to \infty$.*
2. *If $\mu_1 < 0$, then* (8.6) *has a unique positive equilibrium, denoted as θ. Moreover, $\|u(\cdot,t) - \theta\|_{C([0,L])} \to 0$ as $t \to \infty$ provided that $u(x,0)$ is nonnegative and nontrivial.*

Remark 8.2.6 If $d < g(\mathrm{e}^{-k_0 L})$, then (8.6) has a unique positive solution θ. Moreover, $\|u(\cdot,t) - \theta\|_{C([0,L])} \to 0$ as $t \to \infty$ provided that $u(x,0)$ is nonnegative and nontrivial. Indeed, let μ_1 be the principal eigenvalue of (8.19). Integrating (8.19) in $[0, L]$ and using the no-flux boundary conditions, we have

$$\mu_1 \int_0^L \varphi \,\mathrm{d}x = \int_0^L [d - g(\mathrm{e}^{-k_0 x})]\varphi \,\mathrm{d}x \le [d - g(\mathrm{e}^{-k_0 L})] \int_0^L \varphi \,\mathrm{d}x < 0.$$

Hence $\mu_1 < 0$, and we may conclude by Theorem 8.2.5. Indeed, it was shown in [14] (see also [8]) that there exists a number $d_* = d_*(D, \alpha, L) > 0$ such that $\mu_1 < 0$ if and only if $d < d_*$; that is, d_* is the critical death rate for the persistence of a single phytoplankton species. In particular, the above argument shows that $d_* \ge g(\mathrm{e}^{-k_0 L})$ holds for any D, α, L.

Thanks to Proposition 8.1.1, Proposition 8.2.2 and Lemma 8.2.7, the single equation (8.6) generates a monotone dynamical system with precompact trajectories. Part 2 of Theorem 8.2.5 can then be proved by constructing super- and sub-solutions, and arguing similarly as in Chapter 5. Here, we present an alternative argument based on the concept of subhomogeneity (see e.g. [40, Chap. 2]).

First, we prove that (8.6) generates a dissipative system.

Lemma 8.2.7 *There exists an $M_0 > 0$, independent of initial data, such that*

$$\limsup_{t\to\infty} \|u(\cdot,t)\|_{C^1([0,L])} \le M_0,$$

for all nonnegative solutions u of (8.6).

Proof Define

$$u_*(t) = \inf_{0<x<L} u(x,t) \quad \text{and} \quad u^*(t) = \sup_{0<x<L} u(x,t).$$

By the fact that $0 \le g(\exp(-k_0 x - \int_0^x u(y,t)\,\mathrm{d}y)) \le g(1)$ is uniformly bounded, the Harnack principle (Theorem 1.2.6) says that

$$u^*(t) \le C' u_*(t) \quad \text{for } t \ge 1 \tag{8.20}$$

for some constant C' that is independent of u.

Choose $M_1 > 0$ such that $g(\mathrm{e}^{-M_1}) - d < 0$ (using $g(0) = 0$), and then choose $\delta_1 > 0$ such that

$$C'\delta_1 |g(1) - d| + (L - \delta_1)(g(\mathrm{e}^{-M_1}) - d) < 0. \tag{8.21}$$

Let u be a nonnegative solution of (8.6) with initial data u_0. We first claim that

$$\limsup_{t\to\infty} \|u(\cdot,t)\|_{L^1([0,L])} \le M_0', \tag{8.22}$$

where $M_0' := \max\{M_1, C'LM_1/\delta_1\}$ is independent of initial data.

To this end, it is enough to show that there exists a $\delta_2 > 0$ such that the differential inequality

$$\frac{\mathrm{d}}{\mathrm{d}t}\int_0^L u(x,t)\,\mathrm{d}x \le -\delta_2 \int_0^L u(x,t)\,\mathrm{d}x \tag{8.23}$$

holds whenever $\int_0^L u(x,t)\,\mathrm{d}x > M_0' := \max\{M_1, C'LM_1/\delta_1\}$.

Now,

$$M_1 < \frac{\delta_1}{C'L}\int_0^L u(x,t)\,\mathrm{d}x \le \frac{\delta_1}{C'}u^*(t) \le \delta_1 u_*(t). \tag{8.24}$$

Integrating (8.6) over $[0, L]$, we obtain

$$\begin{aligned}
&\frac{\mathrm{d}}{\mathrm{d}t}\int_0^L u(x,t)\,\mathrm{d}x \\
&\le \int_0^L [g(\mathrm{e}^{-\int_0^x u(y,t)\,\mathrm{d}y}) - d]u(x,t)\,\mathrm{d}x \\
&\le \int_0^L [g(\mathrm{e}^{-xu_*(t)}) - d]u(x,t)\,\mathrm{d}x \\
&\le \int_0^{\delta_1} (g(1) - d)u(x,t)\,\mathrm{d}x + \int_{\delta_1}^L [g(\mathrm{e}^{-M_1}) - d]u(x,t)\,\mathrm{d}x \quad \text{(by (8.21), (8.24))} \\
&\le u^*(t)\delta_1(g(1) - d) + u_*(t)\int_{\delta_1}^L (g(\mathrm{e}^{-M_1}) - d)\,\mathrm{d}x \\
&\le u_*(t)\left[C'\delta_1(g(1) - d) + (L - \delta_1)(g(\mathrm{e}^{-M_1}) - d)\right] \\
&\le \left(\frac{1}{C'L}\int_0^L u(x,t)\,\mathrm{d}x\right)\left[C'\delta_1(g(1) - d) + (L - \delta_1)(g(\mathrm{e}^{-M_1}) - d)\right].
\end{aligned}$$

This proves (8.23) holds whenever $\int_0^L u(x,t)\,\mathrm{d}x > \max\{M_1, C'LM_1/\delta_1\}$. We obtain (8.22).

It then follows from (8.22) and (8.20) that

$$\limsup_{t\to\infty} \|u(\cdot,t)\|_{C([0,L])} \le M_0'', \tag{8.25}$$

for some M_0'' independent of u.

Finally, the uniform eventual boundedness in $C^1([0, L])$ follows from parabolic L^p estimates and Sobolev embeddings. □

Next, we observe that (8.6) generates a subhomogeneous dynamical system.

Lemma 8.2.8 *Let u be a nonnegative solution of* (8.6) *with initial data u_0, and let u_λ be the solution of* (8.6) *with initial data λu_0. If $0 < \lambda < 1$ and $u_0 \ge, \not\equiv 0$, then*

$$\lambda u(\cdot,t) \ll_{\mathcal{K}} u_\lambda(\cdot,t) \quad \textit{for } t > 0. \tag{8.26}$$

In particular, (8.6) *has at most one positive equilibrium.*

Proof It suffices to observe that λu satisfies

$$\begin{cases} \partial_t(\lambda u) < D\partial_{xx}(\lambda u) - \alpha\partial_x(\lambda u) + [g(\mathrm{e}^{-k_0 x - \int_0^x (\lambda u)}) - d](\lambda u) & \text{in } (0,L)\times(0,\infty), \\ D\partial_x(\lambda u) - \alpha(\lambda u) = 0 & \text{in } \{0,L\}\times(0,\infty). \end{cases}$$

One can then repeat the proof of Proposition 8.2.2 (and Lemma 8.2.3) to prove (8.26).

It remains to show the uniqueness of positive equilibrium. Suppose to the contrary that there exist two distinct positive equilibrium solutions, θ_1 and θ_2, of (8.6). By interchanging θ_1, θ_2 if necessary, we assume that $\theta_1 - \theta_2 \notin \mathcal{K}$. Then

$$\lambda^* := \sup\{\lambda > 0 : \lambda\theta_2 \le_{\mathcal{K}} \theta_1\} \in (0,1).$$

However, (8.26) implies that

$$\lambda^*\theta_2 \ll_{\mathcal{K}} \theta_1.$$

This contradicts the maximality of λ^*. □

Proof (of Theorem 8.2.5) Let $X = C([0,L])$, $C = C([0,L];\mathbb{R}_+)$ and $\mathcal{K}$ be given by (8.7). Let $\Phi_t : C \to C$ be the semiflow generated by (8.6). We will apply Corollary C.2.2. To this end, we need to verify hypotheses (S1')–(S3') and (S5') from Subsection C.2. First, it is clear that $\Phi_t(0) = 0$ for all $t > 0$. Therefore (S1') follows from Proposition 8.2.2. (S2') and (S3') are consequences of Lemmas 8.2.7 and 8.2.8 respectively. Finally, observe that for each $t > 0$, Φ_t is compact and its Fréchet derivative at zero, denoted by $D\Phi_t(0) : C \to C$, is given by $D\Phi_t(0)[\phi_0] = \phi(\cdot, t)$, where ϕ is the unique solution of

$$\begin{cases} \partial_t\phi = D\partial_{xx}\phi - \alpha\partial_x\phi + [g(\mathrm{e}^{-k_0 x}) - d]\phi & \text{in } [0,L]\times(0,\infty), \\ D\partial_x\phi - \alpha\phi = 0 & \text{on } \{0,L\}\times(0,\infty), \\ \phi(x,0) = \phi_0(x) & \text{in } [0,L]. \end{cases}$$

This verifies (H5'). Now, let μ_1 and ϕ_1 be the principal eigenvalue and eigenfunction of (8.19). Since $D\Phi_t(0)[\phi_1] = \mathrm{e}^{-\mu_1 t}\phi_1$ and $\phi_1 \in C \setminus \{0\}$, we deduce that

$$r(D\Phi_t) = \mathrm{e}^{-\mu_1 t}. \tag{8.27}$$

The conclusion thus follows from (8.27) and Theorem C.2.2. □

8.3 Dynamics for Two Competing Phytoplankton Species

Consider the competition model for two phytoplankton species

$$\begin{cases} \partial_t u_i = D_i \partial_{xx} u_i - \alpha_i \partial_x u_i + u_i\,[g_i(I(x,t)) - d_i] & \text{for } 0 < x < L,\ t > 0,\ i = 1,2, \\ D_i \partial_x u_i - \alpha_i u_i = 0 & \text{for } x = 0, L,\ t > 0,\ i = 1,2, \\ u_i(x,0) = u_{i,0}(x) & \text{for } 0 \le x \le L,\ i = 1,2, \\ I(x,t) = \exp(-k_0 x - \sum_{i=1}^2 \int_0^x u_i(y,t)\,\mathrm{d}y) & \text{for } 0 \le x \le L,\ t > 0. \end{cases} \tag{8.28}$$

Note that we are assuming without loss of generality that $I_0 = 1$ and $k_1 = k_2 = 1$. The two phytoplankton populations may differ in their dispersal rates, sinking rates, growth rates and death rates.

Suppose (8.28) has two semitrivial equilibria $E_1 = (\hat{u}, 0)$ and $E_2 = (0, \hat{v})$. According to Theorem 8.2.5, E_1 and E_2 exist if and only if, for $i = 1, 2$, the principal eigenvalue of

$$\begin{cases} D_i \varphi'' - \alpha_i \varphi' + [g_i(\mathrm{e}^{-k_0 x}) - d_i]\varphi + \mu\varphi = 0 & \text{in } [0, L], \\ D_i \varphi' - \alpha_i \varphi = 0 & \text{for } x = 0, L, \end{cases} \tag{8.29}$$

is negative. (By Remark 8.2.6, this is guaranteed if $d_i < g_i(\mathrm{e}^{-k_0 L})$ for $i = 1, 2$.)

Definition 8.3.1 We say that $X = C([0, L]) \times C([0, L])$ is a Banach space with order induced by the solid cone $K = \mathcal{K} \times (-\mathcal{K})$, such that

$$\begin{cases} (u_0, v_0) \le_K (\bar{u}_0, \bar{v}_0) & \text{iff} \quad \bar{u}_0 - u_0 \in \mathcal{K} \text{ and } v_0 - \bar{v}_0 \in \mathcal{K}; \\ (u_0, v_0) <_K (\bar{u}_0, \bar{v}_0) & \text{iff} \quad (u_0, v_0) \le_K (\bar{u}_0, \bar{v}_0) \text{ and } (u_0, v_0) \ne (\bar{u}_0, \bar{v}_0); \\ (u_0, v_0) \ll_K (\bar{u}_0, \bar{v}_0) & \text{iff} \quad \bar{u}_0 - u_0 \in \operatorname{Int}\mathcal{K} \text{ and } v_0 - \bar{v}_0 \in \operatorname{Int}\mathcal{K}. \end{cases}$$

It is proved in [25] that (8.28) generates a strongly monotone dynamical system in the ordered Banach space X with the order induced by the cone K.

Theorem 8.3.2 *Let (u_1, u_2), $(\bar{u}_1, \bar{u}_2)$ be solutions of* (8.28). *Then the following hold:*

1. *If $(u_1(\cdot, 0), u_2(\cdot, 0)) <_K (\bar{u}_1(\cdot, 0), \bar{u}_2(\cdot, 0))$, then*

$$(u_1(\cdot, t), u_2(\cdot, t)) <_K (\bar{u}_1(\cdot, t), \bar{u}_2(\cdot, t)) \quad \textit{for each } t > 0.$$

2. *If $(u_1(\cdot, 0), u_2(\cdot, 0)) <_K (\bar{u}_1(\cdot, 0), \bar{u}_2(\cdot, 0))$, and at least one of them belongs to $\operatorname{Int}\mathcal{K} \times \operatorname{Int}\mathcal{K}$, then*

$$(u_1(\cdot, t), u_2(\cdot, t)) \ll_K (\bar{u}_1(\cdot, t), \bar{u}_2(\cdot, t)) \quad \textit{for } t > 0.$$

Proof The proof follows the ideas in Section 8.2.1. We refer the readers to [25, Corollary 3.4] for details. □

Define $X = C([0, L]) \times C([0, L])$, and

$$C = C_1 \times C_2 = C([0, L]; \mathbb{R}_+) \times C([0, L]; \mathbb{R}_+), \quad K = \mathcal{K} \times (-\mathcal{K}),$$

where $\mathcal{K}$ is given in (8.7). We recall the hypotheses (H1')–(H5') from Appendix E.

Lemma 8.3.3 *The semiflow $\Phi_t : C \to C$ generated by* (7.1) *is continuously differentiable, and satisfies the conditions* (H1')–(H5').

Proof (of Lemma 8.3.3) Now, (H1') and (H4') follow directly from Theorem 8.3.2.

Next, notice that the semiflow operator is continuously differentiable. Furthermore, if $(\tilde{u}, \tilde{v})$ is an equilibrium of Φ_t, then $D\Phi_t(\tilde{u}, \tilde{v}) : X \to X$ is given by $D\Phi_t(\tilde{u}, \tilde{v})[p_0, q_0] = (p(\cdot, t), q(\cdot, t))$, where (p, q) is the unique solution of the linear parabolic system

$$\begin{cases} p_t - D_1 p_{xx} + \alpha_1 p_x = [g_1(\tilde{I}) - d_1]p + \tilde{u} g_1'(\tilde{I})\tilde{I}[-\int_0^x (p(y,t) + q(y,t))\,\mathrm{d}y] \\ q_t - D_2 q_{xx} + \alpha_2 q_x = [g_2(\tilde{I}) - d_2]q + \tilde{v} g_2'(\tilde{I})\tilde{I}[-\int_0^x (p(y,t) + q(y,t))\,\mathrm{d}y] \end{cases} \tag{8.30}$$

in $[0, L] \times \mathbb{R}_+$ and

$$\begin{cases} \tilde{I}(x) = \mathrm{e}^{-k_0 x - \int_0^x \tilde{u}(y)\,\mathrm{d}y - \int_0^x \tilde{v}(y)\,\mathrm{d}y} & \text{in } [0, L], \\ D_1 p_x - \alpha_1 p = D_2 q_x - \alpha_2 q = 0 & \text{on } \{0, L\} \times \mathbb{R}_+, \\ p(x, 0) = p_0(x), \quad q(x, 0) = q_0(x) & \text{in } \Omega. \end{cases} \tag{8.31}$$

To verify condition (H2'), set $(\tilde{u}, \tilde{v}) = (0, 0)$ in (8.30)–(8.31) so that the system decouples. Next, observe that

$$D\Phi_t(0, 0)[p_0, q_0] = (f_1(p_0), f_2(q_0)) = (\mathrm{e}^{tL_1}[p_0], \mathrm{e}^{tL_2}[q_0]),$$

where e^{tL_i} denotes the analytic semigroup generated by

$$L_i = D_i \partial_{xx} - \alpha_i \partial_x + g_i(\mathrm{e}^{-k_0 x}) - d_i$$

in the space $C(\overline{\Omega})$, such that the solution respects the no-flux boundary condition in positive time. For $i = 1, 2$, let $\tilde{\mu}_i$ and $\tilde{\phi}_i(x) > 0$ be the principal eigenvalue and eigenfunctions of (8.29). Then $f_i(\tilde{\phi}_i) = \mathrm{e}^{-\tilde{\mu}_i t}\tilde{\phi}_i$. Since f_i is linear and strongly positive in $C(\overline{\Omega})$ and since we enforce the condition $\mu_i < 0$ in this chapter, it follows from the Krein-Rutman Theorem (Theorem B.3.2) that $r(f_i) = \mathrm{e}^{-\tilde{\mu}_i t} > 1$. This verifies (H2'), in view of Remark E.2.1. (H3') follows from Theorem 8.2.5 and the fact that $\mu_i < 0$.

We will verify (H5') using Lemma E.2.10. For that purpose, observe that $D_1\Phi_t^{(1)}(E_2)$ is an analytic semigroup generated by certain linear elliptic operators, so they are automatically compact and strongly monotone with respect to $C_1 = C([0, L]; \mathbb{R}_+)$. On the other hand, $D_2\Phi_t^{(2)}(E_2) : X \to X$ is generated by the solution to the linear parabolic equation with nonlocal term:

$$q_t - D_2 q_{xx} + \alpha_2 q_x = [g_2(\tilde{I}_2(x)) - d_2]q + \tilde{v}g_2'(\tilde{I}(x))\tilde{I}(x)[-\int_0^x q(y,t)\,\mathrm{d}y]$$

on $[0, L] \times (0, \infty)$ with no-flux boundary conditions, where

$$\tilde{I}_2(x) = \exp(-k_0 x - \int_0^x \hat{v}(y)\,\mathrm{d}y) \text{ and } E_2 = (0, \hat{v}).$$

It is not hard to see that $D_2\Phi_t^{(2)}(E_2)$ is compact. Furthermore, by the arguments in Section 8.2.1, we can once again show that the linear operator $D_2\Phi_t^{(2)}(E_2)$ is strongly monotone with respect to $\mathcal{K}$. Moreover, $r(D_2\Phi_t^{(2)}(E_2)) = \mathrm{e}^{-\tilde{\mu}t}$, where $\tilde{\mu} \in \mathbb{R}$ and $\phi \in \mathcal{K}$ are, respectively, the principal eigenvalue and principal eigenfunction of

$$\begin{cases} -D_2\phi_{xx} + \alpha_2\phi_x = [g_2(\tilde{I}_2) - d_2]\phi + \tilde{v}g_2'(\tilde{I}_2)\tilde{I}_2[-\int_0^x \phi(y,t)\,\mathrm{d}y] + \tilde{\mu}\phi, & 0 < x < L, \\ D_2\phi_x - \alpha_2\phi = 0, & x = 0, L. \end{cases} \tag{8.32}$$

For details of the existence of ϕ and $\tilde{\mu}$, we refer to [25, Theorem 4.4].

We claim that $\tilde{\mu} > 0$. Suppose to the contrary that $\tilde{\mu} \le 0$, then we use the fact that $\phi \in \operatorname{Int}\mathcal{K}$ to get

$$\tilde{v}g_2'(\tilde{I}_2(x))\tilde{I}_2(x)\int_0^x \phi(y,t)\,\mathrm{d}y > 0 \quad \text{for } 0 < x < L.$$

Then (8.32) yields that

$$-D_2\phi_{xx} + \alpha_2\phi_x < [g_2(\tilde{I}_2(x)) - d_2]\phi + \tilde{\mu}\phi \quad \text{for } 0 < x < L.$$

Next, we use the facts that $\int_0^x \phi(y)\,\mathrm{d}y > 0$ for $x \in (0, L]$ and $\inf_{[0,L]} \hat{v} > 0$ to obtain a constant $c > 0$ such that $\min_{[0,L]}(c\hat{v} - \phi) = 0$. Then $\varphi = c\hat{v} - \phi$ satisfies

$$\begin{cases} D_2\varphi_{xx} - \alpha_2\varphi_x + [g_2(\tilde{I}_2(x)) - d_2]\varphi + \tilde{\mu}\varphi < \tilde{\mu}c\hat{v} \le 0 & \text{in } [0, L], \\ D_2\varphi_x - \alpha_2\varphi = 0 & \text{for } x = 0, L, \\ \min_{[0,L]} \varphi = 0. \end{cases}$$

By the strict differential inequality and nonnegativity of φ, we must have $\varphi > 0$ in $(0, L)$ and that $\varphi(x_0) = 0$ for some $x_0 \in \{0, L\}$. But the Hopf boundary point lemma says that $\varphi_x(x_0) \ne 0$, which contradicts the boundary condition $D_2\varphi_x(x_0) - \alpha_2\varphi(x_0) = 0$. Hence, we have established that $\tilde{\mu} > 0$, i.e. $r(D_2\Phi_t^{(2)}(E_2)) = \mathrm{e}^{-\tilde{\mu}t} < 1$. The last condition, which translates to $D_1\Phi_t^{(2)}(E_2)[p_0] \ne 0$ for $p_0 \in C(\overline{\Omega})$, $p_0 > 0$ in $\overline{\Omega}$, also holds by virtue of (8.30)–(8.31).

Now, we can invoke Lemma E.2.10 to conclude that (H5') holds. □

Hence, the results in Appendix E can be applied to establish the following results.

Theorem 8.3.4 *If E_1 and E_2 are both linearly unstable, then* (8.28) *has at least one stable positive equilibrium $E^* = (u^*, v^*)$.*

Proof This is a consequence of Theorem E.2.15. □

Theorem 8.3.5 *Suppose* (8.28) *has no positive equilibria, then exactly one of the following statements holds.*

- E_1 *is globally asymptotically stable among all solutions of* (8.28) *with nonnegative, nontrivial initial data.*
- E_2 *is globally asymptotically stable among all solutions of* (8.28) *with nonnegative, nontrivial initial data.*

Proof This is a consequence of Theorem E.2.13. □

Selection for more buoyant phytoplankton species

As an application, we state and prove a theorem which says that, in a eutrophic water column in which competition is only limited by light, if both species have the same diffusion rate, then the more buoyant (or less prone to sinking) phytoplankton species is selected. Before doing so, we first prove two eigenvalue lemmas. For $D > 0$, $\alpha \in \mathbb{R}$ and $h \in C([0, L])$, let $\lambda_1(D, \alpha, h)$ and ψ be the principal eigenvalue and positive eigenfunction of

$$\begin{cases} D\psi_{xx} + \alpha\psi_x + h(x)\psi + \lambda_1\psi = 0, & 0 < x < L, \\ \psi_x = 0, & x = 0, L. \end{cases} \tag{8.33}$$

Lemma 8.3.6 *If* $h \in C([0, L])$ *is strictly decreasing, then* $\psi_x < 0$ *for* $0 < x < L$.

Proof Rewrite (8.33) as

$$\begin{cases} D(\mathrm{e}^{\alpha x/D}\psi_x)_x + \mathrm{e}^{\alpha x/D}[h(x) + \lambda_1]\psi = 0, & 0 < x < L, \\ \psi_x = 0, & x = 0, L. \end{cases} \tag{8.34}$$

Integrating (8.34) over $(0, L)$, we have

$$\int_0^L \mathrm{e}^{\alpha x/D}\psi[h(x) + \lambda_1]\,\mathrm{d}x = 0,$$

which implies that $h(x) + \lambda_1$ changes sign in $(0, L)$. Since $h(x)$ is strictly decreasing in $(0, L)$, there exists a unique $x_0 \in (0, L)$ such that $h(x) + \lambda_1 > 0$ for $0 < x < x_0$ and $h(x) + \lambda_1 < 0$ for $x_0 < x < L$. Hence, by (8.34) we see that $(\mathrm{e}^{\alpha x/D}\psi_x)_x < 0$ for $0 < x < x_0$ and $(\mathrm{e}^{\alpha x/D}\psi_x)_x > 0$ for $x_0 < x < L$. That is, $\mathrm{e}^{\alpha x/D}\psi_x$ is strictly decreasing in $(0, x_0)$ and strictly increasing in (x_0, L). Since $\psi_x(0) = \psi_x(L) = 0$, we have $\psi_x < 0$ in $(0, L)$. □

Lemma 8.3.7 *If* $h \in C([0, L])$ *is strictly decreasing, then*

$$\partial_\alpha\lambda_1(D, \alpha, h) > 0 \quad \textit{for any } D > 0 \textit{ and } \alpha \in \mathbb{R}.$$

Proof For definiteness, normalize the positive eigenfunction ψ by $\max_{[0,L]}\psi = 1$, so that it is uniquely determined. By arguing as in Proposition 1.3.15, it follows that $\lambda_1(D,\alpha,h)$ is smooth in $D > 0$ and $\alpha \in \mathbb{R}$. For simplicity of notation, we denote $\partial_\alpha \psi$ by ψ', etc. Differentiating (8.34) with respect to α, we have

$$\begin{cases} D\psi'_{xx} + \alpha\psi'_x + \psi_x + [h(x) + \lambda_1]\psi' = -\lambda'_1\psi, & 0 < x < L, \\ \psi'_x = 0, & x = 0, L. \end{cases} \tag{8.35}$$

Multiplying (8.35) by $e^{\alpha x/D}$, we can write the first equation as

$$D(e^{\alpha x/D}\psi'_x)_x + e^{\alpha x/D}[h(x) + \lambda_1]\psi' = -e^{\alpha x/D}\psi_x - e^{\alpha x/D}\lambda'_1\psi. \tag{8.36}$$

Multiplying (8.36) by ψ and integrating by parts, we have

$$\begin{aligned} &-D\int_0^L e^{\alpha x/D}\psi_x\psi'_x + \int_0^L e^{\alpha x/D}\psi'\psi[h(x) + \lambda_1] \\ &= -\int_0^L e^{\alpha x/D}\psi\psi_x - \lambda'_1\int_0^L e^{\alpha x/D}\psi^2. \end{aligned}$$

Integrating by parts again, we have

$$\begin{aligned} 0 &= \int_0^L \psi'[D(e^{\alpha x/D}\psi_x)_x + e^{\alpha x/D}\psi[h(x) + \lambda_1] \\ &= -\int_0^L e^{\alpha x/D}\psi\psi_x - \lambda'_1\int_0^L e^{\alpha x/D}\psi^2, \end{aligned}$$

where the first equality follows from (8.34). Hence, we have

$$\lambda'_1 = -\frac{\int_0^L e^{\alpha x/D}\psi\psi_x}{\int_0^L e^{\alpha x/D}\psi^2} > 0,$$

where the last inequality follows from $\psi_x < 0$, which is proved in Lemma 8.3.6. □

Now we state and prove the selection for buoyant species in a eutrophic water column [25, Theorem 2.2], i.e. the buoyant species has the advantage provided other factors are held constant. For this purpose, we set $D_1 = D_2$, $\alpha_1 < \alpha_2$, $g_1 = g_2$, $d_1 = d_2 := d$ in (8.28) and consider the following competition system.

$$\begin{cases} \partial_t u_i = D\partial_{xx}u_i - \alpha_i\partial_x u_i + u_i\,[g(I(x,t)) - d] & \text{for } 0 < x < L,\ t > 0,\ i = 1, 2, \\ D\partial_x u_i - \alpha_i u_i = 0 & \text{for } x = 0, L,\ t > 0,\ i = 1, 2, \\ u_i(x, 0) = u_{i,0}(x) & \text{for } 0 \le x \le L,\ i = 1, 2, \\ I(x,t) = \exp(-k_0 x - \sum_{i=1}^2 \int_0^x u_i(y,t)\,dy) & \text{for } 0 \le x \le L,\ t > 0. \end{cases} \tag{8.37}$$

Theorem 8.3.8 *If $\alpha_1 < \alpha_2$, and both semi-trivial equilibria exist, then the more buoyant species u_1 drives the less buoyant species u_2 to extinction, as long as the initial data $u_{i,0}$ is nonnegative and nontrivial for $i = 1, 2$.*

Proof By Theorem 8.3.5, it suffices to establish, for system (8.37), the linear instability of E_2 and the nonexistence of positive equilibria.[1] Recall that $E_2 = (0, \tilde{v})$, where $\tilde{v}$ is the unique positive solution of

$$\begin{cases} D\tilde{v}_{xx} - \alpha_2 \tilde{v}_x + \tilde{v}[g(\sigma_2) - d] = 0, & 0 < x < L, \\ D\tilde{v}_x - \alpha_2 \tilde{v} = 0, & x = 0, L, \end{cases} \tag{8.38}$$

where $\sigma_2(x) = \exp(-k_0 x - \int_0^x \tilde{v}(s)\,\mathrm{d}s)$.

Step 1. We claim that $\lambda_1(D, \alpha_2, g(\sigma_2) - d) = 0$. Indeed, one can easily verify that zero is an eigenvalue of (8.33) with the positive eigenfunction being $\mathrm{e}^{-\alpha_2 x/D}\tilde{v}$. By the characterization of the principal eigenvalue (Theorem 1.3.6, assertion 4), it follows that $\lambda_1(D, \alpha_2, g(\sigma_2) - d) = 0$.

Step 2. To show that $E_2 = (0, \tilde{v})$ is linearly unstable, it suffices to show that $\lambda_1(D, \alpha_1, g(\sigma_2) - d) < 0$, where $\lambda_1(D, \alpha, h)$ is the principal eigenvalue of (8.33); see [25, Proposition 4.5]. Indeed, the equilibrium E_2 is unstable if and only if the first species can invade when rare, which is in turn equivalent to the negativity of the principal eigenvalue $\tilde{\lambda}$ of

$$\begin{cases} D\tilde{\phi}_{xx} - \alpha_1 \tilde{\phi}_x + [g(\sigma_2) - d]\tilde{\phi} + \tilde{\lambda}\tilde{\phi} = 0 & \text{for } 0 < x < L, \\ D\tilde{\phi}_x - \alpha_1 \tilde{\phi} = 0 & \text{for } x = 0, L. \end{cases} \tag{8.39}$$

By the transformation $\psi = \mathrm{e}^{-\alpha_1 x/D}\tilde{\phi}$, we deduce that $\tilde{\lambda} = \lambda_1(D, \alpha_1, g(\sigma_2) - d)$. Hence, it remains to show that $\lambda_1(D, \alpha_1, g(\sigma_2) - d) < 0$. Since $s \mapsto g(s)$ is strictly increasing and $x \mapsto \sigma_2$ is strictly decreasing, it follows from Lemma 8.3.7 that $\lambda_1(D, \alpha, g(\sigma_2) - d)$ is strictly increasing in α. Combining with Step 1, we have

$$\lambda_1(D, \alpha_1, g(\sigma_2) - d) < \lambda_1(D, \alpha_2, g(\sigma_2) - d) = 0.$$

This shows the linear instability of E_2.

Step 3. We claim that (8.37) has no positive equilibrium. Assume to the contrary that (8.37) has a positive equilibrium (u_1, v_1). Then it follows again by assertion 4 of Theorem 1.3.6 that

$$\lambda_1(D, \alpha_1, g(\sigma^*) - d) = 0 = \lambda_1(D, \alpha_2, g(\sigma^*) - d), \tag{8.40}$$

[1] Thanks to Theorem 8.3.5, one can alternatively check that E_1 is linearly stable and the nonexistence of positive equilibrium. This is a special feature of diffusive Lotka–Volterra-like systems, which satisfy the additional condition (H5'). For general competitively systems satisfying only (H1')–(H4'), it is necessary to show the linear instability of a semitrivial equilibrium, together with nonexistence of positive equilibria; see Proposition E.2.7 and also [13, 16, 30].

where $\sigma^*(x) = \exp(-k_0x - \int_0^x [u_1(s)+v_1(s)]\,\mathrm{d}s)$. However, since $g(\sigma^*)-d$ is strictly decreasing, Lemma 8.3.7 implies that $\lambda_1(D, \alpha, g(\sigma^*) - d)$ is strictly increasing in α. This is in contradiction with (8.40). □

8.4 The N-Species Model – Application of the Principal Floquet Bundle

Let $N \geq 2$ and consider the N-species competition model.

$$\begin{cases} \partial_t u_i = D\partial_{xx}u_i - \alpha_i\partial_x u_i + u_i\,[g(I(x,t)) - d] & \text{for } 0 < x < L,\ t > 0,\ 1 \leq i \leq N, \\ D\partial_x u_i - \alpha_i u_i = 0 & \text{for } x = 0, L,\ t > 0,\ 1 \leq i \leq N, \\ u_i(x,0) = u_{i,0}(x) & \text{for } 0 \leq x \leq L,\ 1 \leq i \leq N, \\ I(x,t) = \exp(-k_0x - \sum_{i=1}^N \int_0^x u_i(y,t)\,\mathrm{d}y) & \text{for } 0 \leq x \leq L,\ t > 0, \end{cases} \tag{8.41}$$

where $D, d > 0$ are constants independent of i, and $\alpha_1 < \alpha_2 < \cdots < \alpha_N$. Also, we assume for simplicity that all species have the same growth response function g. Note also that we are assuming without loss of generality that $I_0 = 1$ and $k_i = 1$ for all i.

We also assume that

$$d < g(\mathrm{e}^{-k_0L}). \tag{8.42}$$

Under the assumption (8.42), it follows from Remark 8.2.6 that, for each $\alpha \in \mathbb{R}$, the single species problem (8.6) has a unique positive equilibrium $\theta_\alpha(x)$ that is globally asymptotically stable among all positive solutions of (8.6). In particular,

$$E_i = (0, ..., 0, \theta_{\alpha_i}, 0, ..., 0)$$

is an equilibrium of (8.41), for each i.

We will prove that, in general, competition exclusion takes place when the N species are similar, i.e. α_i lies in a small interval.

Theorem 8.4.1 *For each $\hat{\alpha} \in \mathbb{R}$, there exists an $\epsilon > 0$ such that for arbitrary N and arbitrary increasing sequence $(\alpha_i)_{i=1}^N \subset (\hat{\alpha} - \epsilon, \hat{\alpha} + \epsilon)$, every positive solution $(u_i)_{i=1}^N$ of (8.41) converges to the equilibrium $E_1 = (\theta_{\alpha_1}, 0, ..., 0)$ as $t \to \infty$.*

The rest of this section is organized as follows: In Subsection 8.4.1, we derive some uniform bounds for positive solutions to the time-dependent problem (8.41). In Subsection 8.4.2, we use the smallness of ϵ to show that for any positive solutions $(u_i)_{i=1}^N$ of (8.41), the total population $\sum_{i=1}^N u_i$ eventually enters a neighborhood of the positive equilibrium of the single species problem. In Subsection 8.4.3, we introduce the notion of normalized principal bundle, which is a generalized notion of principal eigenvalue for elliptic or periodic-parabolic operators. In Subsection 8.4.4, we prove a general exclusion criterion and then give the proof of Theorem 8.4.1.

8.4.1 A priori estimates

Define

$$G(s) = \int_0^s g(\mathrm{e}^{-\tau})\,\mathrm{d}\tau - sd. \tag{8.43}$$

Then $G(0) = 0$, $G'(s) = g(\mathrm{e}^{-s}) - d$, and, since $G'(+\infty) = -d < 0$, there exists an $M_1 > 0$ such that

$$G(s) < 0 \quad \text{for } s \geq M_1. \tag{8.44}$$

In the following we will also use the notation

$$\hat{u}(x,t) = \sum_{i=1}^{N} u_i(x,t), \quad U_i(x,t) := \int_0^x u_i(y,t)\,\mathrm{d}y \ \text{ and } \ \hat{U}(x,t) = \sum_{i=1}^{N} U_i(x,t).$$

Lemma 8.4.2 *Let* $(u_i)_{i=1}^N$ *be a nonnegative solution of* (8.41) *such that*

$$\sum_{i=1}^{N} \|u_i(x,0)\|_{L^1([0,L])} \leq M,$$

then

$$\begin{cases} \sup\limits_{t\geq 0} \sum_{i=1}^N \|u_i(x,t)\|_{L^1([0,L])} \leq \max\{M, M_1\}, \\ \limsup\limits_{t\to\infty} \sum_{i=1}^N \|u_i(x,t)\|_{L^1([0,L])} \leq M_1. \end{cases} \tag{8.45}$$

Proof Integrating (8.41) with respect to x from 0 to L, and adding i from 1 to N, we obtain

$$\begin{aligned} \frac{\mathrm{d}}{\mathrm{d}t}\hat{U}(L,t) &= \int_0^L \sum_{i=1}^{N} [g(\exp(-k_0 x - \hat{U}(x,t))) - d]u_i(x,t)\,\mathrm{d}x \\ &\leq \int_0^L [g(\exp(-\hat{U}(x,t)) - d]\partial_x \hat{U}(x,t)\,\mathrm{d}x \\ &= \int_0^L \partial_x[G(\hat{U}(x,t))]\,\mathrm{d}x = G(\hat{U}(L,t)), \end{aligned}$$

where we used $G(0) = 0$ and $\hat{U}(0,t) = 0$ in the last equality. Since $G(s) < 0$ for $s \geq M_1$, it is not difficult to deduce (8.45) from the above differential inequality. □

Lemma 8.4.3 *There exists a* C_1 *such that for any* N *and any* $(\alpha_i)_{i=1}^N \subset \mathbb{R}$, *every positive solution* $(u_i)_{i=1}^N$ *of* (8.41) *satisfies*

$$\limsup_{t\to\infty} \sum_{i=1}^{N} \|u_i(x,t)\|_{C^{2+\beta,1+\beta/2}([0,L]\times[t,t+1))} \leq C_1, \tag{8.46}$$

where C_1 depends on $\max\{|\alpha_i|\}$*, but does not depend on the number N and the initial data.*

Proof Fix an arbitrary positive solution $(u_i)_{i=1}^N$ of (8.41). By Lemma 8.4.2,

$$\limsup_{t\to\infty} \sum_{i=1}^N \|u_i\|_{L^1([0,L]\times[t-4,t])} \le 4M_1$$

so there exists a $T_0 > 0$ such that

$$\sum_{j=1}^N \|u_i\|_{L^1([0,L]\times[t-4,t])} \le 5M_1 \quad \text{for } t \ge T_0.$$

Observe that the equation of u_i can be regarded as a linear parabolic equation with non-autonomous coefficients:

$$\begin{cases} \partial_t u_i - D\partial_{xx}u_i - \alpha_i\partial_x u_i = \tilde{\sigma}(x,t)u_i & \text{in } [0,L]\times(0,\infty), \\ D\partial_x u_i - \alpha_i u_i = 0 & \text{on } \{0,L\}\times(0,\infty), \end{cases} \tag{8.47}$$

where

$$\tilde{\sigma}(x,t) = g(\exp(-k_0x - \hat{U}(x,t))) - d$$

is uniformly bounded in $L^\infty([0,L]\times[0,\infty))$. We can apply the Harnack principle (Theorem 1.2.6) to deduce that

$$\sup_{0<x<L} u_i(x,t) \le C_H \inf_{0<x<L} u_i(x,t) \quad \text{for } t \ge 1, \tag{8.48}$$

where C_H does not depend on i or the initial data. Then, we apply the local maximum principle (Theorem 1.2.7) to yield

$$\|u_i\|_{L^\infty([0,L]\times[t-3,t])} \le C\|u_i\|_{L^1([0,L]\times[t-4,t])} \quad \text{for } t \ge 5. \tag{8.49}$$

Next, we apply the Sobolev embedding theorem and the parabolic L^p estimate to the linear parabolic equation to improve the above estimate to (C' denotes a generic constant)

$$\begin{aligned} \|u_i\|_{C^{\beta,\beta/2}([0,L]\times[t-2,t])} &\le C'\|u_i\|_{W^{2+p,1+p}([0,L]\times[t-2,t])} \\ &\le C'\|u_i\|_{L^\infty([0,L]\times[t-3,t])} \le C'\|u_i\|_{L^1([0,L]\times[t-4,t])}. \end{aligned} \tag{8.50}$$

Then $\tilde{\sigma}(x,t)$ in (8.47) is Hölder continuous, so that by a parabolic Schauder estimate, the above can then be improved to

$$\|u_i\|_{C^{2+\beta,1+\beta/2}([0,L]\times[t-1,t])} \le C'\|u_i\|_{L^1([0,L]\times[t-4,t])}. \tag{8.51}$$

The desired conclusion follows by summing i from 1 to N, and taking the supremum for $t \geq T_0$ to obtain

$$\sum_{j=1}^{N} \|u_i\|_{C^{2+\beta,1+\beta/2}([0,L]\times[T_0-1,\infty))} \leq C' \sup_{t\geq T_0} \sum_{j=1}^{N} \|u_i\|_{L^1([0,L]\times[t-4,t])} \leq 5CM_1.$$

Note that the constants depend on $\max\{|\alpha_i|\}$ but are independent of N. This completes the proof. □

Lemma 8.4.4 *There exists a constant* $\delta_0 > 0$ *such that for any positive solution* $(u_i)_{i=1}^N$ *of* (8.41), *we have*

$$\liminf_{t\to\infty}\left[\inf_{0<x<L}\sum_{i=1}^{N} u_i(x,t)\right] \geq \delta_0.$$

Proof Let $(u_i)_{i=1}^N$ be a positive solution of (8.41). Integrating (8.41) with respect to $x \in [0, L]$ and summing i from 1 to N, we have

$$\begin{aligned}\frac{\mathrm{d}}{\mathrm{d}t}\hat{U}(L,t) &= \int_0^L [g(\mathrm{e}^{-k_0x-\hat{U}(x,t)}) - d] \sum_{i=1}^{N} u_i(x,t)\,\mathrm{d}x \\ &= \int_0^L \partial_x \left[G(k_0x + \hat{U}(x,t)) - G(k_0x)\right] \mathrm{d}x \\ &\quad + k_0 \int_0^L \left[g(\mathrm{e}^{-k_0x}) - g(\mathrm{e}^{-k_0x-\hat{U}(x,t)})\right] \mathrm{d}x \\ &\geq G(k_0L + \hat{U}(L,t)) - G(k_0L), \end{aligned} \tag{8.52}$$

where $G(s) = \int_0^s [g(\mathrm{e}^{-\tau}) - d]\,\mathrm{d}\tau$. Note that the last inequality follows from the fact that $g'(s) > 0$ for $s > 0$.

Observe now that, by (8.42),

$$G'(k_0L) = g(\mathrm{e}^{-k_0L}) - d > 0.$$

Hence there exists a $\delta_1 > 0$ such that $G(k_0L + s) - G(k_0L) > 0$ for $s \in (0, \delta_1]$. Since the mapping $t \mapsto \hat{U}(L,t)$ satisfies the differential inequality (8.52), it follows that

$$\liminf_{t\to\infty} \sum_{i=1}^{N} \|u_i\|_{L^1[0,L]} = \liminf_{t\to\infty} \hat{U}(L,t) \geq \delta_1.$$

By applying Harnack's inequality (8.48) once again, we can convert the above lower estimate of the L^1 integral to the desired pointwise estimate. This proves the lemma. □

8.4.2 A rough estimate

Proposition 8.4.5 *For each $\eta > 0$ and $\hat{\alpha} \in \mathbb{R}$, there exists an $\epsilon > 0$ such that for any $N \in \mathbb{N}$ and $(\alpha_i)_{i=1}^N \subset (\hat{\alpha} - \epsilon, \hat{\alpha} + \epsilon)$, any positive solution of* (8.41) *satisfies*

$$\limsup_{t\to\infty} \Big\| \sum_{i=1}^N u_i(x,t) - \theta_{\hat{\alpha}}(x) \Big\|_{C^{\beta,\beta/2}([0,L]\times[t,\infty))} < \eta. \tag{8.53}$$

Proof Let a positive solution $(u_i)_{i=1}^N$ of the time-dependent problem (8.41) be given. By interpolation and the *a priori* estimate (8.46), it is enough to show

$$\limsup_{t\to\infty} \Big\| \sum_{i=1}^N u_i(\cdot,t) - \theta_{\hat{\alpha}} \Big\|_{C([0,L])} < \eta. \tag{8.54}$$

Suppose to the contrary that there is an $\eta_0 > 0$ such that for $k \in \mathbb{N}$, there exist $N_k \in \mathbb{N}$, and a sequence $\{\alpha_i^k\}_{i=1}^{N_k}$ and a positive solution $(u_i^k)_{i=1}^{N_k}$ such that

$$\sup_i |\alpha_i^k - \hat{\alpha}| < \frac{1}{k}, \quad \limsup_{t\to\infty} \big\| \hat{U}^k(x,t) - \theta_{\hat{\alpha}}(x) \big\|_{C([0,L])} \geq \eta_0,$$

where $\hat{U}^k(x,t) = \sum_{i=1}^{N_k} u_i^k(x,t)$. We can infer that for each k, there exists $\{t_j^k\}_{j=1}^\infty$ such that

$$\lim_{j\to\infty} t_j^k = +\infty, \quad \text{and} \quad \|\hat{U}^k(\cdot,t_j^k) - \theta_{\hat{\alpha}}\|_{C([0,L])} \geq \eta_0, \quad \text{for all } j \geq 1.$$

By the estimate established in Lemma 8.4.3, we can pass to a subsequence so that

$$u_i^k(x,t+t_j^k) \to \check{u}_i^k(x,t) \quad \text{as } j \to \infty, \text{ in } C_{loc}([0,L]\times\mathbb{R}),$$

where $(\check{u}_i^k)_i$ is some entire solution of (8.41) such that $\check{U}^k = \sum_i \check{u}_i^k$ satisfies

$$\|\check{U}^k(\cdot,0) - \theta_{\hat{\alpha}}\|_{C([0,L])} \geq \eta_0 \quad \text{for all } k. \tag{8.55}$$

By Lemma 8.4.4 and by possibly taking a smaller η_0, we may also assume that

$$\inf_{[0,L]\times\mathbb{R}} \check{U}^k(x,t) \geq \eta_0 \quad \text{for all } k \geq 1. \tag{8.56}$$

Now, since the estimate of Lemma 8.4.3 is independent of N, there is a C_1, independent of k, such that

$$\|\check{U}^k\|_{C^{2+\beta,1+\beta/2}([0,L]\times\mathbb{R})} \leq C_1. \tag{8.57}$$

Hence we can again pass to the limit to assume that, as $k \to \infty$, the sequence $\{\check{U}^k\}_k$ converges in $C_{loc}([0,L]\times\mathbb{R})$ to some bounded entire solution $\check{U}$ of the single species equation (8.6) with $\alpha = \hat{\alpha}$. Moreover, by (8.56) and (8.57), $\check{U}$ is bounded and satisfies

$$\inf_{[0,L]\times\mathbb{R}} \check{U}(x,t) \geq \eta_0. \tag{8.58}$$

By the global attractivity of $\theta_{\hat{\alpha}}$, we must have $\check{U}(x,t) \equiv \theta_{\hat{\alpha}}(x)$ in $[0,L]\times\mathbb{R}$. But this is in contradiction with (8.55). □

8.4.3 The normalized principal bundle

Given two constants, $D > 0$, $\alpha \in \mathbb{R}$ and a function $h(x,t) \in C^{\beta,\beta/2}([0,L]\times\mathbb{R})$, consider

$$\begin{cases} \partial_t\Psi_1(x,t) - D\partial_{xx}\Psi_1(x,t) + \alpha\partial_x\Psi_1(x,t) - h(x,t)\Psi_1(x,t) + d\Psi_1(x,t) & \\ \qquad = H_1(t)\Psi_1(x,t) & \text{for } 0 < x < L,\ t \in \mathbb{R}, \\ D\partial_x\Psi_1(x,t) - \alpha\Psi_1(x,t) = 0 & \text{for } x \in \{0,L\},\ t \in \mathbb{R}, \\ \int_0^L e^{-\alpha x/D}\Psi_1(x,t)\,dx = 1 & \text{for } t \in \mathbb{R}, \\ \Psi_1(x,t) > 0 & \text{for } x \in [0,L],\ t \in \mathbb{R}. \end{cases} \tag{8.59}$$

We call the pair $(\Psi_1(x,t), H_1(t))$ the normalized principal bundle corresponding to (8.59).

Proposition 8.4.6 *The problem* (8.59) *has a unique solution*

$$(\Psi_1, H_1) \in C^{2+\beta,1+\beta/2}([0,L]\times\mathbb{R}) \times C^{1+\beta/2}(\mathbb{R}).$$

Moreover, the following statements hold.

1. *For each $M > 1$, there exists a constant C_M such that*

$$\frac{1}{C_M} \leq \Psi_1(x,t) \leq C_M \quad in\ \overline{\Omega}\times\mathbb{R} \tag{8.60}$$

   provided $(D,\alpha) \in \left[\frac{1}{M}, M\right] \times [-M,M]$.
2. *The normalized principal bundle as a mapping*

$$\begin{gathered} (D,\alpha,h) \mapsto (\Psi_1, H_1) \\ \mathbb{R}_+ \times \mathbb{R} \times C^{\beta,\beta/2}([0,L]\times\mathbb{R}) \to C^{2+\beta,1+\beta/2}([0,L]\times\mathbb{R}) \times C^{1+\beta/2}(\mathbb{R}), \end{gathered}$$

   is smooth.

Proof Letting $\psi_1(x,t) := e^{-\alpha x/D}\Psi_1(x,t)$, the above problem can be transformed to

$$\begin{cases} \partial_t\psi_1(x,t) - D\partial_{xx}\psi_1(x,t) - \alpha\partial_x\psi_1(x,t) - h(x,t)\psi_1(x,t) + d\psi_1(x,t) & \\ \qquad = H_1(t)\psi_1(x,t) & \text{for } 0 < x < L,\ t \in \mathbb{R}, \\ \partial_x\psi_1(x,t) = 0 & \text{for } x \in \{0,L\},\ t \in \mathbb{R}, \\ \int_0^L \psi_1(x,t)\,dx = 1 & \text{for } t \in \mathbb{R}, \\ \psi_1(x,t) > 0 & \text{for } x \in [0,L],\ t \in \mathbb{R}. \end{cases} \tag{8.61}$$

The existence and uniqueness of $(\psi_1(x,t), H_1(t))$ are proved in Theorem 4.1.4. The bounds in (8.60) follow from the Harnack principle (Theorem 1.2.6) and the fact that $\int_0^L e^{-\alpha x/D}\Psi(x,t)\,dx = 1$. (See, e.g. the proof of Theorem 4.2.2 for details.) The smooth dependence on coefficients is proved in Theorem 4.3.4. □

Definition 8.4.7 For given $\hat{\alpha}, \alpha \in \mathbb{R}$, let $\lambda(\alpha, \hat{\alpha})$ and $\phi(x; \alpha, \hat{\alpha})$ be the principal eigenvalue and eigenfunction of

$$\begin{cases} D\phi''(x) + \alpha\phi'(x) + \hat{h}(x)\phi(x) + \lambda\phi(x) = 0 & \text{for } 0 < x < L, \\ \phi'(x) = 0 & \text{for } x = 0, L, \end{cases} \tag{8.62}$$

where

$$\hat{h}(x) = g\left(\exp\left(-k_0 x - \int_0^x \theta_{\hat{\alpha}}(y)\,dy\right)\right) - d. \tag{8.63}$$

By Lemma 8.3.7, we have

$$\partial_\alpha \lambda(\alpha, \hat{\alpha}) > 0 \quad \text{for } \alpha, \hat{\alpha} \in \mathbb{R}. \tag{8.64}$$

Corollary 8.4.8 *There exists $\eta' > 0$ such that for any $\alpha \in \mathbb{R}$ and any function $h(x,t) \in C^{\beta,\beta/2}([0,L] \times \mathbb{R})$, if*

$$|\alpha - \hat{\alpha}| < \eta', \quad \text{and} \quad \|h(x,t) - \hat{h}(x)\|_{C^{\beta,\beta/2}([0,L]\times\mathbb{R})} < \eta', \tag{8.65}$$

then

$$\partial_\alpha H_1(t;\alpha) \geq \eta' \quad \text{for all } t \in \mathbb{R},\ \alpha \in (\hat{\alpha} - \eta', \hat{\alpha} + \eta'),$$

where $\partial_\alpha H_1(t;\alpha,h)$ is the partial derivative of $H_1(t;\alpha,h)$ with respect to the scalar parameter α.

Proof Taking $h(x,t) = \hat{h}(x)$, then by uniqueness of the principal Floquet bundle,

$$(\Psi_1(x,t;\alpha,\hat{h}), H_1(t;\alpha,\hat{h})) = (\phi(x), \lambda(\alpha,\hat{\alpha})),$$

where $(\phi(x), \lambda(\alpha,\hat{\alpha}))$ is the principal eigenpair of (8.62). By (8.64) and continuous dependence, there is an $\epsilon_1 > 0$ such that

$$\eta_0 := \inf_{\alpha \in [\hat{\alpha}-\epsilon_1, \hat{\alpha}+\epsilon_1]} \frac{\partial}{\partial \alpha}\lambda(\alpha,\hat{\alpha}) > 0.$$

Now it follows from the smooth dependence of (Ψ_1, H_1) on (α, h) that there exists $\eta' \in (0, \eta_0/2)$ such that if (8.65) holds, then

$$\begin{aligned} &\sup_{\alpha \in [\hat{\alpha}-\epsilon_1, \hat{\alpha}+\epsilon_1]} \|\partial_\alpha H_1(\cdot;\alpha,h) - \partial_\alpha H_1(\cdot;\alpha,\hat{h})\|_{C^{\beta,\beta/2}([0,L]\times\mathbb{R})} \\ &= \left\|\partial_\alpha H_1(\cdot;\alpha,h) - \partial_\alpha \lambda(\alpha,\hat{\alpha})|_{\alpha=\hat{\alpha}}\right\|_{C^{\beta,\beta/2}([0,L]\times\mathbb{R})} < \frac{\eta_0}{2}. \end{aligned}$$

Hence, for $\alpha \in [\hat{\alpha} - \epsilon_1, \hat{\alpha} + \epsilon_1]$,

$$\partial_\alpha H_1(t; \alpha, h) > \partial_\alpha \lambda(\alpha, \hat{\alpha})|_{\alpha=\hat{\alpha}} - \frac{\eta_0}{2} \geq \frac{\eta_0}{2} > \eta' \quad \text{for } t \in \mathbb{R}.$$

This proves the corollary. □

8.4.4 A general exclusion criterion

Proposition 8.4.9 *There exists an $\eta > 0$ such that if* (8.53) *holds, then for any N and any increasing sequence $(\alpha_i)_{i=1}^N \subset (\hat{\alpha} - \eta, \hat{\alpha} + \eta)$, every positive solution $(u_i)_{i=1}^N$ of* (8.41) *converges to the equilibrium solution $E_1 = (\theta_{\alpha_1}, 0, ..., 0)$ as $t \to \infty$; i.e. the equilibrium E_1 is globally asymptotically stable among all positive solutions.*

Proof Let $\eta' > 0$ be as given in Corollary 8.4.8. It follows from Proposition 8.4.5 that there is an $\epsilon \in (0, \eta')$ such that for any N and any $(\alpha_i)_{i=1}^N \subset (\hat{\alpha} - \epsilon, \hat{\alpha} + \epsilon)$,

$$\limsup_{t\to\infty} \left\| h(x,t) - \hat{h}(x) \right\|_{C^{\beta,\beta/2}([0,L]\times[t,t+1])} < \eta', \tag{8.66}$$

where

$$\begin{aligned} h(x,t) &= g(\exp(-k_0 x - \hat{U}(x,t))) - d \\ &= g(\exp(-k_0 x - \sum_{j=1}^N \int_0^x u_j(y,t)\,dy)) - d, \end{aligned} \tag{8.67}$$

and $\hat{h}(x)$ is given in (8.63).

By (8.66), after possibly a translation in time, we may assume without loss of generality that

$$\|h(\cdot,t) - \hat{h}\|_{C^{\beta,\beta/2}([0,L]\times[0,\infty))} < \eta'. \tag{8.68}$$

Extend $h(x,t)$ evenly in t, so that it is defined for $(x,t) \in [0,L] \times \mathbb{R}$. Since $(\hat{\alpha} - \epsilon, \hat{\alpha} + \epsilon) \subset (\hat{\alpha} - \eta', \hat{\alpha} + \eta')$, we have verified (8.65).

Let $\Psi_1(x,t;\alpha,h)$ and $H_1(t;\alpha,h)$ be the normalized principal bundle considered in the statement of Corollary 8.4.8. We have, for any $\alpha \in [\hat{\alpha} - \epsilon, \hat{\alpha} + \epsilon]$,

$$\inf_{t\in\mathbb{R}} \partial_\alpha H_1(t;\alpha,h) \geq \eta' > 0. \tag{8.69}$$

For each i, we claim that there are $\overline{c}_i > \underline{c}_i > 0$ such that

$$\underline{c}_i e^{-\int_0^t H_1(s;\alpha_i,h)\,ds} \Psi_1(x,t;\alpha_i,h) \leq u_i(x,t) \leq \overline{c}_i e^{-\int_0^t H_1(s;\alpha_i,h)\,ds} \Psi_1(x,t;\alpha_i,h) \tag{8.70}$$

for $(x,t) \in [0,L] \times [0,\infty)$.

Indeed, the left- and right-hand sides of (8.70) are positive solutions to the same linear parabolic equation as u_i. Thanks to (8.60), $\Psi_i(\cdot, 0)$ and the positive solution $u_i(\cdot, 0)$ are bounded uniformly from above and below by positive constants at time $t = 0$. Hence we can choose $\overline{c}_i$ large enough and $\underline{c}_i$ small enough (both being independent of $t \geq 0$) to deduce (8.70) from the classical comparison theorem of linear parabolic equations. This proves (8.70).

By (8.69), we have

$$H_1(t; \alpha_i, h) - H_1(t; \alpha_1, h) \geq (\alpha_i - \alpha_1)\eta' > 0 \quad \text{for all } i > 1, \text{ and all } t \in \mathbb{R}.$$

Hence, we derive from (8.70) that, for $i > 1$,

$$\begin{aligned} \frac{u_i(x,t)}{u_1(x,t)} &\leq C \exp\left(-\int_0^t (H_1(s;\alpha_i) - H_1(s;\alpha_1))\, ds\right) \frac{\Psi_1(x,t;\alpha_i,h)}{\Psi_1(x,t;\alpha_1,h)} \\ &\leq C' \exp\left(-(\alpha_i - \alpha_1)\eta' t\right) \to 0 \quad \text{as } t \to \infty. \end{aligned}$$

Note that we have used (8.60) again in the second inequality. Since we also have $\limsup_{t\to\infty} \sum_{i=1}^N \|u_i\| \leq C_1$ (by Lemmas 8.4.2 and 8.4.3), we deduce that $u_i \to 0$ uniformly for $i = 2, ..., N$. Hence the semiflow generated by (8.41) is asymptotic to the single species model consisting of only the first species u_1. By the compactness of the semiflow, there is a sequence $t_k \to \infty$ such that

$$(\tilde{u}_i^k(x,t))_{i=1}^N = (u_i(x, t - t_k))_{i=1}^N \to (\tilde{u}_i(x,t))_{i=1}^N \quad \text{in } C^{2,1}_{loc}([0, L] \times \mathbb{R}),$$

where $(\tilde{u}_i)_{i=1}^N$ is an entire solution that is contained in the omega limit set. By the above discussion, $\tilde{u}_i \equiv 0$ for all $i = 2, .., N$. Hence, the omega limit set takes the form $\omega_1 \times \{(0, ..., 0)\}$.

By Lemmas 8.4.3 and Lemma 8.4.4, there exists a $\delta_0 > 0$ such that for any $\tilde{u}_1 \in \omega_1$,

$$\delta_0 \leq \tilde{u}_1(x,t) \leq \frac{1}{\delta_0} \quad \text{in } \Omega \times \mathbb{R}.$$

Now, since $\omega = \{(\tilde{u}_1, 0, ..., 0) : \tilde{u}_1 \in \omega_1\}$ is the omega limit set of a trajectory, it follows that ω is internally chain transitive[2] with respect to the semiflow generated by (8.41). It follows then that ω_1 is internally chain transitive with respect to the semiflow generated by the single species problem. Since every nonnegative nontrivial solution of the single species equation converges to θ_{α_1}, it follows that either $\omega_1 = \{0\}$

[2] Let $\Phi_t : X \to X$ be a semiflow in a Banach space X. A subset ω of X is said to be internally chain transitive with respect to the semiflow Φ_t if, for any two points $u_0, v_0 \in \omega$, and any positive numbers $\delta > 0$, $T > 0$, there is a finite sequence

$$C_{\delta,T} = \{u^{(1)} = u_0, u^{(2)}, ..., u^{(m)} = v_0;\ t_1, ..., t_{m-1}\}$$

such that $u^{(j)} \in \omega$ and $t_j \geq T$, and that $\|\Phi_{t_j}(u^{(j)}) - u^{(j+1)}\| < \delta$ for all $1 \leq j \leq m - 1$. The sequence $C_{\delta,T}$ is called a (δ, T)-chain connecting u_0 and v_0. In particular every point u_0 within an internally chain transitive set is chain recurrent, i.e. for any $\delta, T > 0$, there is a (δ, T)-chain connecting u_0 to itself.

or $\omega_1 = \{\theta_{\alpha_1}\}$. Using Lemma 8.4.4, we deduce that $\omega_1 = \{\theta_{\alpha_1}\}$, i.e. $u_1 \rightarrow \theta_{\alpha_1}$ uniformly as $t \longrightarrow \infty$. □

Proof (of Theorem 8.4.1) Let η be given by Proposition 8.4.9. We can then choose $\epsilon \in (0, \eta)$ by Proposition 8.4.5 such that for any N and $(\alpha_i)_{i=1}^N \in (\hat{\alpha} - \epsilon, \hat{\alpha} + \epsilon)$, any positive solution $(u_i)_{i=1}^N$ of (8.41) satisfies (8.53). It then follows from the choice of η above and Proposition 8.4.9 that $E_1 = (\theta_{\alpha_1}, 0..., 0)$ is globally asymptotically stable among all positive solutions of (8.41). □

8.5 Further Reading

In this chapter, we limit ourselves to the relatively simple situation of a eutrophic water column where light is the only limiting factor of population growth. In other situations, factors such as temperature [9], nutrient availability [2] and predation [10] can also be important drivers of the population dynamics.

Much of the existing mathematical literature on phytoplankton is focused on a single species. The single species model was considered in [36] for the self-shading case (i.e., $k_0 = 0$) in infinitely long water columns ($L = \infty$). The existence, uniqueness, and global stability of the steady state are established in [22, 23, 36]. It is shown in [28] that the self-shading model with any finite water column depth has a stable positive steady state, which means that the self-shading model has no critical water column depth beyond which the phytoplankton cannot persist.

For the case with background turbidity ($k_0 > 0$), it is illustrated in [8] that the condition for phytoplankton bloom development can be characterized by critical water column depth and some critical values of the vertical turbulent diffusion coefficient. Du and Hsu [5] studied both single and multiple species competing for light with no advection. For the single-species model, the existence, uniqueness, and global attractivity of a positive equilibrium was established. Hsu and Lou [14] analyzed the critical death rate, critical water column depth, critical sinking or buoyant coefficient, and critical turbulent diffusion rate. Du and Mei [7] investigated the global dynamics of the single species model for the case $D = D(x)$, $\alpha = \alpha(x)$ and the asymptotic profiles of the positive steady states for small or large diffusion and deep water column when D, α are constants. Peng and Zhao [35] considered the effect of time-periodic light intensity I_0 at the surface, due to the diurnal light cycle and seasonal changes. Ma and Ou [32] further studied the model in [35] and assumed that $D(t), \alpha(t)$ were time-periodic functions. They obtained the uniqueness and the global attractivity of the positive periodic solution of the single-species model, when it exists. Du et al. [6] studied the effect of photoinhibition on the single phytoplankton species, and they found that, in contrast to the case of no photoinhibition, where at most one positive steady state can exist, the model with photoinhibition possesses at least two positive steady states in certain parameter ranges. Hsu et al. [15] examined the dynamics of a single species under the assumption that the amount of light absorbed by individuals is proportional to cell size, which varies for populations that reproduced by simple cell division into two equal-sized daughter cells, whereas Pang et al. [34] considered the crowding effect.

Although many mathematical theories have been developed for the single-species phytoplankton model, there are few results for two or more phytoplankton species competing for light. The existence of positive steady state and uniform persistence for two-species model were proved in [5], where there is no sinking or buoyancy; see also [33]. The global dynamics was obtained only recently; see [24, 25] for some classification results based on establishing a generalized comparison principle and the application of the theory of monotone dynamical systems.

Unlike the two-species Lotka–Volterra competition model with diffusion, one main difficulty for the system (8.1)–(8.4) is the lack of a comparison principle, i.e., if $N = 2$ and (u_1, u_2) and $(\tilde{u}_1, \tilde{u}_2)$ are solutions to (8.1)–(8.4), then

$$
\begin{aligned}
&u_1(x,0) \le \tilde{u}_1(x,0), \quad u_2(x,0) \ge \tilde{u}_2(x,0) \quad \text{pointwise in } [0,L],\\
&\not\Longrightarrow u_1(x,t) \le \tilde{u}_1(x,t), \quad u_2(x,t) \ge \tilde{u}_2(x,t) \quad \text{pointwise in } [0,L]\times(0,\infty).
\end{aligned}
$$

For order-preserving properties in the single-species model, Shigesada and Okubo [33] observed that, in the self-shading case (with $k_0 = 0$), the cumulative distribution function $U(x,t) := \int_0^x u(s,t)\,\mathrm{d}s$ satisfies a single reaction-diffusion equation without nonlocal terms. Subsequently, Ishii and Takagi [22, 23] showed that the flow retains the natural order in U. For a related model with a water column of infinite depth, they made use of this fact to obtain a complete classification of the long-time behavior of the population. This fact was used again in Du and Hsu [5] to determine the long-time dynamics for a single-species model with finite water depth. Subsequently, the weak maximum principle for U in the single-species model was established in [32]. The strong maximum principle for U and the generalization to the two-species model is due to [25].

For $N \ge 3$, the competitive system no longer admits a comparison principle. Recently, a sufficient condition leading to competitive exclusion was obtained in [3], which uses the theory of normalized principal Floquet bundles (see Chapter 4). Indeed, when $N = 2$, we have shown that the phytoplankton competition system is a monotone dynamical system, so that the trajectory generically converges to some equilibrium. In such case, the reduction principle [1] concerning the linearized problem *within the set of equilibria* is enough to ascertain the large time limit of the dynamical system. However, for nonmonotone systems, it is not sufficient to only classify the stability of the set of equilibria. In fact, a generalized reduction principle for the nonautonomous, nonperiodic situation is needed, and the principal Floquet bundle seems to be the appropriate notion to study. See [4, 29], where monotonicity results for principal Floquet bundles are established, with application to the evolution of slow dispersal.

The paradox of the plankton proposed by Hutchinson [21] raised the lack of explanation of the diversity of phytoplankton species, despite the small number of limiting factors, such as light, nitrogen, carbon, phosphorus, etc. Based on our results concerning competitive exclusion in eutrophic water columns, it seems that other factors (e.g. nutrient availability and predation) have to be included to account for the coexistence of multiple phytoplankton species. Recently, mathematical models considering the full spectrum of light have been proposed, which are based on the fact that different phytoplankton species have differential preference among different colors of light. We mention [11] which introduces a reaction-diffusion model that generalizes the model in [37] to the spatial context. Finally, in real situations the turbulent diffusion rate is likely to vary with water depth. In this direction, we mention the works [38, 39] investigating the situation of a stratified lake with a well mixed top layer and a poorly mixed bottom layer.

Problems

8.5.1 Let $\mathcal{K}$ be given by (8.7). Show that $\mathcal{K}$ has nonempty interior with respect to the topology of $C([0, L])$, and that (8.8) holds.

8.5.2 Prove part 2 of Theorem 8.2.5 in the following steps.

1. Find $\rho \in C^2([0, L])$ such that

$$\begin{cases} D\rho'' - \alpha\rho' + (g(\exp(-k_0 x - \int_0^x \rho(y)\,dy)) - d)\rho < 0 & \text{for } 0 \le x \le L, \\ D\rho'(0) - \alpha\rho(0) \le 0 \le D\rho'(L) - \alpha\rho(L). \end{cases}$$

2. Let $\tilde{\mu}_1$ and $\tilde{\varphi}$ be the principal eigenvalue and positive eigenfunction of

$$\begin{cases} D\tilde{\varphi}'' - \alpha\tilde{\varphi}' + [g(e^{-k_0 x - \delta_0 x}) - d]\tilde{\varphi} + \tilde{\mu}\tilde{\varphi} = 0 & \text{in } [0, L], \\ D\tilde{\varphi}' - \alpha\tilde{\varphi} = 0 & \text{for } x = 0, L. \end{cases} \tag{8.71}$$

   Assume that $\tilde{\mu}_1 < 0$. Let $\underline{u}_\delta$ and $\overline{u}_M$ be respectively solutions of (8.6) with initial data $\delta\tilde{\varphi}$ and $M\rho$. Deduce that, for $0 < \delta \ll 1$ and $M \gg 1$, we have

$$\underline{u}_\delta(\cdot, t_1) \le_{\mathcal{K}} \underline{u}_\delta(\cdot, t_2), \quad \overline{u}_M(\cdot, t_1) \ge_{\mathcal{K}} \overline{u}_M(\cdot, t_2) \quad \text{for } 0 \le t_1 < t_2.$$

3. Show that $\underline{u}_\delta(x, t) \to \underline{\theta}(x)$ and $\overline{u}_M \to \overline{\theta}$, where $0 < \underline{\theta} \le_K \overline{\theta}$ are positive equilibria of (8.6).
4. Show that $\underline{\theta} = \overline{\theta}$, and conclude that (8.6) has a unique positive equilibrium which attracts all nonnegative, nontrivial solutions.

8.5.3 We say that $\overline{u}$ is a super-solution of (8.6) provided

$$D\partial_x\overline{u}(0, t) - \alpha\overline{u}(0, t) \le 0 \le D\partial_x\overline{u}(L, t) - \alpha\overline{u}(L, t)$$

and

$$\partial_t\overline{u} \ge_{\mathcal{K}} D\partial_{xx}\overline{u} - \alpha\partial_x\overline{u} + [g(\exp(-k_0 x - \int_0^x \overline{u}(y, t)\,dy)) - d]\overline{u}$$

holds for each $t > 0$. Define similarly the notion of sub-solution of (8.6) by reversing the inequalities. Here we recall that for $\phi, \tilde{\phi} \in C([0, L])$, $\phi \leq_{\mathcal{K}} \tilde{\phi}$ if and only if $\int_0^x (\tilde{\phi} - \phi)\, dy \geq 0$ for all $0 \leq x \leq L$.

Prove that if $\underline{u}$ and $\overline{u}$ is a pair of sub- and super-solutions of (8.6) in the above sense, and if $\underline{u}(\cdot, 0) \leq_{\mathcal{K}} \overline{u}(\cdot, 0)$, then

$$\underline{u}(\cdot, t) \leq_{\mathcal{K}} \overline{u}(\cdot, t) \qquad \text{for } t > 0.$$

References

1. L. Altenberg, *Resolvent positive linear operators exhibit the reduction phenomenon*, Proc. Natl. Acad. Sci. USA, 109 (2012), pp. 3705–3710.
2. A. Burson, M. Stomp, E. Greenwell, J. Grosse, and J. Huisman, *Competition for nutrients and light: testing advances in resource competition with a natural phytoplankton community*, Ecology, 99 (2018), pp. 1108–1118.
3. R. S. Cantrell and K.-Y. Lam, *Competitive exclusion in phytoplankton communities in a eutrophic water column*, Discrete Contin. Dyn. Syst. Ser. B, 26 (2021), pp. 1783–1795.
4. ———, *On the evolution of slow dispersal in multispecies communities*, SIAM J. Math. Anal., 53 (2021), pp. 4933–4964.
5. Y. Du and S.-B. Hsu, *On a nonlocal reaction-diffusion problem arising from the modeling of phytoplankton growth*, SIAM J. Math. Anal., 42 (2010), pp. 1305–1333.
6. Y. Du, S.-B. Hsu, and Y. Lou, *Multiple steady-states in phytoplankton population induced by photoinhibition*, J. Differential Equations, 258 (2015), pp. 2408–2434.
7. Y. Du and L. Mei, *On a nonlocal reaction-diffusion-advection equation modelling phytoplankton dynamics*, Nonlinearity, 24 (2011), pp. 319–349.
8. U. Ebert, M. Arrayás, N. Temme, B. Sommeijer, and J. Huisman, *Critical conditions for phytoplankton blooms*, Bulletin of Mathematical Biology, 63 (2001), pp. 1095–1124.
9. K. F. Edwards, M. K. Thomas, C. A. Klausmeier, and E. Litchman, *Phytoplankton growth and the interaction of light and temperature: A synthesis at the species and community level*, Limnology and Oceanography, 61 (2016), pp. 1232–1244.
10. J. J. Elser and R. P. Hassett, *A stoichiometric analysis of the zooplankton–phytoplankton interaction in marine and freshwater ecosystems*, Nature, 370 (1994), pp. 211–213.
11. C. M. Heggerud, K.-Y. Lam, and H. Wang, *Niche differentiation in the light spectrum promotes coexistence of phytoplankton species: a spatial modelling approach*, 2021. arXiv:2109.02634 [math.AP].
12. D. Henry, *Geometric theory of semilinear parabolic equations*, vol. 840 of Lecture Notes in Mathematics, Springer-Verlag, Berlin-New York, 1981.
13. P. Hess and A. C. Lazer, *On an abstract competition model and applications*, Nonlinear Anal., 16 (1991), pp. 917–940.
14. S.-B. Hsu and Y. Lou, *Single phytoplankton species growth with light and advection in a water column*, SIAM J. Appl. Math., 70 (2010), pp. 2942–2974.
15. S.-B. Hsu, L. Mei, and F.-B. Wang, *On a nonlocal reaction-diffusion-advection system modelling the growth of phytoplankton with cell quota structure*, J. Differential Equations, 259 (2015), pp. 5353–5378.
16. S. B. Hsu, H. L. Smith, and P. Waltman, *Competitive exclusion and coexistence for competitive systems on ordered Banach spaces*, Trans. Amer. Math. Soc., 348 (1996), pp. 4083–4094.
17. J. Huisman, P. van Oostveen, and F. J. Weissing, *Critical depth and critical turbulence: Two different mechanisms for the development of phytoplankton blooms*, Limnology and Oceanography, 44 (1999), pp. 1781–1787.

18. ———, *Species dynamics in phytoplankton blooms: Incomplete mixing and competition for light.*, The American Naturalist, 154 (1999), pp. 46–68.
19. J. Huisman and F. J. Weissing, *Light-limited growth and competition for light in well-mixed aquatic environments: An elementary model*, Ecology, 75 (1994), pp. 570–520.
20. ———, *Competition for nutrients and light in a mixed water column: A theoretical analysis*, The American Naturalist, 146 (1995), pp. 536–564.
21. G. E. Hutchinson, *The paradox of the plankton*, Am. Nat., 95 (1961), pp. 137–145.
22. H. Ishii and I. Takagi, *Global stability of stationary solutions to a nonlinear diffusion equation in phytoplankton dynamics*, J. Math. Biol., 16 (1982/83), pp. 1–24.
23. H. Ishii and I. Takagi, *A nonlinear diffusion equation in phytoplankton dynamics with self-shading effect*, in Mathematics in biology and medicine (Bari, 1983), vol. 57 of Lecture Notes in Biomath., Springer, Berlin, 1985, pp. 66–71.
24. D. Jiang, K.-Y. Lam, and Y. Lou, *Competitive exclusion in a nonlocal reaction-diffusion-advection model of phytoplankton populations*, Nonlinear Anal. Real World Appl., 61 (2021), p. 15. Paper No. 103350.
25. D. Jiang, K.-Y. Lam, Y. Lou, and Z.-C. Wang, *Monotonicity and global dynamics of a nonlocal two-species phytoplankton model*, SIAM J. Appl. Math., 79 (2019), pp. 716–742.
26. C. A. Klausmeier and E. Litchman, *Algal games: The vertical distribution of phytoplankton in poorly mixed water columns*, Limnology and Oceanography, 46 (2001), pp. 1998–2007.
27. C. A. Klausmeier, E. Litchman, and S. A. Levin, *Phytoplankton growth and stoichiometry under multiple nutrient limitation*, Limnology and Oceanography, 49 (2004), pp. 1463–1470.
28. T. Kolokolnikov, C. Ou, and Y. Yuan, *Phytoplankton depth profiles and their transitions near the critical sinking velocity*, J. Math. Biol., 59 (2009), pp. 105–122.
29. K.-Y. Lam and Y. Lou, *The principal floquet bundle and the dynamics of fast diffusing communities*. Manuscript submitted for publication.
30. K.-Y. Lam and D. Munther, *A remark on the global dynamics of competitive systems on ordered Banach spaces*, Proc. Amer. Math. Soc., 144 (2016), pp. 1153–1159.
31. E. Litchman, C. A. Klausmeier, J. R. Miller, O. M. Schofield, and P. G. Falkowski, *Multi-nutrient, multi-group model of present and future oceanic phytoplankton communities*, Biogeosciences, 3 (2006), pp. 585–606.
32. M. Ma and C. Ou, *Existence, uniqueness, stability and bifurcation of periodic patterns for a seasonal single phytoplankton model with self-shading effect*, J. Differential Equations, 263 (2017), pp. 5630–5655.
33. L. Mei and X. Zhang, *Existence and nonexistence of positive steady states in multi-species phytoplankton dynamics*, J. Differential Equations, 253 (2012), pp. 2025–2063.
34. D. Pang, H. Nie, and J. Wu, *Single phytoplankton species growth with light and crowding effect in a water column*, Discrete Contin. Dyn. Syst., 39 (2019), pp. 41–74.
35. R. Peng and X.-Q. Zhao, *A nonlocal and periodic reaction-diffusion-advection model of a single phytoplankton species*, J. Math. Biol., 72 (2016), pp. 755–791.
36. N. Shigesada and A. Okubo, *Analysis of the self-shading effect on algal vertical distribution in natural waters*, J. Math. Biol., 12 (1981), pp. 311–326.
37. M. Stomp, J. Huisman, L. J. Stal, and H. C. P. Matthijs, *Colorful niches of phototrophic microorganisms shaped by vibrations of the water molecule*, ISME J., (2007), pp. 271–282.
38. K. Yoshiyama, J. P. Mellard, E. Litchman, and C. A. Klausmeier, *Phytoplankton competition for nutrients and light in a stratified water column.*, The American Naturalist, 174 (2009), pp. 190–203.
39. J. Zhang, J. D. Kong, J. Shi, and H. Wang, *Phytoplankton competition for nutrients and light in a stratified lake: a mathematical model connecting epilimnion and hypolimnion*, J. Nonlinear Sci., 31 (2021), p. 42. Paper No. 35.
40. X.-Q. Zhao, *Dynamical systems in population biology*, CMS Books in Mathematics/Ouvrages de Mathématiques de la SMC, Springer, Cham, second ed., 2017.

Part III
Evolutionary Dynamics

Chapter 9
Elements of Adaptive Dynamics

Abstract Adaptive dynamics is a conceptual framework enabling the exploration of evolutionary questions using ecological models. We discuss the framework of adaptive dynamics in the context of a river population model, hoping to offer to the readers a PDE viewpoint of the theory. Specifically, we introduce the key notions of adaptive dynamics, including invasion exponent, selection gradient, singular strategy, convergence stable strategy, evolutionary stable strategy (ESS), continuously stable strategy, neighborhood invader strategy, dimorphism (coexistence of strategies) and evolutionary branching point, in concrete terms for a reaction-diffusion-advection model, supplemented by analytical results and examples.

9.1 Introduction

Adaptive dynamics is a conceptual framework which focuses on phenotypic evolution in replication, while ignoring the effect of genes and sex, and is a part of the theory of evolution [23, 24, 35]. It is based on the assumption of the separation of time scales between ecological and evolutionary dynamics; the selection principle which favors the individuals with the best adapted trait operates at a fast time scale, while mutational effects due to small variations of the offspring from the parent, of each generation, accumulate over a much longer (evolutionary) time scale.

The simplest situation concerns a single species, which exists initially with a dominant phenotype that is inherited at the time of birth. Since mutation is rare, we assume that the population maintains at an ecological equilibrium, as determined by a single species model, until the rare event that a mutant phenotype is introduced (or arises by mutation). The main questions are: will it coexist with the resident phenotype or competitively exclude the resident? These questions can then be answered by analyzing the corresponding two-phenotype model. In the case of a one-dimensional phenotypic space, one can then record the invasion outcome of a rare phenotype (denoted as $\beta \in \mathbb{R}$) into the equilibrium population of phenotype (denoted as $\alpha \in \mathbb{R}$) by way of a plot over the (α, β) parameter space, and observe the trend of the invasion

K. -Y. Lam, Y. Lou, *Introduction to Reaction-Diffusion Equations*, Lecture Notes on Mathematical Modelling in the Life Sciences, https://doi.org/10.1007/978-3-031-20422-7_9

dynamics. Here the goal of the game is not to maximize the population size, but to minimize the chance of being invaded by exotic phenotypes. This latter consideration leads to the notion of invasion fitness that indicates whether a given phenotype near equilibrium can be invaded by a second phenotype when rare.

The evolution of dispersal can be studied in the above framework. Indeed, A. Hastings [41] considered the mutual invasibility of two species in a spatially heterogeneous but temporally constant environment. Assuming these two species are identical except for their dispersal rates, it was shown that an invasion by a rare mutant is successful if and only if the mutant is the slower diffuser. Regarded as a trait, diffusion is selected against in temporally constant environments. (In contrast, diffusion may be selected for in an advective environment where dispersal is conditional, or in some annual plants in an environment that varies in both space and time.) Later, more specific results were obtained by Dockery et al. [26] for the following reaction-diffusion model:

$$\begin{cases} \partial_t u - \alpha \Delta u = u(m(x) - u - v) & \text{in } \Omega \times (0,\infty), \\ \partial_t v - \beta \Delta v = v(m(x) - u - v) & \text{in } \Omega \times (0,\infty), \\ \mathbf{n} \cdot \nabla u = \mathbf{n} \cdot \nabla v = 0 & \text{on } \partial\Omega \times (0,\infty), \\ u(x,0) = u_0(x), \quad v(x,0) = v_0(x) & \text{in } \Omega, \end{cases} \tag{9.1}$$

where Ω is a bounded domain in $\mathbb{R}^N$ with smooth boundary $\partial\Omega$ and outer unit normal vector $\mathbf{n}$; the constants $\alpha, \beta > 0$ are the dispersal rates; $m(x) \in C^2(\overline{\Omega})$ is the intrinsic growth rate. Assuming $m(x) \geq 0$ and is nonconstant, it follows from Theorem 5.2.4 that (9.1) has at least three equilibria:

$$E_0 = (0,0), \quad E_1 = (\tilde{u}, 0), \quad E_2 = (0, \tilde{v}).$$

Regarding the long-time dynamics, Theorem 7.3.6 says that then the slower diffuser can invade the faster diffuser when rare, but not vice versa. In fact, when $\alpha < \beta$, $E_1 = (\tilde{u}, 0)$ is globally asymptotically stable with respect to all nonnegative, nontrivial solutions.

The above invasion dynamics can be recorded by a pairwise invasibility plot (PIP); see Figure 9.1. It follows from the PIP that only mutations bearing a trait less than the resident trait will succeed. Moreover, it was shown (Theorem 7.3.6) that the slower mutant goes to fixation every time, i.e. the slower mutant phenotype with smaller trait value replaces the original resident phenotype, and becomes the new dominant resident phenotype. This suggests that, with successive mutations, the dominant trait generally decreases, i.e. evolution selects for slower diffusion rate.

Remark 9.1.1 It can be proved that the mutant phenotype, upon successful invasion, can generally go to fixation, provided that the two phenotypes are away from the so-called "singular strategy"; see [32, 34] for the results in finite-dimensional systems, and [11] for the results in infinite-dimensional systems.

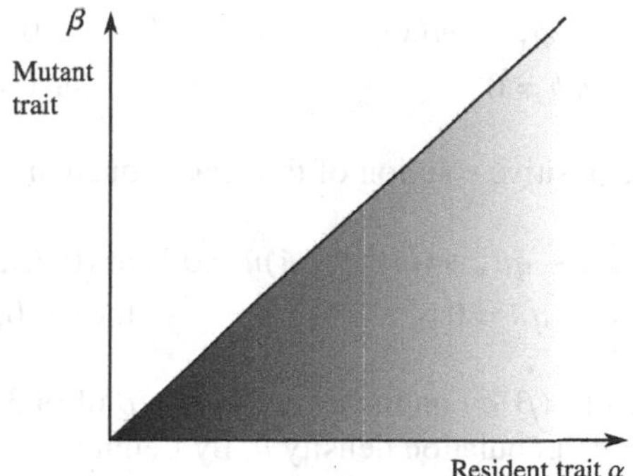

Fig. 9.1: Pairwise-invasibility-plot of (9.1). The shaded area $\{(\alpha, \beta) : 0 < \beta < \alpha\}$ indicates the parameter space where the mutant with trait β can invade the resident population (with trait α) when rare, i.e. $E_1 = (\tilde{u}, 0)$ is unstable.

9.2 Evolution of Dispersal in Advective Environments

We will introduce key concepts in adaptive dynamics by applying them to a specific reaction-diffusion model concerning two competing phenotypes in an advective environment, which is used to model organisms residing in a water column or a river. In the former case, the individuals are pulled by gravity, and in the latter case, the individuals are washed downstream by water flow. Let $\Omega = (0, L)$ be a bounded interval on $\mathbb{R}$, and consider

$$\begin{cases} \partial_t u - \alpha \partial_{xx} u + q \partial_x u - u(m(x) - u - v) = 0 & \text{in } (0, L) \times (0, \infty), \\ \partial_t v - \beta \partial_{xx} v + q \partial_x v - v(m(x) - u - v) = 0 & \text{in } (0, L) \times (0, \infty), \\ \alpha \partial_x u - qu = 0 = \beta \partial_x v - qv = 0 & \text{on } \{0, L\} \times (0, \infty), \\ u(x, 0) = u_0(x), \quad v(x, 0) = v_0(x) & \text{in } [0, L], \end{cases} \tag{9.2}$$

where $\alpha, \beta > 0$ are the unconditional dispersal rates; $m(x)$ is the intrinsic growth rate of the population, which we assume to be continuous and positive in $[0, L]$. Both species are subject to the drift in the direction from $x = 0$ towards $x = L$, with a constant rate $q > 0$. No-flux boundary conditions are imposed at both upstream ($x = 0$) and downstream ends ($x = L$). In such a case, by examining the stability of the trivial solution, it can be proved (via results in Chapter 5) that for any $\alpha, \beta > 0$ and $q \geq 0$, the system (9.2) has at least three equilibria $E_0 = (0, 0)$, $E_1 = (\tilde{u}, 0)$ and $E_2 = (0, \tilde{v})$.

The invasion exponent

We define the invasion exponent, denoted as $\lambda(\beta, \alpha)$, to be the principal eigenvalue of the linear problem

$$\begin{cases} \beta\phi_{xx} - q\phi_x + (m(x) - \tilde{u})\phi + \lambda\phi = 0 & \text{in } (0, L), \\ \beta\phi_x - q\phi = 0 & \text{for } x = 0, L, \end{cases} \tag{9.3}$$

where $\tilde{u}$ is the unique positive solution of the scalar equation

$$\begin{cases} \alpha\tilde{u}_{xx} - q\tilde{u}_x + (m(x) - \tilde{u})\tilde{u} = 0 & \text{in } (0, L), \\ \alpha\tilde{u}_x - q\tilde{u} = 0 & \text{for } x = 0, L. \end{cases} \tag{9.4}$$

Note the dependence of $\lambda(\beta, \alpha)$ on the diffusion rate β of (9.3), and its dependence on α via the equilibrium population density $\tilde{u}$. By Lemma 7.1.1, the sign of $\lambda(\beta, \alpha)$ characterizes the linear stability of $(\tilde{u}, 0)$. Specifically,

1. If $\lambda(\beta, \alpha) < 0$, then the mutant population with trait β, when rare, can invade the resident population with trait α.
2. If $\lambda(\beta, \alpha) > 0$, then the mutant population with trait β, when rare, fails to invade the resident population with trait α.

Observe that

$$\lambda(\beta, \alpha) = 0 \quad \text{when} \quad \beta = \alpha.$$

Indeed, in such a case zero is an eigenvalue of (9.3) with a positive eigenfunction $\tilde{u}$. By the characterization of the principal eigenvalue, we deduce $\lambda(\alpha, \alpha) = 0$. Differentiating the identity once yields

$$\partial_\alpha\lambda(\alpha, \alpha) + \partial_\beta\lambda(\alpha, \alpha) = 0 \quad \text{for all } \alpha. \tag{9.5}$$

Differentiating (9.5) with respect to α yields

$$\partial^2_{\alpha\alpha}\lambda(\alpha, \alpha) + 2\partial^2_{\beta\alpha}\lambda(\alpha, \alpha) + \partial^2_{\beta\beta}\lambda(\alpha, \alpha) = 0 \quad \text{for all } \alpha. \tag{9.6}$$

The selection gradient

Next, let us fix a resident trait with value α, and let $\beta \neq \alpha$ be a nearby mutant trait. Since $\lambda(\alpha, \alpha) = 0$, the sign of the *selection gradient*, given by

$$\lambda_\beta(\alpha, \alpha) := \left[\partial_\beta\lambda(\beta, \alpha)\right]_{\beta=\alpha}, \tag{9.7}$$

tells us in which direction the mutant trait is able to invade the current resident trait α. On the one hand, if the sign is positive then $\lambda(\beta, \alpha) < 0$ for $\beta < \alpha$ and $\beta \approx \alpha$, i.e. β slightly smaller than α is favored. On the other hand, if it is negative then the slightly greater trait is favored. By (9.5), the selection gradient is also equal to $-\lambda_\alpha(\alpha, \alpha)$.

Lemma 9.2.1 *Let $\alpha > 0$ be fixed, and let $\lambda_\beta(\alpha, \alpha)$ be the invasion exponent given in* (9.7), *then*

$$\lambda_\beta(\alpha, \alpha) = -\frac{\int_0^L (e^{-\alpha x/q}\tilde{u})_x \tilde{u}_x \, dx}{\int_0^L e^{-\alpha x/q}\tilde{u}^2 \, dx}, \tag{9.8}$$

where $\tilde{u}$ is the unique positive solution of (9.4).

Proof Differentiating (9.3) with respect to β (denoting by $'$ the derivative with respect to β), we obtain

$$\begin{cases} \beta\phi'_{xx} - q\phi'_x + (m(x) - \tilde{u})\phi' + \lambda\phi' = -\phi_{xx} - \lambda'\phi & \text{in } (0, L), \\ \beta\phi'_x - q\phi' = -\phi_x & \text{for } x = 0, L. \end{cases} \tag{9.9}$$

Next, note that when $\beta = \alpha$, 0 is an eigenvalue of (9.3) with the positive eigenfunction $\tilde{u}$. By the characterization of the principal eigenvalue (Theorem 1.3.6), we have

$$\lambda = 0 \quad \text{and} \quad \phi = \tilde{u} \quad \text{when } \beta = \alpha.$$

Setting $\beta = \alpha$ in (9.9), we obtain

$$\begin{cases} \beta\left(e^{qx/\beta}(e^{-qx/\beta}\phi')_x\right)_x + (m(x) - \tilde{u})\phi' = -\tilde{u}_{xx} - \lambda'\tilde{u} & \text{in } (0, L), \\ \beta\phi'_x - q\phi' = -\tilde{u}_x & \text{for } x = 0, L, \end{cases} \tag{9.10}$$

where we have used the identity

$$\beta\phi'_{xx} - q\phi'_x = \beta\left(e^{qx/\beta}(e^{-qx/\beta}\phi')_x\right)_x.$$

Multiplying (9.4) by $e^{-qx/\beta}\phi'$ and integrating, we get

$$\beta\int_0^L e^{-qx/\beta}\phi'\left(e^{qx/\beta}(e^{-qx/\beta}\tilde{u})_x\right)_x dx + \int_0^L e^{-qx/\beta}\phi'(m(x) - \tilde{u})\tilde{u}\, dx = 0. \tag{9.11}$$

Similarly, multiply (9.10) by $e^{-qx/\beta}\tilde{u}$ to get

$$\begin{aligned} &\beta\int_0^L e^{-qx/\beta}\tilde{u}\left(e^{qx/\beta}(e^{-qx/\beta}\phi')_x\right)_x dx + \int_0^L e^{-qx/\beta}\phi'(m - \tilde{u})\tilde{u}\, dx \\ &= -\int_0^L e^{-qx/\beta}\tilde{u}\tilde{u}_{xx}\, dx - \lambda'\int_0^L e^{-qx/\beta}\tilde{u}^2\, dx. \end{aligned} \tag{9.12}$$

Integrating by parts and subtracting (9.12) from (9.11), we get

$$\begin{aligned} &\beta\left[\phi'(e^{-qx/\beta}\tilde{u})_x\Big|_0^L - \tilde{u}(e^{-qx/\beta}\phi')_x\Big|_0^L\right] - \int_0^L e^{-qx/\beta}\tilde{u}\tilde{u}_{xx}\, dx \\ &= -\lambda'\int_0^L e^{-qx/\beta}\tilde{u}^2\, dx. \end{aligned} \tag{9.13}$$

Next, by the boundary condition in (9.4) and (9.10), we have

$$\beta\,\phi'(\mathrm{e}^{-qx/\beta}\tilde{u})_x\Big|_0^L = \mathrm{e}^{-qx/\beta}\phi'(\beta\tilde{u}_x - q\tilde{u})\Big|_0^L = 0 \tag{9.14}$$

and

$$\beta\,\tilde{u}(\mathrm{e}^{-qx/\beta}\phi')_x\Big|_0^L = \mathrm{e}^{-qx/\beta}\tilde{u}(\beta\tilde{\phi}'_x - q\tilde{\phi}')\Big|_0^L = -\,\mathrm{e}^{-qx/\beta}\tilde{u}\tilde{u}_x\Big|_0^L. \tag{9.15}$$

Substituting (9.14) and (9.15) into (9.13), we have

$$\mathrm{e}^{-qx/\beta}\tilde{u}\tilde{u}_x\Big|_0^L - \int_0^L \mathrm{e}^{-qx/\beta}\tilde{u}\tilde{u}_{xx} = -\lambda' \int_0^L \mathrm{e}^{-qx/\beta}\tilde{u}^2\,\mathrm{d}x.$$

Integrating the left-hand side by parts again, we obtain (9.8). □

Remark 9.2.2 With a minor modification of the proof, one can show that the formula (9.8) holds for a general Robin type boundary condition; see [65, Lemma 5.1].

Singular strategy

A trait $\hat{\alpha}$ is called a *singular strategy* if it is a value at which the selection gradient vanishes, i.e.

$$\left[\partial_\beta \lambda(\beta,\alpha)\right]_{(\beta,\alpha)=(\hat{\alpha},\hat{\alpha})} = 0,$$

that is, $\lambda_\beta(\hat{\alpha},\hat{\alpha}) = 0$. By (9.5), a singular strategy $\hat{\alpha}$ is also characterized by the equation $\lambda_\alpha(\hat{\alpha},\hat{\alpha}) = 0$.

Example. Assume $m(x) = a\mathrm{e}^{bx}$ for some $a, b > 0$. Take $\hat{\alpha} := q/b$, then $\tilde{u}(x;\hat{\alpha}) = m(x)$, so

$$\begin{cases} \beta\phi_{xx} - q\phi_x + \lambda(\beta,\hat{\alpha})\phi = 0 & \text{in } [0,L], \\ \beta\phi_x - q\phi = 0 & \text{for } x = 0, L. \end{cases}$$

This implies that $\phi = \mathrm{e}^{qx/\beta}$ and $\lambda(\beta,\hat{\alpha}) \equiv 0$ for all β. In particular, $\lambda_\beta(\hat{\alpha},\hat{\alpha}) = 0$. That is, $\hat{\alpha} = q/b$ is a singular strategy.

Convergence stable strategy

As we have seen, for a monomorphic resident population, the direction of advantageous mutation is determined by the sign of the selection gradient. By successive mutation, it can be argued that the endpoint $\hat{\alpha}$ of adaptive dynamics is a singular strategy satisfying, for some $\delta > 0$,

$$\left[\partial_\beta \lambda(\beta,\alpha)\right]_{\beta=\alpha} \begin{cases} < 0 & \text{when } \alpha \in (\hat{\alpha}-\delta, \hat{\alpha}), \\ = 0 & \text{when } \alpha = \hat{\alpha}, \\ > 0 & \text{when } \alpha \in (\hat{\alpha}, \hat{\alpha}+\delta). \end{cases} \tag{9.16}$$

We say that $\hat{\alpha}$ is a *convergence stable strategy* if it is a singular strategy and (9.16) holds; see Figure 9.2.

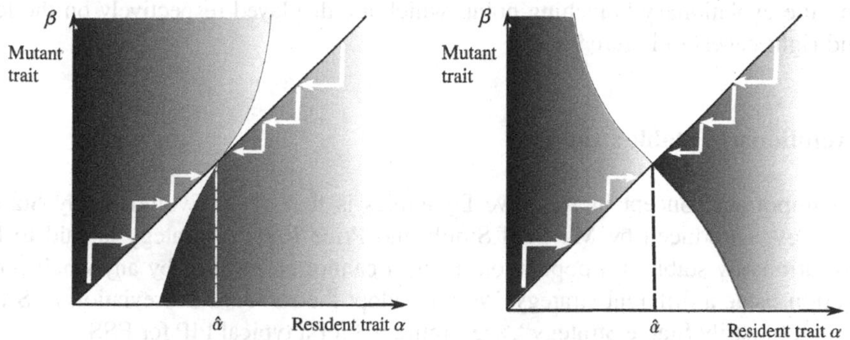

Fig. 9.2: Pairwise-invasibility-plot of (9.2). The dominant trait of the population evolves towards a convergence stable strategy upon successive mutations. The shaded areas indicate the parameter space where the mutant with trait β can invade the resident population (with trait α) when rare, i.e. $E_1 = (\tilde{u}, 0)$ is unstable. If we denote the nullcline different from the diagonal by $\beta = g(\alpha)$, then the left panel has β as an increasing function of α, while in the right panel β is a decreasing function of α. In both cases the singular strategy $\hat{\alpha}$ is convergence stable, but $\hat{\alpha}$ is an evolutionary branching point (hence not an evolutionary endpoint) on the left panel, while it is an ESS (hence an evolutionary endpoint) in the right panel.

Proposition 9.2.3 *Suppose that $m_x, m > 0$ in $[0, L]$, and m_x/m is monotone in $[0, L]$, then there exists a $q_1 > 0$ such that for any $q \in (0, q_1]$, there exists an $\hat{\alpha} = \hat{\alpha}(q)$ such that*

$$\partial_\beta \lambda(\hat{\alpha}, \hat{\alpha}) = 0 \quad \textit{and} \quad \frac{\mathrm{d}}{\mathrm{d}s}\left[\partial_\beta \lambda(s,s)\right]_{s=\hat{\alpha}} > 0.$$

In fact, we have

$$\left[\partial_\beta \lambda(\beta,\alpha)\right]_{\beta=\alpha} = \begin{cases} < 0 & \textit{when } \alpha \in (0, \hat{\alpha}), \\ = 0 & \textit{when } \alpha = \hat{\alpha}, \\ > 0 & \textit{when } \alpha \in (\hat{\alpha}, \infty). \end{cases}$$

In particular, for each $q \in (0, q_1]$, there is a unique singular strategy $\hat{\alpha}(q)$, which is also a convergence stable strategy.

Proof See [60, Corollary 5.10]. □

Example. Assume $m(x) = ae^{bx}$ for some $a, b > 0$. Take $\hat{\alpha} := q/b$. By Proposition 9.2.3, $\hat{\alpha}$ is a convergence stable strategy. Note that strict monotonicity of m_x/m is not needed therein.

What happens if the population is near or has reached the convergence stable strategy? Next, we discuss two important situations: the evolutionarily stable strategy and the evolutionary branching point, which are displayed respectively on the left and right panels in Figure 9.2.

Evolutionarily stable strategy

An important concept in Adaptive Dynamics is that of an Evolutionarily Stable Strategy, introduced by Maynard Smith and Price [67]. A strategy is said to be evolutionarily stable if a population using it cannot be invaded by any small population using a different strategy. We will adopt the standard abbreviation ESS for "Evolutionarily Stable Strategy"; see Figure 9.3 for a typical PIP for ESS.

Definition 9.2.4 Suppose there is an open interval I such that (9.4) has a unique positive solution for all $\alpha \in I$.

- $\hat{\alpha} \in I$ is said to be an *ESS* if

$$\lambda(\beta, \hat{\alpha}) > 0 \quad \text{for all } \beta \in I \setminus \{\hat{\alpha}\}.$$

- $\hat{\alpha} \in I$ is said to be a *local ESS* if

$$\lambda(\beta, \hat{\alpha}) > 0 \quad \text{for all } \beta \in I \setminus \{\hat{\alpha}\} \text{ such that } \beta \approx \hat{\alpha}.$$

Theorem 9.2.5 *Suppose $m, m_x > 0$ in $[0, L]$ and*

$$2 \inf_{[0,L]} \frac{m_x}{m} > \sup_{[0,L]} \frac{m_x}{m}. \tag{9.17}$$

If m_x/m is nondecreasing and nonconstant, then for each $0 < q \ll 1$, the unique singular strategy given in Proposition 9.2.3 is a local ESS.

Proof It is enough to show that

$$\lim_{q \to 0} \left[\partial_{\beta\beta}\lambda(\beta, \alpha)\right]_{(\beta,\alpha)=(\hat{\alpha}(q),\hat{\alpha}(q))} > 0.$$

See [60, Corollary 6.6] for the details. □

Example. Assume $L = 1$ and $m(x) = e^{x+ax^2}$ for some $a \in \mathbb{R}$. It is easy to check that if $a < 0$, then $m(x)$ satisfies all the assumptions in Theorem 9.2.5. Note that when $a = 0$, $m(x)$ satisfies all other assumptions except m_x/m being nonconstant.

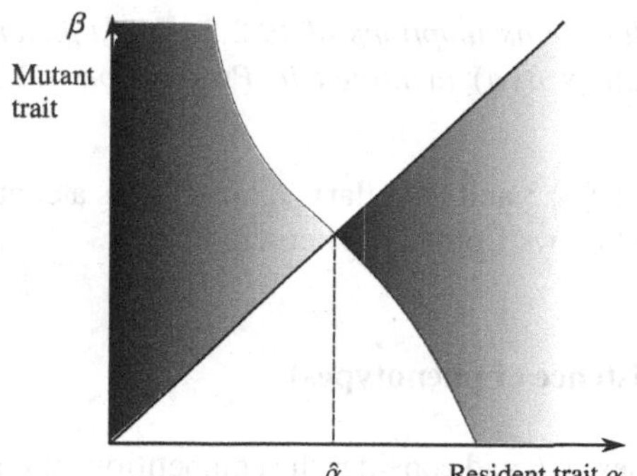

Fig. 9.3: Pairwise-invasibility-plot in the case where there is an ESS. The shaded area indicates the parameter region where the mutant with trait β can invade the resident population (with trait α) when rare, i.e. $E_1 = (\tilde{u}, 0)$ is unstable. The strategy $\hat{\alpha}$, as identified as the point of intersection $(\hat{\alpha}, \hat{\alpha})$ on the diagonal, is an ESS.

By definition, a strategy $\hat{\alpha}$ is an ESS (resp. local ESS) if it can resist invasion by all other strategies (resp. all nearby strategies), i.e. any kind of mutation does not alter the evolutionary dynamics once the population adopts the ESS. A stronger concept is needed to ensure that a monomorphic population with a different initial strategy will converge towards it in the long run.

Continuously stable strategy

We say that $\hat{\alpha}$ is a *continuously stable strategy* [28, 29] if it is a convergence stable strategy as well as an ESS. In fact, the ESS depicted in Figure 9.3 is also a continuously stable strategy. Combining Proposition 9.2.3 and Theorem 9.2.5, we obtain the following.

Corollary 9.2.6 *Under the assumptions of* (9.2.5), *for sufficiently small* $q > 0$, *the unique singular strategy* $\hat{\alpha}(q)$ *identified in Proposition* 9.2.3 *is a continuously stable strategy.*

Neighborhood invader strategy

We say that $\hat{\alpha}$ is a *neighborhood invader strategy* [1] if there exists a $\delta > 0$ such that the equilibrium $E_1 = (\tilde{u}, 0)$ of the system (9.2) is globally asymptotically stable provided

$$\alpha = \hat{\alpha}, \quad \text{and} \quad 0 < |\beta - \hat{\alpha}| < \delta.$$

In particular, if $\hat{\alpha}$ is a neighborhood invader strategy, then it is an ESS.

In [11], it is proved that for a large class of models including reaction-diffusion models, a continuously stable strategy must be a neighborhood invader strategy.

Theorem 9.2.7 *Under the assumptions of* (9.2.5), *for sufficiently small* $q > 0$, *the unique singular strategy* $\hat{\alpha}(q)$ *identified in Proposition* 9.2.3 *is a neighborhood invader strategy.*

Proof By Proposition 9.2.3 and Corollary 9.2.6, $\hat{\alpha}(q)$ is a continuously stable strategy. The conclusion follows from [11, Theorem 6.2]. □

Dimorphism (coexistence of phenotypes)

Fix the two trait values α, β and consider the competition system (9.2). Suppose that the two phenotypes are mutually invasible, i.e.

$$\lambda(\beta, \alpha) < 0 \quad \text{and} \quad \lambda(\alpha, \beta) < 0,$$

then the two species coexist by a general theorem in monotone dynamical systems; see [44, Corollary 1], or Corollary E.2.8. One can therefore use the PIP to explore the parameter space in the $\alpha - \beta$ plane for coexistence, by superimposing the PIP with its reflection along the diagonal. In [85], these so-called *mutual invasibility plots* were introduced, and the parameter region of dimorphism was studied via singularity and unfolding theory; see [36] for its application to reaction-diffusion models. Coexistence regions can typically be found around a singular strategy $\hat{\alpha}$. See Figure 9.4 for an illustration. A generic condition is given in the following lemma.

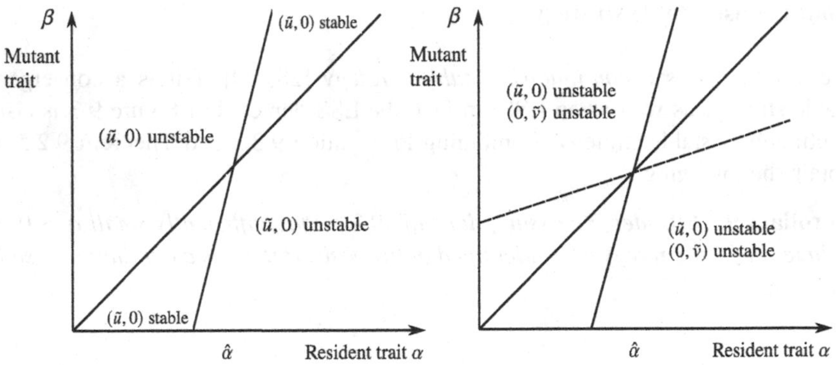

Fig. 9.4: The left panel is a PIP of (9.2) indicating the local stability of $E_1 = (\tilde{u}, 0)$. The right panel is the corresponding mutual invasibility plot, which is the superposition of the PIP with its reflection across the diagonal (that indicates the local stability of $E_2 = (0, \tilde{v})$). The regions of mutual invasibility, i.e. where both E_1 and E_2 are unstable, are indicated.

Lemma 9.2.8 *Let $\hat{\alpha}$ be a singular strategy. If*

$$\left[\partial^2_{\beta\beta}\lambda + \partial^2_{\alpha\alpha}\lambda\right]_{(\beta,\alpha)=(\hat{\alpha},\hat{\alpha})} < 0, \tag{9.18}$$

then there exists a $\delta > 0$ such that

$$\lambda(\hat{\alpha} + s, \hat{\alpha} - s) < 0 \quad \textit{and} \quad \lambda(\hat{\alpha} - s, \hat{\alpha} + s) < 0 \quad \textit{for all } s \in (0, \delta].$$

In particular, the two phenotypes $\hat{\alpha} \pm s$ are mutually invasible for all $0 < s \ll 1$.

Proof It is enough to show that $g(s) := \lambda(\hat{\alpha} + s, \hat{\alpha} - s)$ has a strict local maximum at $s = 0$. Indeed,

$$g'(0) = (\partial_\alpha\lambda - \lambda_\beta\lambda) = 0 \quad \text{at } (\beta, \alpha) = (\hat{\alpha}, \hat{\alpha}),$$

as $\hat{\alpha}$ is a singular strategy, and that

$$g''(0) = \partial^2_{\alpha\alpha}\lambda - 2\partial^2_{\beta\alpha}\lambda + \partial^2_{\beta\beta}\lambda = 2\left(\partial^2_{\alpha\alpha}\lambda + \partial^2_{\beta\beta}\lambda\right) < 0 \quad \text{at } (\beta, \alpha) = (\hat{\alpha}, \hat{\alpha}),$$

where the second equality follows from (9.6). This completes the proof. □

Example. Assume $L = 1$ and $m(x) = \exp(b_1 x) + \exp(b_2 x)$ for some $b_1 > b_2 > 0$. Then for $\alpha = q/b_1$ and $\beta = q/b_2$, (9.2) has a positive steady state given by $(e^{b_1 x}, e^{b_2 x})$, which suggests that these two phenotypes are mutually invasible. We conjecture that for this example, there exists a unique singular strategy, which is an evolutionary branching point defined below.

For the choice of m in the previous example, it seems that there is a singular strategy which is a convergent stable strategy but it is not an ESS. This brings up the concept of evolutionary branching point, which we will discuss next.

Evolutionary branching point

We say that $\hat{\alpha}$ is an *evolutionary branching point* if it is a singular strategy and it satisfies

$$\partial^2_{\alpha\alpha}\lambda < \partial^2_{\beta\beta}\lambda < 0 \quad \text{at } (\beta, \alpha) = (\hat{\alpha}, \hat{\alpha}).$$

Note that an evolutionary branching point is locally convergence stable, since it is a singular strategy and

$$\begin{aligned}\frac{\mathrm{d}}{\mathrm{d}s}\left[\partial_\beta\lambda(s,s)\right]_{s=\hat{\alpha}} &= \partial^2_{\beta\alpha}\lambda(\hat{\alpha},\hat{\alpha}) + \partial^2_{\beta\beta}\lambda(\hat{\alpha},\hat{\alpha}) \\ &= \frac{1}{2}\left[\partial^2_{\beta\beta}\lambda(\hat{\alpha},\hat{\alpha}) - \partial^2_{\alpha\alpha}\lambda(\hat{\alpha},\hat{\alpha})\right] > 0,\end{aligned}$$

where we used (9.6) in the second equality. Intuitively, if one starts with a monomorphic population with trait α_0, then the adaptive dynamics framework suggests that the monomorphic population will evolve towards the evolutionary branching point $\hat{\alpha}$. But what happens after that is very different from our previous discussion of ESS or continuously stable strategies. Indeed, note that

$$\partial_\beta \lambda(\hat{\alpha}, \hat{\alpha}) = 0 \quad \text{and} \quad \partial^2_{\beta\beta}\lambda(\hat{\alpha}, \hat{\alpha}) < 0,$$

which mean that $\hat{\alpha}$ is a fitness minimum, i.e. all nearby mutant phenotypes can invade the resident phenotype with trait $\hat{\alpha}$. In this case, the monomorphic population is not an evolutionary endpoint, so we are led outside of the monomorphic framework and we have to extend the formalism. In fact, we expect that the resident population will become dimorphic, i.e. consisting of two distinct phenotypes (β, α) belonging to the parameter space of dimorphism; see [20, 63, 70].

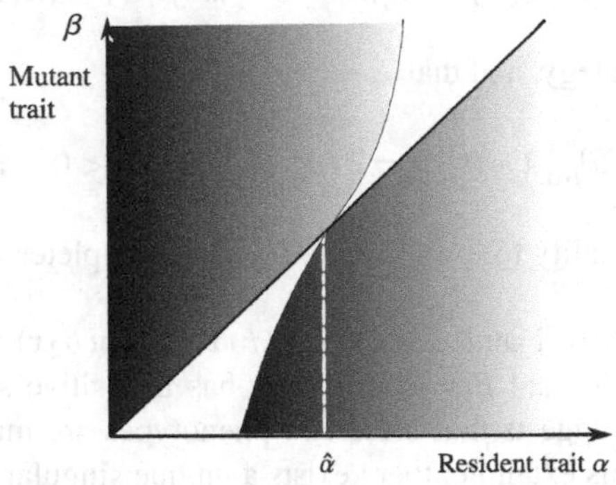

Fig. 9.5: Pairwise-invasibility-plot in the case where there is an evolutionary branching point. The shaded area indicates the parameter space where the mutant with trait β can invade the resident population (with trait α) when rare, i.e. $E_1 = (\tilde{u}, 0)$ is unstable. The strategy $\hat{\alpha}$, as identified as the point of intersection $(\hat{\alpha}, \hat{\alpha})$ on the diagonal, is an evolutionary branching point.

In [60], we proved that spatial heterogeneity alone is enough to cause evolutionary branching in the model (9.2).

Theorem 9.2.9 *Suppose* $m, m_x > 0$ *in* $[0, L]$ *and*

$$2 \inf_{[0,L]} \frac{m_x}{m} > \sup_{[0,L]} \frac{m_x}{m}. \tag{9.19}$$

If m_x/m *is nonincreasing and nonconstant, then for each* $0 < q \ll 1$, *the unique singular strategy given in Proposition* 9.2.3 *is an evolutionary branching point.*

Proof By Proposition 9.2.3, we have

$$\partial^2_{\alpha\alpha}\lambda < \partial^2_{\beta\beta}\lambda \quad \text{at } (\beta, \alpha) = (\hat{\alpha}, \hat{\alpha}).$$

Next, the inequality $\partial^2_{\beta\beta}(\hat{\alpha},\hat{\alpha}) < 0$ follows from the proof of [60, Theorem 6.5], specifically, using equation (57) therein and the sentence that follows. □

Example. Assume $L = 1$ and $m(x) = e^{x+ax^2}$ for some $a \in \mathbb{R}$. It is easy to check that if $a > 0$, then $m(x)$ satisfies all assumptions in Theorem 9.2.5. Note that when $a = 0$, $m(x)$ satisfies all assumptions except m_x/m being nonconstant. Indeed, Theorem 9.2.5 does not hold for the case $a = 0$.

Example. Assume $L = 1$ and $m(x) = \exp(b_1 x) + \exp(b_2 x)$ for some $b_1 > b_2 > 0$. It is easy to check that all assumptions in Theorem 9.2.9 are satisfied, including

$$\left(\frac{m_x}{m}\right)_x = \frac{(b_1 - b_2)^2 e^{(b_1+b_2)x}}{(e^{b_1 x} + e^{b_2 x})^2} > 0.$$

9.3 Further Reading

The application of game theory to the study of evolution flourished after Maynard Smith and Price introduced the pioneering concept of the evolutionarily stable strategy (ESS) [67, 82], which is closely related to the concept of Nash equilibrium [75] in noncooperative games. Among the first applications to theoretical biology are the so-called matrix models with discrete trait space [42, 43]. Also, Auslander et al. [2] applied the Nash equilibrium concept to evolutionary games with continuous traits. Several researchers merged Maynard Smith's evolutionary game theory with more conventional game theory [5, 6, 81], allowing evolutionary biologists to solve specific problems of biological interest. Later, a group of scientists advanced the dynamics aspect of evolutionary game theory by formalizing the framework which is called adaptive dynamics [23, 35]. Adaptive dynamics thus provides a unified and coherent framework encompassing the concepts of ESS, convergence stable strategy, evolutionary branching points, and more. See [68] for a more comprehensive survey article, as well as the monographs [22, 33] and references therein. We also recommend the interested readers to read the excellent introductory lecture notes by Odo Diekmann [24]. See also [50] for a database of research papers in adaptive dynamics.

The framework of adaptive dynamics has been applied to many specific biological problems. However, in most cases the underlying biological model is given by a set of ordinary differential equations. In particular, studies of the evolution of dispersal using spatially explicit models are relatively rare. In this direction, we first mention the paper of Hamilton and May [38], who developed a discrete-time model of seed dispersal; see also [37].

In this chapter, we give a quick exposition of various concepts in adaptive dynamics in terms of a concrete reaction-diffusion-advection model, with applications to the evolution of dispersal in mind. For continuous-time models, in the absence of spatial drift or directed movement, the evolution of unconditional dispersal was considered in [26, 41] for spatially varying and temporally constant environments. Here the faster diffusion rate is selected against, and zero diffusion rate is convergence stable. For time-periodic environments, this is no longer the case [45, 62, 69]. See also [4] for more precise results, and [59] for general time-varying environments, which are not necessarily time-periodic.

In the above works, movement of the individuals are assumed to be unconditional, i.e. equal in all directions. Selection acts differently on the dispersal behavior in the presence of environmental drift or when directed movement is involved. In [56, 57], the following model was considered

$$
\begin{cases}
\partial_t u - \operatorname{div}(\mu_1 \nabla u - \alpha_1 u \nabla m) = u(m - u - v) & \text{for } x \in \Omega,\ t > 0, \\
\partial_t v - \operatorname{div}(\mu_2 \nabla v - \alpha_2 v \nabla m) = v(m - u - v) & \text{for } x \in \Omega,\ t > 0, \\
\mathbf{n} \cdot (\mu_1 \nabla u - \alpha_1 u \nabla m) = \mathbf{n} \cdot (\mu_2 \nabla v - \alpha_2 v \nabla m) = 0 & \text{for } x \in \partial\Omega,\ t > 0,
\end{cases}
\tag{9.20}
$$

with nonnegative, nontrivial initial data. Here the individuals of the i-th species adopt a combination of unconditional dispersal with rate μ_i and a directed movement up the gradient of $m(x)$, which is a nonconstant function of space representing the habitat quality. In contrast to the results in [26, 41], it is shown that intermediate value of dispersal can evolve, and both very slow and very fast movements are selected against. Indeed, when α_i/μ_i is large, the species tends to specialize on the region around local maximum points of $m(x)$, while it adopts a generalist strategy when α_i/μ_i is small. Sufficient conditions are given for the existence of ESS. Intuitively, the magnitude of the ESS in diffusion rate remains proportional to the magnitude of the drift rate.

Next, consider the following model of river populations:

$$
\begin{cases}
\partial_t u - \mu_1 \partial_{xx}^2 u - q \partial_x u = u(r - u - v) & \text{for } [0, L] \times \mathbb{R}, \\
\partial_t v - \mu_2 \partial_{xx}^2 v - q \partial_x v = v(r - u - v) & \text{for } [0, L] \times \mathbb{R}, \\
\mu_1 \partial_x u - qu = \mu_2 \partial_x v - qv = 0 & \text{for } x = 0,\ t > 0, \\
\mu_1 \partial_x u - qu = -bqu, \quad \mu_2 \partial_x v - qv = -bqv & \text{for } x = L,\ t > 0,
\end{cases}
\tag{9.21}
$$

with nonnegative, nontrivial initial data. Here μ_i, q, r, b are positive constants. A net loss of individuals with rate bq is imposed at the downstream end ($x = L$), while the no-flux condition is imposed on the upstream end ($x = 0$).This model was previously considered in [83, 84] for other ecological questions. In [60, 65, 66], the evolution of diffusion rate was studied. Depending on the values of the model parameters, we find up to 9 qualitatively different PIPs, and that for a given environment there sometimes exist up to three distinct evolutionarily stable strategies; see [40]. We also mention the recent work [46, 47] wherein the influence of network topology on the resulting PIP is investigated for a patch model.

One of the contributions of adaptive dynamics is to demonstrate that evolution of a monomorphic population can bring the focal species to a fitness minima, from which evolutionary branching initiates. To explain this fact, suppose the trait of the focal species is given by $\bar{\alpha}(t)$, as a function of the evolutionary time t. Then the invasion fitness at time t is given by $\lambda(\beta, \bar{\alpha}(t))$ and coevolves according to the trait

$\bar{\alpha}(t)$ of the focal species, while the selection gradient $\partial_\beta \lambda(\bar{\alpha}(t), \bar{\alpha}(t))$ determines the direction of evolution. The evolution towards a fitness minima is thus possible since the invasion fitness "landscape" can change as the trait $\bar{\alpha}(t)$ of the focal species changes. Mechanisms causing evolutionary branching include predatorial trade-off between two different preys [86], temporally and spatially varying carrying capacities [69], cyclic or chaotic (in time) local dynamics [27], and temporal variation caused by catastrophes [78]. In [60], it is shown that spatial heterogeneity alone in Lotka–Volterra systems can lead to evolutionary branching.

The evolution of dispersal is closely connected to the theory of habitat choice. Ideal free distribution is a special kind of habitat choice which was introduced by Fretwell and Lucas [30, 31]. In our modeling setting, it corresponds to the perfect alignment between the population distribution and the carrying capacity. It has been shown, robustly across a range of mathematical modeling frameworks, that a dispersal strategy leading to ideal free distribution is both an ESS and a neighborhood invader strategy in spatially heterogeneous but temporally constant environments [3, 13, 14, 15, 16, 17, 21, 51, 52, 53]. See also [10, 12] for the recent extension of ideal free distribution theory in time-periodic environments. In particular, a novel generalized notion of ideal free distribution is introduced in [12] in case resource matching is no longer feasible. Roughly speaking, IFD is attainable by some dispersal strategy provided the environment has no generalized sinks.

In a series of papers, Champagnat et al. [18, 19, 20] introduced a probabilistic model describing a population of discrete individuals, whose phenotypes are described by a vector of trait values. By combining various scalings in the size of population, birth/death rates, the size and rate of mutation, this microscopic model is shown to lead to a range of different macroscopic models. In the limit of large population size and rare mutations, the so-called trait substitution sequence, in which the population is monomorphic with the dominant trait modeled by a jump process in the direction dictated by the selection gradient, is rigorously justified. If one further lets the step size of mutation tend to zero, one then recovers the so-called canonical equation of adaptive dynamics. This provides a rigorous foundation of the "large population, small and rare mutation" heuristics in the framework of adaptive dynamics.

Alternatively, one may start from the probabilistic model of discrete individuals to first pass to the limit of large population size alone (without the rare mutation limit), which leads to deterministic selection-mutation models for the density of traits [19]. Under a further limit of small mutations, and with a proper time scaling, these integro-PDE models exhibits dynamics on the set of finite sums of Dirac masses on the trait space. This latter program is partly studied in [25, 64] for specific models, and an alternative version of canonical equation is derived. Note that the mutation here is small but not necessarily rare, and there remains a trace of the phenotypic distribution in the dynamics. Precisely, in the small mutation limit, the phenotypic distribution is approximately a sum of Dirac masses, and takes the form of $\exp\left(-\frac{u(x,t)+o(1)}{\epsilon}\right)$, where ϵ is the small mutation rate, and $u(x,t)$ satisfies a class of Hamilton–Jacobi equations with a special nonnegativity constraint; see [9, 48, 74, 79] for recent progress on the uniqueness of the solutions to such equations.

More generally, the equilibrium solutions of selection-mutation models tend to select ESS values, under the small mutation limit. For selection-mutation equations involving a continuous trait variable [49], Bürger proved the existence and uniqueness of globally stable equilibrium distributions in L^1 [7]. Later, selection-mutation equations with ecological background were introduced; see [8] for competitive and prey-predator interactions, and [76] which treats age of maturity as an evolutionary trait. In these works, the ESS is often unique and globally attractive. See also [55] where, depending on parameters and initial data, the selection-mutation model of competitive interaction may select between two alternative ESSs (two alternative evolutionary endpoints); or form a Dirac measure supported at two distinct traits, thus forming an ESS coalition.

The incorporation of migration and spatial structure to the selection-mutation models brings new mathematical challenges; see [39, 54, 58, 71, 72, 80] for the convergence of equilibria to the stationary Dirac mass supported at an ESS trait, and [61, 73, 77] for the moving Dirac concentrations in the small mutation limit of the time-dependent problem. The topic of selection-mutation models will be further discussed in Chapter 10.

Problems

9.3.1 Consider a species with phenotypic space $[-1, 1]$. For given $\alpha, \beta \in [-1, 1]$, let $u(t)$ (resp. $v(t)$) be the population of the phenotype α (resp. β). Suppose that (u, v) satisfies the following ODE system

$$\begin{cases} \frac{\mathrm{d}}{\mathrm{d}t}u = u(r(\alpha) - K(\alpha,\alpha)u - K(\alpha,\beta)v), \\ \frac{\mathrm{d}}{\mathrm{d}t}v = v(r(\beta) - K(\beta,\alpha)u - K(\beta,\beta)v), \end{cases}$$

where r, K are smooth and have positive infimum over $[-1, 1]$ and $[-1, 1]^2$ respectively. Show that the invasion exponent is given by

$$\lambda(\beta, \alpha) = r(\beta) - \frac{K(\beta,\alpha)r(\alpha)}{K(\alpha,\alpha)}.$$

9.3.2 In the setting of Problem 9.3.1 suppose, in addition, that

$$r \equiv 1, \quad \text{and} \quad K(\beta,\alpha) = 1 - \eta(\beta - \alpha)(\beta + k\alpha),$$

where η is small enough so that $\inf K > 0$. Show that the invasion exponent is given by

$$\lambda(\beta, \alpha) = \eta(\beta - \alpha)(\beta - k\alpha).$$

Show the following.

1. 0 is a singular strategy.
2. If $\eta(1-k) < 0$, then 0 is a convergence stable strategy.
3. If $\eta < 0$, then 0 is an ESS.
4. If $\eta > 0$ and $k > 1$, then 0 is an evolutionary branching point.
5. If $\eta > 0$ and $k < 0$, then ± 1 are local ESS.
6. If $\eta > 0$ and $k < -1$, show that ± 1 are global ESS.

Here local and global ESS are considered within the class of strategies $[-1, 1]$.

References

1. J. Apaloo, *Revisiting strategic models of evolution: The concept of neighborhood invader strategies*, Theor. Pop. Biol., 52 (1997), pp. 71–77.
2. D. Auslander, J. Guckenheimer, and G. Oster, *Random evolutionarily stable strategies*, Theor. Pop. Biol., 13 (1978), pp. 276–293.
3. I. Averill, Y. Lou, and D. Munther, *On several conjectures from evolution of dispersal*, J. Biol. Dyn., 6 (2012), pp. 117–130.
4. X. Bai, X. He, and W.-M. Ni, *Dynamics of a periodic-parabolic Lotka–Volterra competition-diffusion system in heterogeneous environments*. J. Eur. Math. Soc. (2022), in press, 2022.
5. J. S. Brown and T. L. Vincent, *Coevolution as an evolutionary game*, Evolution, 41 (1987), pp. 66–79.
6. J. S. Brown and T. L. Vincent, *A theory for the evolutionary game*, Theor. Pop. Biol., 31 (1987), pp. 140–166.
7. R. Bürger, *Perturbations of positive semigroups and applications to population genetics*, Math. Z., 197 (1988), pp. 259–272.
8. A. Calsina and C. Perelló, *Equations for biological evolution*, Proc. Roy. Soc. Edinburgh Sect. A, 125 (1995), pp. 939–958.
9. V. Calvez and K.-Y. Lam, *Uniqueness of the viscosity solution of a constrained Hamilton–Jacobi equation*, Calc. Var. Partial Differential Equations, 59 (2020), p. 46. Paper No. 163.
10. R. S. Cantrell and C. Cosner, *Evolutionary stability of ideal free dispersal under spatial heterogeneity and time periodicity*, Math. Biosci., 305 (2018), pp. 71–76.
11. R. S. Cantrell, C. Cosner, and K.-Y. Lam, *Resident-invader dynamics in infinite dimensional systems*, J. Differential Equations, 263 (2017), pp. 4565–4616.
12. ———, *Ideal free dispersal under general spatial heterogeneity and time periodicity*, SIAM J. Appl. Math., 81 (2021), pp. 789–813.
13. R. S. Cantrell, C. Cosner, and Y. Lou, *Evolution of dispersal and the ideal free distribution*, Math. Biosci. Eng., 7 (2010), pp. 17–36.
14. ———, *Evolutionary stability of ideal free dispersal strategies in patchy environments*, J. Math. Biol., 65 (2012), pp. 943–965.
15. R. S. Cantrell, C. Cosner, Y. Lou, and D. Ryan, *Evolutionary stability of ideal free dispersal strategies: a nonlocal dispersal model*, Can. Appl. Math. Q., 20 (2012), pp. 15–38.
16. R. S. Cantrell, C. Cosner, Y. Lou, and S. J. Schreiber, *Evolution of natal dispersal in spatially heterogenous environments*, Math. Biosci., 283 (2017), pp. 136–144.
17. R. S. Cantrell, C. Cosner, and Y. Zhou, *Ideal free dispersal in integrodifference models*, J. Math. Biol., 85 (2022). Paper No. 6.
18. N. Champagnat, R. Ferrière, and S. Méléard, *From individual stochastic processes to macroscopic models in adaptive evolution*, Stoch. Models, 24 (2008), pp. 2–44.
19. N. Champagnat, R. Ferrière, and S. Méléard, *Unifying evolutionary dynamics: From individual stochastic processes to macroscopic models*, Theor. Pop. Biol., 69 (2006), pp. 297–321.

20. N. Champagnat and S. Méléard, *Polymorphic evolution sequence and evolutionary branching*, Probab. Theory Related Fields, 151 (2011), pp. 45–94.
21. C. Cosner, *Reaction-diffusion-advection models for the effects and evolution of dispersal*, Discrete Contin. Dyn. Syst., 34 (2014), pp. 1701–1745.
22. F. Dercole and S. Rinaldi, *Analysis of Evolutionary Processes: The Adaptive Dynamics Approach and Its Applications: The Adaptive Dynamics Approach and Its Applications*, Princeton University Press, 2008.
23. U. Dieckmann and R. Law, *The dynamical theory of coevolution: a derivation from stochastic ecological processes*, J. Math. Biol., 34 (1996), pp. 579–612.
24. O. Diekmann, *A beginner's guide to adaptive dynamics*, in Mathematical modelling of population dynamics, vol. 63 of Banach Center Publ., Polish Acad. Sci. Inst. Math., Warsaw, 2004, pp. 47–86.
25. O. Diekmann, P.-E. Jabin, S. Mischler, and B. Perthame, *The dynamics of adaptation: An illuminating example and a Hamilton—Jacobi approach*, Theor. Pop. Biol., 67 (2005), pp. 257–271.
26. J. Dockery, V. Hutson, K. Mischaikow, and M. Pernarowski, *The evolution of slow dispersal rates: a reaction diffusion model*, J. Math. Biol., 37 (1998), pp. 61–83.
27. M. Doebeli and G. D. Ruxton, *Evolution of dispersal rates in metapopulation models: Branching and cyclic dynamics in phenotype space*, Evolution, 51 (1997), pp. 1730–1741.
28. I. Eshel, *Evolutionary and continuous stability*, J. Theoret. Biol., 103 (1983), pp. 99–111.
29. I. Eshel and U. Motro, *Kin selection and strong evolutionary stability of mutual help*, Theor. Pop. Biol., 19 (1981), pp. 420–433.
30. S. D. Fretwell, *Populations in a Seasonal Environment*, Princeton University Press, 1972.
31. S. D. Fretwell and H. L. Lucas, *On territorial behavior and other factors influencing habitat distribution in birds*, Acta Biotheor., 19 (1969), pp. 16–36.
32. S. A. H. Geritz, *Resident-invader dynamics and the coexistence of similar strategies*, J. Math. Biol., 50 (2005), pp. 67–82.
33. S. A. H. Geritz and M. Gyllenberg, *The Mathematical Theory of Adaptive Dynamics*, Cambridge University Press, 2008.
34. S. A. H. Geritz, M. Gyllenberg, F. J. A. Jacobs, and K. Parvinen, *Invasion dynamics and attractor inheritance*, J. Math. Biol., 44 (2002), pp. 548–560.
35. S. A. H. Geritz, E. Kisdi, J. A. J. Metz, et al., *Evolutionarily singular strategies and the adaptive growth and branching of the evolutionary tree*, Evolutionary ecology, 12 (1998), pp. 35–57.
36. M. Golubitsky, W. Hao, K.-Y. Lam, and Y. Lou, *Dimorphism by singularity theory in a model for river ecology*, Bull. Math. Biol., 79 (2017), pp. 1051–1069.
37. M. Gyllenberg, E. Kisdi, and M. Utz, *Evolution of condition-dependent dispersal under kin competition*, J. Math. Biol., 57 (2008), pp. 285–307.
38. W. D. Hamilton and R. M. May, *Dispersal in stable habitats*, Nature, 269 (1977), pp. 578–581.
39. W. Hao, K.-Y. Lam, and Y. Lou, *Concentration phenomena in an integro-PDE model for evolution of conditional dispersal*, Indiana Univ. Math. J., 68 (2019), pp. 881–923.
40. ———, *Ecological and evolutionary dynamics in advective environments: critical domain size and boundary conditions*, Discrete Contin. Dyn. Syst. Ser. B, 26 (2021), pp. 367–400.
41. A. Hastings, *Can spatial variation alone lead to selection for dispersal?*, Theor. Pop. Biol., 24 (1983), pp. 244–251.
42. J. Hofbauer and K. Sigmund, *The theory of evolution and dynamical systems. Mathematical aspects of selection*, vol. 7 of London Mathematical Society Student Texts, Cambridge University Press, Cambridge, 1988. Translated from the German.
43. ———, *Evolutionary game dynamics*, Bull. Amer. Math. Soc. (N.S.), 40 (2003), pp. 479–519.
44. S. B. Hsu, H. L. Smith, and P. Waltman, *Competitive exclusion and coexistence for competitive systems on ordered Banach spaces*, Trans. Amer. Math. Soc., 348 (1996), pp. 4083–4094.
45. V. Hutson, K. Mischaikow, and P. Poláčik, *The evolution of dispersal rates in a heterogeneous time-periodic environment*, J. Math. Biol., 43 (2001), pp. 501–533.

46. H. Jiang, K.-Y. Lam, and Y. Lou, *Are two-patch models sufficient? The evolution of dispersal and topology of river network modules*, Bull. Math. Biol., 82 (2020), p. 42. Paper No. 131.
47. ———, *Three-patch models for the evolution of dispersal in advective environments: varying drift and network topology*, Bull. Math. Biol., 83 (2021), p. 46. Paper No. 109.
48. Y. Kim, *On the uniqueness of solutions to one-dimensional constrained Hamilton–Jacobi equations*, Min. Theor. Appl., 6 (2021), pp. 145–154.
49. M. Kimura, *A stochastic model concerning the maintenance of genetic variability in quantitative characters*, Proceedings of the National Academy of Sciences, 54 (1965), pp. 731–736.
50. E. Kisdi, *A list of papers in adaptive dynamics*. https://www.mv.helsinki.fi/home/kisdi/addyn.htm. Accessed: 2022-05-18.
51. L. Korobenko and E. Braverman, *A logistic model with a carrying capacity driven diffusion*, Can. Appl. Math. Q., 17 (2009), pp. 85–104.
52. ———, *On logistic models with a carrying capacity dependent diffusion: stability of equilibria and coexistence with a regularly diffusing population*, Nonlinear Anal. Real World Appl., 13 (2012), pp. 2648–2658.
53. ———, *On evolutionary stability of carrying capacity driven dispersal in competition with regularly diffusing populations*, J. Math. Biol., 69 (2014), pp. 1181–1206.
54. K.-Y. Lam, *Stability of Dirac concentrations in an integro-PDE model for evolution of dispersal*, Calc. Var. Partial Differential Equations, 56 (2017), p. 46. Paper No. 79.
55. ———, *Dirac-concentrations in an integro-PDE model from evolutionary game theory*, Discrete Contin. Dyn. Syst. Ser. B, 24 (2019), pp. 737–754.
56. K.-Y. Lam and Y. Lou, *Evolution of conditional dispersal: evolutionarily stable strategies in spatial models*, J. Math. Biol., 68 (2014), pp. 851–877.
57. ———, *Evolutionarily stable and convergent stable strategies in reaction-diffusion models for conditional dispersal*, Bull. Math. Biol., 76 (2014), pp. 261–291.
58. ———, *An integro-PDE model for evolution of random dispersal*, J. Funct. Anal., 272 (2017), pp. 1755–1790.
59. ———, *The principal floquet bundle and the dynamics of fast diffusing communities*. Manuscript submitted for publication, 2022.
60. K.-Y. Lam, Y. Lou, and F. Lutscher, *Evolution of dispersal in closed advective environments*, J. Biol. Dyn., 9 (2015), pp. 188–212.
61. K.-Y. Lam, Y. Lou, and B. Perthame, *A Hamilton–Jacobi approach to evolution of dispersal*, 2022. arXiv:2205.05534 [math.AP].
62. S. A. Levin, D. Cohen, and A. Hastings, *Dispersal strategies in patchy environments*, Theor. Pop. Biol., 26 (1984), pp. 165–191.
63. S. Lion, *Theoretical approaches in evolutionary ecology: Environmental feedback as a unifying perspective*, Am. Nat., 191 (2018), pp. 21–44.
64. A. Lorz, S. Mirrahimi, and B. Perthame, *Dirac mass dynamics in multidimensional nonlocal parabolic equations*, Comm. Partial Differential Equations, 36 (2011), pp. 1071–1098.
65. Y. Lou and F. Lutscher, *Evolution of dispersal in open advective environments*, J. Math. Biol., 69 (2014), pp. 1319–1342.
66. Y. Lou and P. Zhou, *Evolution of dispersal in advective homogeneous environment: the effect of boundary conditions*, J. Differential Equations, 259 (2015), pp. 141–171.
67. J. F. Maynard Smith and G. R. Price, *The logic of animal conflict*, Nature, 246 (1973), pp. 15–18.
68. B. J. McGill and J. S. Brown, *Evolutionary game theory and adaptive dynamics of continuous traits*, Annual Review of Ecology, Evolution, and Systematics, 38 (2007), pp. 403–435.
69. M. A. McPeek and R. D. Holt, *The evolution of dispersal in spatially and temporally varying environments*, The American Naturalist, 140 (1992), pp. 1010–1027.
70. J. A. J. Metz, S. A. H. Geritz, G. Meszéna, F. J. A. Jacobs, and J. S. Van Heerwaarden, *Adaptive dynamics: a geometrical study of the consequences of nearly faithful reproduction*, North-Holland, Amsterdam, 1996, pp. 183–231.
71. S. Mirrahimi, *A Hamilton–Jacobi approach to characterize the evolutionary equilibria in heterogeneous environments*, Math. Models Methods Appl. Sci., 27 (2017), pp. 2425–2460.

72. S. Mirrahimi and S. Gandon, *Evolution of specialization in heterogeneous environments: Equilibrium between selection, mutation and migration*, Genetics, 214 (2020), pp. 479–491.
73. S. Mirrahimi and B. Perthame, *Asymptotic analysis of a selection model with space*, J. Math. Pures Appl. (9), 104 (2015), pp. 1108–1118.
74. S. Mirrahimi and J.-M. Roquejoffre, *A class of Hamilton–Jacobi equations with constraint: uniqueness and constructive approach*, J. Differential Equations, 260 (2016), pp. 4717–4738.
75. J. F. Nash, Jr., *Equilibrium points in n-person games*, Proc. Nat. Acad. Sci. U.S.A., 36 (1950), pp. 48–49.
76. Àngel Calsina and J. M. Palmada, *Steady states of a selection-mutation model for an age structured population*, J. Math. Anal. Appl., 400 (2013), pp. 386–395.
77. S. Nordmann and B. Perthame, *Dynamics of concentration in a population structured by age and a phenotypic trait with mutations. Convergence of the corrector*, J. Differential Equations, 290 (2021), pp. 223–261.
78. K. Parvinen, *Evolutionary branching of dispersal strategies in structured metapopulations*, J. Math. Biol., 45 (2002), pp. 106–124.
79. B. Perthame and G. Barles, *Dirac concentrations in Lotka–Volterra parabolic PDEs*, Indiana Univ. Math. J., 57 (2008), pp. 3275–3301.
80. B. Perthame and P. E. Souganidis, *Rare mutations limit of a steady state dispersal evolution model*, Math. Model. Nat. Phenom., 11 (2016), pp. 154–166.
81. J. Roughgarden, *Resource partitioning among competing species—a coevolutionary approach*, Theor. Pop. Biol., 9 (1976), pp. 388–424.
82. J. M. Smith, *Evolution and the Theory of Games*, Cambridge University Press, 1982.
83. D. C. Speirs and W. S. C. Gurney, *Population persistence in rivers and estuaries*, Ecology, 82 (2001), pp. 1219–1237.
84. O. Vasilyeva and F. Lutscher, *Population dynamics in rivers: analysis of steady states*, Can Appl Math Q, 18 (2011), pp. 439–469.
85. X. Wang and M. Golubitsky, *Singularity theory of fitness functions under dimorphism equivalence*, J. Math. Biol., 73 (2016), pp. 525–573.
86. J. Zu, K. Wang, and M. Mimura, *Evolutionary branching and evolutionarily stable coexistence of predator species: Critical function analysis*, Mathematical Biosciences, 231 (2011), pp. 210–224.

Chapter 10
Selection-Mutation Models

Abstract In this chapter, we discuss the class of selection-mutation models. This class of models describe populations that are structured by, in addition to other variables, continuous (phenotypic) trait variables such as movement rate. The dynamical and equilibrium solutions of these models exhibit the dominant phenotype under the processes of selection and mutation. In Section 10.1, we will discuss a selection-mutation model studied by Magal and Webb, where the population is structured by trait and time only. In Section 10.2, we consider the problem of evolution of dispersal rate by incorporating the spatial variable in the model, and discuss a result of Perthame and Souganidis related to the evolution of slow dispersal in spatially heterogeneous and temporally constant environments.

10.1 Populations Structured by a Phenotypic Trait

Consider a population structured by a phenotypic trait $z \in \Omega$, where Ω is a bounded and smooth domain in $\mathbb{R}^N$. The theory of evolution asserts that phenotypic frequency is being shaped by the processes of mutation and selection. Let $u(z,t)$ be the frequency function, then

$$\partial_t u = \underbrace{M[u]}_{\text{Mutation}} + \underbrace{F(z, u(\cdot,t))u}_{\text{Selection}}$$

where $F(z, u(\cdot,t))$ is the fitness of the trait z when the resident population distribution is $u(\cdot,t)$. The mutation is usually modeled by a second-order operator or an integral operator. See [8] for a more detailed discussion of mutation and selection models.

Magal and Webb [29] considered the following more specific model.

$$\begin{cases} \partial_t u = \epsilon^2 \Delta u + u\left[m(z) - \int_\Omega u(y,t)\,\mathrm{d}y\right] & \text{for } z \in \Omega,\ t > 0, \\ \mathbf{n}\cdot\nabla u = 0 & \text{for } z \in \partial\Omega,\ t > 0, \\ u(z,0) = u_0(z) & \text{for } z \in \Omega, \end{cases} \tag{10.1}$$

K. -Y. Lam, Y. Lou, *Introduction to Reaction-Diffusion Equations*, Lecture Notes on Mathematical Modelling in the Life Sciences, https://doi.org/10.1007/978-3-031-20422-7_10

where mutation is modeled by a diffusion process with rate ϵ^2. The fitness of a particular trait $z \in \Omega$ at time t is given by $m(z) - \int_\Omega u(y,t)\,dy$, where $m(z)$ is a positive continuous function modeling the proliferation rate of phenotype z, and the integral in the square bracket indicates that all phenotypes compete equally. In the following result, the statement concerning the long-time convergence to equilibrium is due to [29, Theorem 3.1].

Theorem 10.1.1 *Let $m = m(z)$ be a continuous function which is strictly positive in $\overline{\Omega}$. Then, for each $\epsilon > 0$, the model* (10.1) *has a unique positive equilibrium θ_ϵ, which attracts all nonnegative, nontrivial solutions of* (10.1). *Furthermore, if m attains its supremum in Ω at a unique interior point $\hat{z} \in \Omega$, then as $\epsilon \to 0$,*

$$\theta_\epsilon(z) \to \left(\sup_\Omega m\right)\delta_0(z - \hat{z}) \quad \text{in distribution.} \tag{10.2}$$

Proof Fix $\epsilon > 0$, and consider the eigenvalue problem

$$\begin{cases} -\epsilon^2\Delta\phi = m(z)\phi + \mu\phi & \text{for } z \in \Omega, \\ \mathbf{n}\cdot\nabla\phi = 0 & \text{for } z \in \partial\Omega. \end{cases} \tag{10.3}$$

By Theorem 1.3.6, the eigenvalue problem (10.3) has a principal eigenvalue $\mu_\epsilon \in \mathbb{R}$ such that μ_ϵ is simple and has a positive eigenfunction ϕ_ϵ, which we normalize by $\int_\Omega \phi_\epsilon\,dy = 1$. Also, by evaluating at the minimum point of ϕ_ϵ, and using $\inf_\Omega m > 0$, we deduce that $\mu_\epsilon \le -\inf m < 0$. Note that $\theta_\epsilon = -\mu_\epsilon\phi_\epsilon$ is a positive equilibrium.

Next, define $\mathcal{L} = -\epsilon^2\Delta - m - \mu_\epsilon$ and let $v(\cdot,t) = e^{-\mathcal{L}t}v_0$ be the unique solution of

$$\begin{cases} \partial_t v - \epsilon^2\Delta v = m(z)v + \mu_\epsilon v & \text{for } z \in \Omega,\ t > 0, \\ \mathbf{n}\cdot\nabla v = 0 & \text{for } z \in \partial\Omega,\ t > 0, \\ v(z,0) = v_0(z) & \text{for } z \in \Omega. \end{cases}$$

By Remark 4.2.3, there exist $C, \gamma > 0$ such that

$$\|e^{-\mathcal{L}t}v_0\|_{C(\overline{\Omega})} \le Ce^{-\gamma t}\|v_0\|_{C(\overline{\Omega})} \quad \text{for } t > 0, \tag{10.4}$$

whenever $v_0 \in C(\overline{\Omega})$ satisfies $\int_\Omega v_0\phi_\epsilon\,dy = 0$.

Given a nonnegative, nontrivial initial data $u_0(z)$. Write

$$u_0(z) = k_0\phi_\epsilon(z) + v_0(z),$$

where we set $k_0 = \int_\Omega \phi_\epsilon u_0\,dy / \int_\Omega |\phi_\epsilon|^2\,dy$. Note that $k_0 > 0$, since u_0 is nonnegative and nontrivial. Also, this means $\int_\Omega v_0\phi_\epsilon\,dy = 0$. It is clear that

$$e^{-\mathcal{L}t}u_0 = k_0\phi_\epsilon + e^{-\mathcal{L}t}v_0.$$

By (10.4), we obtain

$$e^{-\mathcal{L}t}u_0 \to k_0\phi_\epsilon > 0 \quad \text{in } C(\overline{\Omega}) \text{ as } t \to \infty. \tag{10.5}$$

Now let $u(z,t)$ be the solution of (10.1) with initial data u_0, and define

$$\rho(t) = \int_\Omega u(y,t)\,dy \quad \text{and} \quad \sigma(t) = \int_\Omega \phi_\epsilon(y)u(y,t)\,dy.$$

Note that $\rho(t) > 0$ and $\sigma(t) > 0$ for $t \geq 0$.

Multiplying (10.1) by ϕ_ϵ, and integrating in z, we obtain

$$\frac{d}{dt}\sigma(t) = (-\mu_\epsilon - \rho(t))\sigma(t), \tag{10.6}$$

which implies

$$\sigma(t) = \sigma(0)e^{-\mu_\epsilon t - \int_0^t \rho(\tau)\,d\tau}. \tag{10.7}$$

Next, notice that the solution $u(z,t)$ can be written as

$$u(z,t) = e^{-\mu_\epsilon t - \int_0^t \rho(\tau)\,d\tau} e^{-t\mathcal{L}} u_0 = \frac{\sigma(t)}{\sigma(0)}(k_0\phi_\epsilon + o(1)), \tag{10.8}$$

where we used (10.5). Integrating the above in z, we obtain

$$\rho(t) = \frac{\sigma(t)}{\sigma(0)}\left(k_0 \int_\Omega \phi_\epsilon\,dy + o(1)\right) = \frac{\sigma(t)}{\sigma(0)}(k_0 + o(1)). \tag{10.9}$$

Substituting (10.9) into (10.6), we have

$$\frac{d}{dt}\sigma(t) = \left(-\mu_\epsilon - \frac{k_0\sigma(t)}{\sigma(0)} + o(1)\right)\sigma(t) \quad \text{for } t \gg 1.$$

It then follows that $\sigma(t) \to -\frac{\mu_\epsilon}{k_0}\sigma(0)$ as $t \to \infty$. Substituting back into (10.8), we deduce that

$$u(z,t) = \left(-\frac{\mu_\epsilon}{k_0} + o(1)\right)(k_0\phi_\epsilon(z) + o(1)) \to -\mu_\epsilon\phi_\epsilon(z) \quad \text{as } t \to \infty.$$

This proves part 1 concerning the global convergence to the equilibrium $-\mu_\epsilon\phi_\epsilon(z)$.

Finally, we apply Proposition 1.3.17 to deduce that, as $\epsilon \to 0$,

$$\phi_\epsilon(z) \to \delta_0(z - \hat{z}) \quad \text{in distribution.}$$

This completes the proof. □

Remark 10.1.2 When $\Omega = \mathbb{R}$, and $m(z)$ is confining, i.e. $m(z) \to -\infty$ as $|z| \to \infty$, then the convergence to a scalar multiple of the principal eigenfunction is proved in [4]. When $m(z)$ is not confining, then there may no longer be a "fittest phenotype" and convergence to equilibrium is no longer expected. See, e.g. [3].

Remark 10.1.3 The convergence of θ_ϵ to a Dirac measure in (10.2) is based upon a special case of semiclassical analysis. More generally, if the critical points of m are nondegenerate, and m attains the global maximum at more than one point, then the limit measure is supported in a subset of these points, and the exact weights at each point depend on the associated Hessian matrices [20]. In general, we expect the solution to concentrate on a subset of the global maximum points where mutation causes the minimal decrease in fitness, i.e. the biological niche is widest. See also [27].

A way to obtain multi-modal distribution is by symmetry:

Example 1. [One-dimensional domain] Choose

$$\Omega = (-2, 2), \quad \text{and} \quad m(z) = 2 - \cos\left(\frac{z}{\pi}\right),$$

then for small mutation the unique equilibrium is symmetric with respect to $z = 0$ and is supported at the two points $z = \pm 1$.

Example 2. [Higher-dimensional domain] Let $N \geq 2$. If Ω is the unit ball in $\mathbb{C}$, and $m(z)$ is nonconstant and rotationally symmetric, i.e.

$$m(z) = m\left(e^{\frac{2\pi i}{N}} z\right) \quad \text{for } z \in \Omega.$$

Then under mild conditions, one can show that, for small mutation, the support of the equilibrium solution consists of at least N points.

Remark 10.1.4 In fact, several theoretical studies have shown that the phenotypic distributions in continuous trait space often evolve towards single or multiple peaks that represent distinct phenotypes in nature [14, 15, 21, 35, 37].

Remark 10.1.5 In the previous example, the distinct points are the global maximum point(s) of $m(z)$, which form a discrete set in the generic situation. Also, the case of global maximum attained at more than one point is not generic since the maximum values at various peaks are generically distinct. In [23], the following mutation-selection model with nontrivial competition kernel was considered.

$$\begin{cases} \partial_t u = \Delta u + u\left(m(z) - \int_\Omega K(z, z')u(z', t)\, dz'\right) & \text{in } \Omega \times (0, \infty), \\ \mathbf{n} \cdot \nabla u = 0 & \text{on } \partial\Omega \times (0, \infty), \\ u(z, 0) = u_0(z) & \text{in } \Omega. \end{cases}$$

Under various assumptions on $m(z)$ and $K(z, z')$, it is proved that equilibrium solutions exhibiting (i) a single boundary Dirac mass; (ii) a single interior Dirac mass; or (iii) two Dirac masses, exist and are stable with respect to small perturbation of model parameters.

The Case $\Omega = \mathbb{R}^N$

The following nonlocal parabolic equation is considered in [6, 28, 35].

$$\begin{cases} \epsilon\partial_t u_\epsilon = \epsilon^2 \Delta u_\epsilon + u_\epsilon R(z, I_\epsilon(t)) & \text{for } z \in \mathbb{R}^N,\ t > 0, \\ u_\epsilon(z, 0) = u_\epsilon^0(z) & \text{for } z \in \mathbb{R}^N, \end{cases} \tag{10.10}$$

where R is bounded from above, strictly decreasing in the second argument, and

$$I_\epsilon(t) = \int_{\mathbb{R}^N} \psi(z) u_\epsilon(z, t)\, \mathrm{d}z. \tag{10.11}$$

The model (10.10) arises in the theory of adaptive evolution, which describes the evolution of a population structured with phenotypic trait z, where $u_\epsilon(z, t)$ denotes the number of individuals with trait $z \in \mathbb{R}^N$ at time $t > 0$. The term $I_\epsilon(t)$ given in (10.11) describes the population burden, or the competition among different phenotypes, that is weighted by $\psi(z)$. Roughly speaking, a phenotype consumes more resources if $\psi(z)$ is larger. If $\psi \equiv 1$, then we obtain the same population burden in (10.1). The population dynamics is driven by births and deaths, as represented by the net growth rate R. Note that growth rate is in general a nonconstant function of the phenotypic trait, meaning that some individuals are better adapted. In the special case

$$R(z, I) = m(z) - I, \quad \psi(z) \equiv 1,$$

the equation (10.10) is a version of (10.1) with a rescaling in time. Under suitable assumptions, the selection for higher growth rate and the dynamics of adaptation cause the population density to concentrate on a set of dominant traits, meaning that it degenerates to a Dirac mass, or a sum of Dirac masses, located at the dominant trait(s), and the rescaling in time is crucial to capture the dynamics of the Dirac masses. Namely, as $\epsilon \to 0$,

$$u_\epsilon(z, t) \to M(t)\delta(z - \bar{z}(t)) \quad \text{in distribution,} \tag{10.12}$$

where $M(t) > 0$ is the total population and $\bar{z}$ is the dominant trait at time t.

Below, we briefly outline the proof, which is based on the WKB-ansatz

$$v_\epsilon(z, t) = -\epsilon \log u_\epsilon(z, t).$$

Then (10.10) becomes

$$\begin{cases} \partial_t v_\epsilon - \epsilon \Delta v_\epsilon + |\nabla v_\epsilon|^2 + R(z, I_\epsilon(t)) = 0 & \text{for } z \in \mathbb{R}^N,\ t > 0, \\ v_\epsilon(z, 0) = v_\epsilon^0(z) \equiv -\epsilon \log u_\epsilon^0(z) & \text{for } z \in \mathbb{R}^N. \end{cases} \tag{10.13}$$

By establishing the uniform Lipschitz estimate for v_ϵ on compact subsets of $\mathbb{R}^N \times \mathbb{R}_+$, and the BV estimate for I_ϵ, one can then pass to a subsequence to obtain

$$v_\epsilon \to v \quad \text{in } C_{loc}(\mathbb{R}^N \times \mathbb{R}_+) \quad \text{and} \quad I_\epsilon(t) \to I(t) \quad \text{in } L^1_{loc}(\mathbb{R}_+),$$

for some $v \in C_{loc}(\mathbb{R}^N \times \mathbb{R})_+$ and $I \in \mathrm{BV}_{loc}(\mathbb{R}_+)$. See [6] for details. It can then be verified that the limits $(v(z,t), I(t))$ form the viscosity solution of a Hamilton–Jacobi equation with a constraint:

$$\begin{cases} \partial_t v + |\nabla v|^2 + R(z, I(t)) = 0 & \text{for } z \in \mathbb{R}^N,\ t > 0, \\ \inf_{z\in\mathbb{R}^N} v(z,t) = 0 & \text{for } t > 0, \\ v(z,0) = v^0(z). \end{cases} \tag{10.14}$$

(See, e.g. [5] for the definition of viscosity solution.) The positivity constraint $\inf v(\cdot,t) \equiv 0$ comes from a uniform positive upper and lower bounds for the total population $I_\epsilon(t)$.

The solution $(v(z,t), I(t))$ of (10.14) characterizes the Dirac dynamics. In the case when

$$z \mapsto R(z,I) \text{ is strictly concave and } v^0(z) \text{ is strictly convex,} \tag{10.15}$$

it is proved in [28] that there exists a continuous function $\bar{z}(t)$ such that

$$v(z,t) = 0 \quad \text{if and only if} \quad z = \bar{z}(t).$$

Since $v(z,t) > 0$ for $z \neq \bar{z}$, we have $u_\epsilon(z,t) = \exp\left(-\frac{v(z,t)+o(1)}{\epsilon}\right) \to 0$ for $z \neq \bar{z}(t)$. Hence, (10.12) holds. (In fact, it holds with $M(t) = I(t)/\psi(\bar{z}(t))$.)

In general (with R not necessarily concave), $u_\epsilon(z,t)$ converges to a measure supported on the null set of $z \mapsto v(z,t)$. A natural question is whether the limiting equation (10.14) has a unique solution (v, I); if so, then the evolutionary dynamics can be uniquely determined by the first-order equation (10.14). The uniqueness of (10.14) in the class of $\mathrm{Lip}_{loc}(\mathbb{R}^N \times \mathbb{R}_+) \times \mathrm{BV}_{loc}(\mathbb{R}_+)$ was first established in [35] for the case when $R(z,I)$ is separable:

$$R(z,I) = B(z)Q(I) - D(z) \quad \text{or} \quad B(z) - D(z)Q(I)$$

with $Q(I)$ being a monotone function. Subsequently, it was proved in [31] with $R(z,I)$ being strictly decreasing in I, under the convex regime (10.15). The general uniqueness is established in [10], assuming only that $R(z,I)$ is strictly decreasing in I and $v(z) \to \infty$ as $|z| \to \infty$.

In [15], a similar selection-mutation problem is considered for the case of competition for two independent resources $I_1(t)$ and $I_2(t)$, and it can similarly be proved that the local uniform limit of the rate function v_ϵ and the two resources, denoted by $(v(z,t), I_1(t), I_2(t))$, satisfy a similar Hamilton–Jacobi equation with positive constraint. Note that there is one more parameter I_i to be determined from the positivity constraint. It is an interesting open question to prove if, and in what sense, the limit $(v(z,t), I_1(t), I_2(t))$ can be uniquely determined by the constrained Hamilton–Jacobi equation.

10.2 Populations Structured by Space and a Phenotypic Trait

In this section, we discuss the evolution of dispersal in bounded spatial domains. For this purpose we shall introduce a selection-mutation model of a population structured by space and a phenotypic trait. We will discuss a result due to Perthame and Souganidis [36]; see also [24].

To motivate the model, consider first the following multi-species model in [16].

$$\begin{cases} \partial_t u_i = z_i \Delta u_i + u_i(m(x) - \sum_{j=1}^N u_j) + \epsilon^2 \sum_{j=1}^N M_{ij} u_j & \text{for } x \in D,\, t > 0, \\ \mathbf{n} \cdot \nabla u_i = 0 & \text{for } x \in \partial D,\, t > 0, \\ u_i(x, 0) = u_{i,0}(x) & \text{for } x \in D, \end{cases} \tag{10.16}$$

where $N \geq 2$, $1 \leq i \leq N$ and D is a bounded domain in $\mathbb{R}^n$ with smooth boundary ∂D and outer unit normal vector $\mathbf{n}$; z_i is the spatial diffusion rate of the i-th species. In addition to the Lotka–Volterra competition nonlinearity, the mutation process is modeled by the term $\epsilon^2 M_{ij}$, where ϵ^2 is the mutation rate.

If we formally let the number of species $N \to \infty$, we obtain the following integro-PDE studied in [24, 36]. In the following $u = u(x, z, t)$, where $x \in D$ is the spatial variable, $z \in \mathbb{R}$ is the trait variable, and $t > 0$ is time. The mutation-selection model is given by

$$\begin{cases} \partial_t u = \Theta(z) \Delta_x u + u \left(m(x) - \int_0^1 u(x, z, t)\, \mathrm{d}z \right) + \epsilon^2 \partial_{zz} u & \text{for } x \in D,\, z \in \mathbb{R},\, t > 0, \\ \mathbf{n} \cdot \nabla_x u = 0 & \text{for } x \in \partial D,\, z \in \mathbb{R},\, t > 0, \end{cases} \tag{10.17}$$

where $\Theta : \mathbb{R} \to \mathbb{R}$ is a positive, 1-periodic function, and periodic initial data is imposed:

$$u(x, z, 0) = u_0(x, z) \quad \text{is 1-periodic in } z.$$

In Chapter 7, we saw that a slower diffuser is selected when there are two distinct phenotypes with different diffusion rates. This result generalizes to the N-species case in some situations [11, 25]. Here we prove a similar result in the context of the mutation-selection model. We will analyze the positive equilibrium solutions of (10.17). Any such solution, denoted by $u_\epsilon(x, z)$,[1] satisfies the following nonlocal semilinear elliptic equation:

$$\begin{cases} \Theta(z) \Delta_x u_\epsilon + u_\epsilon (m(x) - \rho_\epsilon(x)) + \epsilon^2 \partial_{zz} u_\epsilon = 0 & \text{for } x \in D,\, z \in \mathbb{R}, \\ \mathbf{n} \cdot \nabla_x u_\epsilon = 0 & \text{for } x \in \partial D,\, z \in \mathbb{R}, \\ u_\epsilon(x, z) = u_\epsilon(x, z+1) & \text{for } x \in D,\, z \in \mathbb{R}, \end{cases} \tag{10.18}$$

where $\rho_\epsilon(x)$ is given by

[1] Note that (10.17) does not admit a comparison principle. Nevertheless, it can be proved that there exists at least one positive equilibrium if the trivial solution is unstable; see [22] for details. In particular, there is at least one positive equilibrium if $\int_D m(x)\, \mathrm{d}x > 0$.

$$\rho_\epsilon(x) = \int_0^1 u_\epsilon(x,z)\,\mathrm{d}z. \tag{10.19}$$

In the following, we show that u_ϵ concentrates as a Dirac mass supported at the trait value with the minimal spatial diffusion rate, i.e. the slowest diffuser is selected.

Theorem 10.2.1 *Suppose that*

1. *$m(x)$ is positive and nonconstant in $\overline{D}$;*
2. *$\Theta(z)$ is continuous, positive and 1-periodic;*
3. *there exists a unique $\hat{z} \in (0,1)$ such that $\inf_{\mathbb{R}} \Theta = \Theta(\hat{z})$.*

Let u_ϵ be any positive solution of (10.18), *then as $\epsilon \to 0$, we have*

$$u_\epsilon(x,z) \to \delta_0(z-\hat{z})\hat{\theta}(x) \quad \textit{in distribution in } D\times I, \tag{10.20}$$

where $\hat{\theta}$ is the unique positive solution of

$$\begin{cases} \Theta(\hat{z})\Delta\hat{\theta} + \hat{\theta}(m(x)-\hat{\theta}) = 0 & \textit{in } D,\\ \mathbf{n}\cdot\nabla\hat{\theta} = 0 & \textit{on } \partial D. \end{cases} \tag{10.21}$$

For convenience, we define

$$a = \inf_{\mathbb{R}} \Theta(z) \quad \text{and} \quad b = \sup_{\mathbb{R}} \Theta(z).$$

Since $\Theta(z)$ is positive and 1-periodic, it is clear that $0 < a < b$.

Below we give some an *a priori* estimate of ρ_ϵ. Note that estimates similar to that in (10.20) are possible even if we replace the periodic condition in z by a Neumann or Dirichlet condition; see [19, 24].

Lemma 10.2.2 *For each $p > 1$, there exists a constant C_1 independent of $\epsilon \in (0,1]$ such that for any positive solution $u_\epsilon(x,z)$ of* (10.18), *we have*

$$\|\rho_\epsilon\|_{W^{2,p}(D)} \le C_1,$$

and for $0 < \epsilon \ll 1$,

$$0 < \frac{1}{2}\inf_D m \le \rho_\epsilon(x) \le 2\sup_D m \quad \textit{for } x \in D, \tag{10.22}$$

where $\rho_\epsilon(x) = \int_0^1 u_\epsilon(x,z)\,\mathrm{d}z$.

Proof First, we estimate $\|\rho_\epsilon\|_{C(\overline{D})}$. To this end, divide (10.18) by $\Theta(z)$ and integrate in z over $[0,1]$ to get

$$\begin{cases} \Delta_x\rho_\epsilon + \epsilon^2 P_\epsilon + Q_\epsilon(m(x)-\rho_\epsilon(x)) = 0 & \text{in } D,\\ \mathbf{n}\cdot\nabla\rho_\epsilon = 0 & \text{on } \partial D, \end{cases} \tag{10.23}$$

where, using the periodicity of Θ and u_ϵ in z,

$$P_\epsilon(x) = \int_0^1 \frac{1}{\Theta(z)} \partial_{zz} u_\epsilon \,\mathrm{d}z = \int_0^1 \partial_{zz}\left(\frac{1}{\Theta(z)}\right) u_\epsilon \,\mathrm{d}z,$$

and $Q_\epsilon = \int_0^1 \frac{1}{\Theta(z)} u_\epsilon(x,z)\,\mathrm{d}z$ is a positive function. Using the fact that

$$|P_\epsilon(x)| \le b \left\| \partial_{zz}\left(\frac{1}{\Theta(z)}\right)\right\|_{L^\infty([0,1])} Q_\epsilon(x) = C_0 Q_\epsilon(x) \quad \text{in } D,$$

we can rewrite (10.23) as

$$\Delta_x \rho_\epsilon + Q_\epsilon\left[O(\epsilon^2) + m(x) - \rho_\epsilon(x)\right] = 0 \quad \text{in } D, \quad \mathbf{n}\cdot\nabla\rho_\epsilon = 0 \quad \text{on } \partial D.$$

By the maximum principle, we deduce (10.22) for $0 < \epsilon \ll 1$. Since Q_ϵ also satisfies

$$\frac{1}{b}\rho_\epsilon(x) \le Q_\epsilon(x) \le \frac{1}{a}\rho_\epsilon(x) \quad \text{in } D, \tag{10.24}$$

we deduce that $\|Q_\epsilon\|_{L^\infty(D)}$ is also bounded uniformly in $\epsilon > 0$. Furthermore, by (10.23) we obtain

$$\begin{aligned}|\Delta_x \rho_\epsilon| &\le \epsilon^2 \int_0^1 \left|\partial_{zz}\left(\frac{1}{\Theta(z)}\right)\right| u_\epsilon \,\mathrm{d}z + |Q_\epsilon||m-\rho_\epsilon| \\ &\le \epsilon^2 \left\|\partial_{zz}\left(\frac{1}{\Theta(z)}\right)\right\|_{L^\infty([0,1])} |\rho_\epsilon| + |Q_\epsilon||m-\rho_\epsilon|.\end{aligned} \tag{10.25}$$

Since $\|Q_\epsilon\|_{L^\infty}$ and $\|\rho_\epsilon\|_{L^\infty}$ are bounded uniformly in $\epsilon > 0$, we may apply the elliptic L^p estimate to obtain the desired $W^{2,p}$ bound of ρ_ϵ which is uniform in ϵ.□

By Lemma 10.2.2, one can pass to a sequence and assume that $\lim_{\epsilon\to 0}\rho_\epsilon$ converges in $C^1(\overline{D})$. In the following, we show that $m - \lim_{\epsilon\to 0}\rho_\epsilon$ is nonconstant.

Lemma 10.2.3 *Suppose that there exist $\rho \in C^1(\overline{\Omega})$ and a sequence $\epsilon = \epsilon_n \to 0$ such that $\rho_\epsilon \to \rho$ in $C^1(\overline{\Omega})$. Then $m - \rho$ is nonconstant.*

Proof Suppose to the contrary that, along a sequence, $m - \rho_\epsilon \to C$ for some constant $C \in \mathbb{R}$. We divide our discussion into two cases: $C = 0$ or $C \neq 0$.

Suppose $m - \rho_\epsilon \to 0$, then (10.25) implies $\Delta\rho_\epsilon \to 0$ uniformly in D. This, together with the homogeneous Neumann boundary condition, implies that ρ_ϵ converges to a constant. But this contradicts the fact that $m - \rho_\epsilon \to 0$ and that m is nonconstant in D.

Next, suppose $m - \rho_\epsilon \to C$ for some $C \neq 0$. Then we may assume that $m - \rho_\epsilon$ does not change sign in D. However, this is in contradiction with

$$\int_D \rho_\epsilon(m(x) - \rho_\epsilon)\,\mathrm{d}x = 0, \tag{10.26}$$

which is obtained if we integrate (10.18) over $D \times [0,1]$. □

Lemma 10.2.4 *There exists a C_1 independent of ϵ such that*

$$\epsilon \left\| \frac{\partial_z u_\epsilon}{u_\epsilon} \right\|_{C(\overline{D}\times[0,1])} + \left\| \frac{\nabla_x u_\epsilon}{u_\epsilon} \right\|_{C(\overline{D}\times[0,1])} \le C_1. \tag{10.27}$$

Proof For each $z_0 \in \mathbb{R}$, define

$$U_\epsilon(x,y) := u_\epsilon(x, z_0 + \epsilon y).$$

Then U_ϵ is a positive solution to

$$\mathcal{L}_\epsilon U_\epsilon = A(y,\epsilon)\Delta_x U_\epsilon + \partial_{zz} U_\epsilon + h_\epsilon(x) U_\epsilon = 0 \quad \text{in } D \times \mathbb{R},$$

and satisfies the Neumann boundary condition when $x \in \partial D$. Here $A(y,\epsilon)$ is a Lipschitz continuous function (with Lipschitz constant bounded uniformly in ϵ), such that $a \le A(y,\epsilon) \le b$, i.e. it is uniformly elliptic. We can then apply the Harnack inequality to $D \times [-1,1]$ and deduce

$$\sup_{D\times[-1,1]} U_\epsilon \le C \inf_{D\times[-1,1]} U_\epsilon \tag{10.28}$$

for some constant C that is independent of $z_0 \in D$ and $\epsilon > 0$. Note that the estimate holds up to the boundary of x variable, thanks to the Neumann boundary condition. Next, fix $p > 1$ suitably large to apply the Sobolev embedding to get

$$\sup_{D\times(-1/2,1/2)} \left[|\partial_z U_\epsilon| + |\nabla_x U_\epsilon| \right] \le C \|U_\epsilon\|_{W^{2,p}(D\times(-3/4,3/4))}. \tag{10.29}$$

Next, we recall the elliptic interior L^p estimate

$$\|U_\epsilon\|_{W^{2,p}(D\times(-3/4,3/4))} \le C \|U_\epsilon\|_{L^p(D\times(-1,1))}. \tag{10.30}$$

Note that the constants in (10.29) and (10.30) are independent of z_0 and ϵ. Combining the above, we have

$$\sup_{D\times(-1/2,1/2)} \left[|\partial_z U_\epsilon| + |\nabla_x U_\epsilon| \right] \le C \|U_\epsilon\|_{L^p(D\times(-1,1))} \le C \sup_{D\times[-1,1]} U_\epsilon. \tag{10.31}$$

In view of (10.28), we deduce that

$$|\partial_z U_\epsilon(x,0)| + |\nabla_x U_\epsilon(x,0)| \le C \inf_{D\times[-1,1]} U_\epsilon \le C'' U_\epsilon(x,0) \quad \text{for } x \in D,$$

for some constant C'' that is independent of z_0, ϵ, and $x \in D$. This can be written as

$$\epsilon |\partial_z u_\epsilon(x,z_0)| + |\nabla_x u_\epsilon(x,z_0)| \le C'' u_\epsilon(x,z_0).$$

This proves (10.27). □

Introduce the WKB-ansatz

$$w_\epsilon(x,z) = -\epsilon \log u_\epsilon(x,z). \tag{10.32}$$

Then we have

$$-\epsilon\partial_{zz}w_\epsilon + |\partial_z w_\epsilon|^2 + m(x) - \rho_\epsilon(x) = \Theta(z)\left(\frac{\Delta_x w_\epsilon}{\epsilon} - \frac{|\nabla_x w_\epsilon|^2}{\epsilon^2}\right) \quad \text{in } D\times\mathbb{R}, \tag{10.33}$$

and satisfies the Neumann boundary condition on $\partial D\times[0,1]$ while being 1-periodic in z.

Lemma 10.2.5 *$\{w_\epsilon\}$ is precompact in $C(\overline{D}\times[0,1])$ and satisfies*

$$\lim_{\epsilon\to 0}\inf_{D\times[0,1]} w_\epsilon = 0. \tag{10.34}$$

Proof By Lemma 10.2.4, we have

$$\sup_{D\times[0,1]}\left[|\partial_z w_\epsilon| + \frac{1}{\epsilon}|\nabla_x w_\epsilon|\right] \le C, \tag{10.35}$$

i.e. w_ϵ is equicontinuous. It suffices to show (10.34), since then the precompactness in $C(\overline{D}\times[0,1])$ follows directly from the Arzelà–Ascoli Theorem.

We will show (10.34) in two steps. First, we claim that

$$\limsup_{\epsilon\to 0}\inf_{D\times[0,1]} w_\epsilon \le 0. \tag{10.36}$$

Otherwise, one can pass to a sequence in $\epsilon\to 0$ such that $u_\epsilon = \exp\left(-\frac{w_\epsilon}{\epsilon}\right)\to 0$ in $C(\overline{D}\times[0,1])$, and hence $\rho_\epsilon\to 0$ in $C(\bar{D})$. This is impossible in view of (10.22).

It remains to show that

$$\liminf_{\epsilon\to 0}\inf_{D\times I} w_\epsilon \ge 0. \tag{10.37}$$

Otherwise, there exist a sequence $\epsilon\to 0$ and $(x_\epsilon, z_\epsilon)\in D\times I$ such that

$$\limsup_{\epsilon\to 0} w_\epsilon(x_\epsilon, z_\epsilon) < 0.$$

By the equicontinuity, there exists a $\delta > 0$ independent of ϵ such that

$$\limsup_{\epsilon\to 0}\sup_{B_\delta(x_\epsilon,z_\epsilon)} w_\epsilon \le -\eta_0 \quad \text{for some } \eta_0 > 0.$$

But this implies that

$$\liminf_{\epsilon\to 0}\inf_{B_\delta(x_\epsilon,z_\epsilon)} u_\epsilon \ge \exp\left(\frac{\eta_0}{\epsilon}\right).$$

This is in contradiction with (10.22), which says that $\rho_\epsilon(x) \le 2\sup_D m$. □

By Lemmas 10.2.4 and 10.2.5, we may pass to a sequence $\epsilon = \epsilon_n \to 0$ and assume that

$$w_\epsilon(x,z) \to w(z) \quad \text{and} \quad \rho_\epsilon \to \rho(x) \tag{10.38}$$

respectively in $C(\overline{D} \times I)$ and $C(\overline{D})$. Note that the limit $w(z)$ is independent of x due to (10.27). We will show that the two limit functions w and ρ are related via a Hamilton–Jacobi equation.

Definition 10.2.6 For each $z \in \mathbb{R}$ and $h \in L^\infty(D)$, define $\Lambda_h(z)$ and $\phi_h(x,z)$ to be the principal eigenvalue and eigenfunction of

$$\begin{cases} -\Theta(z)\Delta_x\phi = (m(x) - h(x))\phi + \Lambda\phi & \text{in } D, \\ \mathbf{n} \cdot \nabla_x\phi = 0 & \text{on } \partial D. \end{cases} \tag{10.39}$$

Next, we introduce the notion of a viscosity solution to a first-order Hamilton–Jacobi equation.

Definition 10.2.7 Let $w \in C(\mathbb{R})$ and $\rho \in C(\overline{D})$.

1. We say that $w \in C(\mathbb{R})$ is a viscosity supersolution of

$$|w'|^2 - \Lambda_\rho(z) = 0 \quad \text{in } I, \tag{10.40}$$

if, for any $\varphi \in C^\infty(\mathbb{R})$, if $w - \varphi$ attains a strict local minimum at $z_0 \in \mathbb{R}$, then it holds that

$$|\varphi'(z_0)|^2 - \Lambda_\rho(z_0) \geq 0.$$

2. We say that $w \in C(\mathbb{R})$ is a viscosity subsolution of (10.40) if, for any $\varphi \in C^\infty(\mathbb{R})$, if $w - \varphi$ attains a strict local maximum at $z_0 \in \mathbb{R}$, then it holds that

$$|\varphi'(z_0)|^2 - \Lambda_\rho(z_0) \leq 0.$$

3. We say that $w \in C(\mathbb{R})$ is a viscosity solution of (10.40) if it is both a viscosity subsolution and a viscosity supersolution of (10.40).

Lemma 10.2.8 *Suppose that, along a sequence $\epsilon = \epsilon_n \to 0$, the convergence* (10.38) *holds for some $w \in C(I)$ and $\rho \in C(\overline{D})$. Then w is a viscosity solution of* (10.40).

Proof We apply the method of perturbed test functions [17]. Let $\psi(x,z) = \log \phi_\rho(x,z)$, where $\phi_\rho(\cdot,z)$ is the positive eigenfunction associated with the principal eigenvalue $\Lambda_\rho(z)$ of (10.39). Note that

$$-\Theta(z)(|\nabla_x\psi|^2 + \Delta_x\psi) = m(x) - \rho(x) + \Lambda_\rho(z) \quad \text{in } D \times \mathbb{R}. \tag{10.41}$$

Suppose $w - \varphi$ attains a strict local minimum point at $z_0 \in I$. Then $w_\epsilon(x,z) - \varphi(z) + \epsilon\psi(x,z)$ has a local minimum $(x_\epsilon, z_\epsilon) \in \overline{D} \times I$ such that $z_\epsilon \to z_0$. At the point (x_ϵ, z_ϵ), we have

$$\begin{cases} \Delta_x w_\epsilon(x_\epsilon, z_\epsilon) \geq -\epsilon \Delta_x \psi(x_\epsilon, z_\epsilon), \quad \nabla_x w_\epsilon(x_\epsilon, z_\epsilon) = -\epsilon \nabla_x \psi(x_\epsilon, z_\epsilon), \\ \partial_{zz} w_\epsilon(x_\epsilon, z_\epsilon) \geq -\epsilon \partial_{zz} \psi(x_\epsilon, z_\epsilon) + \varphi''(z_\epsilon), \\ \partial_z w_\epsilon(x_\epsilon, z_\epsilon) = -\epsilon \partial_z \psi(x_\epsilon, z_\epsilon) + \varphi'(z_\epsilon). \end{cases}$$

Note that the above holds even if $x_\epsilon \in \partial D$. (This can be proved by extending the function by reflecting along the boundary, and observing that the extended function also satisfies the same PDE in a neighborhood of $\partial D \times \mathbb{R}$. Here we used the homogeneous Neumann boundary condition.) Substituting the above into (10.33) and using (10.41), we deduce that at the point (x_ϵ, z_ϵ),

$$\begin{aligned} &\epsilon(\epsilon \partial_{zz} \psi(x_\epsilon, z_\epsilon) - \varphi''(z_\epsilon)) + \left| -\epsilon \partial_z \psi(x_\epsilon, z_\epsilon) + \varphi'(z_\epsilon) \right|^2 \\ &\geq \rho_\epsilon(x_\epsilon) - \rho(x_\epsilon) + \Lambda_\rho(z_\epsilon). \end{aligned}$$

Using the fact that ψ and Λ_ρ are smooth in $z \in I$ (see Proposition 1.3.15), we may take $\epsilon \to 0$ in the above inequality, and obtain $|\varphi'(z_0)|^2 \geq \Lambda_\rho(z_0)$. This proves that φ is a viscosity supersolution of (10.40). It follows in a similar manner that φ is also a viscosity subsolution. □

By the property of viscosity solutions, we obtain some qualitative properties of w.

Lemma 10.2.9 *Under the assumptions of Lemma* 10.2.8, *suppose there is a* $\hat{z} \in [0, 1]$ *such that* $\Theta(z) > \Theta(\hat{z})$ *if* $z \in [0, 1] \setminus \{\hat{z}\}$. *Then*

1. $\Lambda_\rho(z) > 0$ *if* $z \in [0, 1] \setminus \{\hat{z}\}$;
2. $w(z) > 0$ *if* $z \in [0, 1] \setminus \{\hat{z}\}$.

Proof We claim that

$$\Lambda_\rho(z) \geq 0 \quad \text{for all } z \in \mathbb{R}. \tag{10.42}$$

Indeed, for each interval $[a', b'] \subset \mathbb{R}$, we can take $\varphi = 1/\tilde{\sigma}$ where $\tilde{\sigma}$ is a test function such that $\tilde{\sigma} > 0$ in (a', b') and $\tilde{\sigma} = 0$ otherwise. Then $w - \varphi$ attains its maximum at some point $z' \in (a', b')$, so that the definition of viscosity subsolution yields

$$\Lambda_\rho(z') \geq |\varphi'(z')|^2 \geq 0.$$

Since $(a', b') \subset I$ is arbitrary, we deduce that $\Lambda_\rho(z) \geq 0$ on a dense subset of I. By continuity, we have $\Lambda_\rho(z) \geq 0$ for all $z \in I$.

By Lemma 10.2.3, $m - \rho$ is nonconstant, so that the principal eigenvalue Λ_ρ of (10.39) (with $h = \rho$) is strictly increasing in diffusion rate (Proposition 1.3.15). Hence, if $z \in [0, 1] \setminus \{\hat{z}\}$, then $\Theta(z) > \Theta(\hat{z})$ and hence

$$\Lambda_\rho(z) > \Lambda_\rho(\hat{z}) \geq 0.$$

This proves the first assertion.

For the second assertion, we note that $\inf_I w = 0$ in I, by Lemma 10.2.5. Suppose $w(z_0) = 0$ for some $z_0 \in [0, 1]$, it remains to show that $z_0 = \hat{z}$. Indeed, $w + |z - z_0|^2$ attains a strict local minimum at z_0. By virtue of w being a viscosity supersolution of (10.40), $-\Lambda_\rho(z_0) \geq 0$. By the first assertion, we must have $z_0 = \hat{z}$. □

***Proof (of Theorem** 10.2.1)* Passing to a sequence, we may assume that $w_\epsilon(x, z) \to w(z)$ in $C(\overline{W} \times [0, 1])$ and $\rho_\epsilon(x) \to \rho(x)$ in $C^1(\overline{D})$ as $\epsilon \to 0$. By Lemma 10.2.9, $w(z) > 0$ in $[0, 1] \setminus \{\hat{z}\}$, where we recall that $\hat{z}$ is the unique point in $[0, 1]$ such that $\Theta(\hat{z}) = \inf \Theta$. It then follows that

$$u_\epsilon(x, z) = \exp\left(-\frac{w_\epsilon(x, z)}{\epsilon}\right) = \exp\left(-\frac{w(z) + o(1)}{\epsilon}\right) \to 0$$

in compact subsets of $\overline{D} \times ([0, 1] \setminus \{\hat{z}\})$.

It remains to show that $\rho_\epsilon \to \hat{\theta}$, where $\hat{\theta}$ is the unique positive solution of (10.21). To this end, rewrite (10.18) as

$$\Theta(\hat{z})\Delta_x u_\epsilon + u_\epsilon(m(x) - \rho_\epsilon) = -(\Theta(z) - \Theta(\hat{z}))\Delta_x u_\epsilon \quad \text{in } D \times \mathbb{R}.$$

Integrate in z from 0 to 1, we obtain

$$\Theta(\hat{z})\Delta_x \rho_\epsilon + \rho_\epsilon(m - \rho_\epsilon) = -\int_0^1 (\Theta(z) - \Theta(\hat{z}))\Delta_x u_\epsilon \, dz.$$

Multiplying by a test function $\varphi(x) \in C^1(\overline{D})$, and integrating by parts,

$$-\Theta(\hat{z}) \int_D \nabla_x \rho_\epsilon \cdot \nabla_x \varphi \, dx + \int_D \varphi \rho_\epsilon (m - \rho_\epsilon) \, dx = \tilde{R}_\epsilon, \tag{10.43}$$

where $\tilde{R}_\epsilon = \int_{D\times[0,1]} (\Theta(z) - \Theta(\hat{z}))\nabla_x u_\epsilon \cdot \nabla_x \varphi \, dz dx$.

We claim that $\tilde{R}_\epsilon \to 0$. Indeed,

$$\begin{aligned}
|\tilde{R}_\epsilon| &\leq \int_{D\times[0,1]} |\Theta(z) - \Theta(\hat{z})| |\nabla_x u_\epsilon| |\nabla_x \varphi| \, dz dx \\
&\leq C \int_{D\times[0,1]} |z - \hat{z}| u_\epsilon(x, z) \, dz dx \\
&\leq C \left[\delta \int_D \rho_\epsilon \, dx + \int_{D\times[0,\hat{z}-\delta]} u_\epsilon(x, z) \, dx dz + \int_{D\times[\hat{z}+\delta,1]} u_\epsilon(x, z) \, dx dz \right]
\end{aligned}$$

where we used (10.27) for the second inequality. Hence,

$$\limsup |\tilde{R}_\epsilon| \leq C\delta.$$

Letting $\delta \to 0$, we deduce that $\tilde{R}_\epsilon \to 0$.

Hence, letting $\epsilon \to 0$ in (10.43), we deduce that

$$-\Theta(\hat{z}) \int_D \nabla_x \rho \cdot \nabla \varphi \, dx + \int_D \varphi \rho (m - \rho) \, dx = 0.$$

Hence, ρ is a weak solution (and thus a classical solution by the elliptic regularity theory) to (10.21). Since ρ is positive (thanks to Lemma 10.2.2), it must coincide with the unique positive solution $\hat{\theta}$ of (10.21). □

Remark 10.2.10 A related problem in bounded regions with the Neumann boundary condition is studied in [24], where more detailed asymptotics of the equilibrium solutions, as $\epsilon \to 0$, are obtained. These asymptotics are used in proving the stability and uniqueness of equilibrium solution in [22].

Remark 10.2.11 It is conjectured in [36] that, for well-prepared initial data, the time-dependent problem (10.17) possesses solutions exhibiting moving Dirac concentrations. See [26] for a recent result.

10.3 Further Reading

In this chapter, we are interested in models of a population which is structured by a continuous phenotypic trait. Such models can be viewed as the analogue of competition models of infinitely many species. Although mutation is modeled by diffusion in the trait variable, we note that other more realistic formulations, such as integral operators, are also possible [6].

The first half of this chapter concerns the selection of the fittest trait in the context of maximizing growth rate in the nonspatial context. We establish the well-known result that such models often admit a unique (globally asymptotically stable) positive equilibria [7], and that the positive equilibria tends to a single or multiple Dirac mass(es) in the small mutation limit. See also [1, 2] for measure-valued selection-mutation models.

More recently, the dynamics of Dirac concentrations of (10.10) has been investigated with a Hamilton–Jacobi approach. This latter body of work started with a paper of Diekmann et al. [15], which formally derived the Hamilton–Jacobi equation with a constraint for a model with two substitutable resources and a consumer population structured by a continuous trait parameterizing the preference towards the two resources. Asymptotic analysis of phenotype-structured population models has been carried out for various situations, see [12, 14, 15, 33, 34]. The particular case of (10.10) is considered in [6, 28, 34, 35]. See also [18] for the case of time-periodic environments.

The selection-mutation model can be considered as a population approach to rigorously derive the evolutionary dynamics as suggested by the formal adaptive dynamics framework [9, 12, 15, 23, 34]. The models considered in this chapter form a special class that can be obtained by taking appropriate limits from stochastic individual based models as the population size becomes large [13]. Other types of models, such as the class of pure selection models [14, 21], are also possible.

The second half of this chapter concerns a population structured by both spatial and trait variables, in which the slowest diffuser is selected, in the sense that the unique equilibria, which is also locally asymptotically stable, converges to a Dirac-concentration at the slowest diffusing phenotype when the mutation rate tends to zero. See also [30]. With the incorporation of spatial structure, it is more challenging to resolve the time-dependent problem concerning the dynamics of Dirac masses. It is an open problem to show that solutions converge to moving Dirac masses in the small mutation limit, under proper rescaling in time. See [26] for some recent progress for the model (10.17), and also [32] for a result with a similar flavor concerning another selection-mutation model with age structure.

Granted, the selection of slowest diffuser is well known for the two species case, assuming unconditional movement and a spatially heterogeneous environment. The significance of the present approach (with a population structured by space and trait) is to suggest an alternative method to identify an ESS in a given ecological situation. Recently, we performed a detailed analysis of equilibrium in the river model where the population is structured by space and the diffusion rate (as the phenotypic trait) [19]. Depending on the environmental conditions, we proved the convergence of the positive equilibrium to a single interior Dirac-concentration, a boundary Dirac-concentration, and two boundary Dirac-concentrations. These represent different endpoints of evolutionary dynamics. Namely, the presence of an ESS (in fact continuously stable strategy), the absence of a singular strategy (so that one of the boundary traits is convergence stable), and the existence of an evolutionary branching point. The third situation demonstrates the flexibility of the continuum framework in handling a nonmonomorphic population. For the corresponding results in a non-spatial setting where the population is structured only by a continuous trait, see also [23], where it is proved in addition that the problem can have bistability, which is caused by two alternative global ESSs in the adaptive dynamics.

Problems

10.3.1 Let $\epsilon > 0$ and $m \in C^1(\overline{D})$ be given.

1. Suppose $m \leq 0$ in D, show that every nonnegative, nontrivial solution of (10.17) satisfies $u \to 0$ as $t \to \infty$.
2. Suppose the principal eigenvalue λ_1 of

$$\begin{cases} -\Theta(z)\Delta_x \phi - \epsilon^2 \partial_{zz}^2 \phi = m(x)\phi + \lambda\phi = 0 & \text{in } D \times [0,1], \\ \mathbf{n} \cdot \nabla_x \phi = 0 & \text{on } \partial D \times [0,1], \\ \phi(x,0) = \phi(x,1) & \text{in } D, \end{cases}$$

is nonnegative, then every nonnegative, nontrivial solution of (10.17) satisfies $u \longrightarrow 0$ as $t \longrightarrow \infty$.

[On the other hand, if $\lambda_1 < 0$, then it is shown in [22] that there is a $\delta > 0$ such that

$$\delta \leq \liminf_{t\to\infty} \inf_{D\times[0,1]} u \leq \limsup_{t\to\infty} \sup_{D\times[0,1]} u \leq \frac{1}{\delta},$$

and that (10.17) has at least one positive equilibrium.]

10.3.2 Suppose $m(x) \equiv 1$.

1. Verify that $\mathbf{1}(x, z) = 1$ is an equilibrium of (10.17).
2. Show that

$$V(u) := \int_{D\times[0,1]} (u - 1 - \log u)$$

   is a Lyapunov function (see Definition 7.2.2), i.e. if $u(x, z, t)$ is a solution of (10.17), then

$$\frac{\mathrm{d}}{\mathrm{d}t} V(u(\cdot, \cdot, t)) \leq 0 \quad \text{for } t \geq 0.$$

3. Show that $u(x, z, t) \to 1$ as $t \to \infty$, for any nonnegative, nontrivial solution u of (10.17), by invoking LaSalle's invariance principle (Theorem 7.2.3).

References

1. A. S. Ackleh, J. Cleveland, and H. R. Thieme, *Population dynamics under selection and mutation: long-time behavior for differential equations in measure spaces*, J. Differential Equations, 261 (2016), pp. 1472–1505.
2. A. S. Ackleh and N. Saintier, *Diffusive limit to a selection-mutation equation with small mutation formulated on the space of measures*, Discrete Contin. Dyn. Syst. Ser. B, 26 (2021), pp. 1469–1497.
3. M. Alfaro and R. Carles, *Explicit solutions for replicator-mutator equations: extinction versus acceleration*, SIAM J. Appl. Math., 74 (2014), pp. 1919–1934.
4. M. Alfaro and M. Veruete, *Evolutionary branching via replicator-mutator equations*, J. Dynam. Differential Equations, 31 (2019), pp. 2029–2052.
5. G. Barles, *An introduction to the theory of viscosity solutions for first-order Hamilton–Jacobi equations and applications*, in Hamilton–Jacobi equations: approximations, numerical analysis and applications, vol. 2074 of Lecture Notes in Math., Springer, Heidelberg, 2013, pp. 49–109.
6. G. Barles, S. Mirrahimi, and B. Perthame, *Concentration in Lotka–Volterra parabolic or integral equations: a general convergence result*, Methods Appl. Anal., 16 (2009), pp. 321–340.
7. R. Bürger, *Perturbations of positive semigroups and applications to population genetics*, Math. Z., 197 (1988), pp. 259–272.
8. ———, *The mathematical theory of selection, recombination, and mutation*, Wiley Series in Mathematical and Computational Biology, John Wiley & Sons, Ltd., Chichester, 2000.
9. A. Calsina, S. Cuadrado, L. Desvillettes, and G. Raoul, *Asymptotics of steady states of a selection-mutation equation for small mutation rate*, Proc. Roy. Soc. Edinburgh Sect. A, 143 (2013), pp. 1123–1146.

10. V. Calvez and K.-Y. Lam, *Uniqueness of the viscosity solution of a constrained Hamilton–Jacobi equation*, Calc. Var. Partial Differential Equations, 59 (2020), p. 22. Paper No. 163.
11. R. S. Cantrell and K.-Y. Lam, *On the evolution of slow dispersal in multispecies communities*, SIAM J. Math. Anal., 53 (2021), pp. 4933–4964.
12. J. A. Carrillo, S. Cuadrado, and B. Perthame, *Adaptive dynamics via Hamilton–Jacobi approach and entropy methods for a juvenile-adult model*, Math. Biosci., 205 (2007), pp. 137–161.
13. N. Champagnat, R. Ferrière, and S. Méléard, *Unifying evolutionary dynamics: From individual stochastic processes to macroscopic models*, Theor. Pop. Biol., 69 (2006), pp. 297–321.
14. L. Desvillettes, P.-E. Jabin, S. Mischler, and G. Raoul, *On selection dynamics for continuous structured populations*, Commun. Math. Sci., 6 (2008), pp. 729–747.
15. O. Diekmann, P.-E. Jabin, S. Mischler, and B. Perthame, *The dynamics of adaptation: An illuminating example and a Hamilton-–Jacobi approach*, Theor. Pop. Biol., 67 (2005), pp. 257–271.
16. J. Dockery, V. Hutson, K. Mischaikow, and M. Pernarowski, *The evolution of slow dispersal rates: a reaction diffusion model*, J. Math. Biol., 37 (1998), pp. 61–83.
17. L. C. Evans, *The perturbed test function method for viscosity solutions of nonlinear PDE*, Proc. Roy. Soc. Edinburgh Sect. A, 111 (1989), pp. 359–375.
18. S. Figueroa Iglesias and S. Mirrahimi, *Long time evolutionary dynamics of phenotypically structured populations in time-periodic environments*, SIAM J. Math. Anal., 50 (2018), pp. 5537–5568.
19. W. Hao, K.-Y. Lam, and Y. Lou, *Concentration phenomena in an integro-PDE model for evolution of conditional dispersal*, Indiana Univ. Math. J., 68 (2019), pp. 881–923.
20. D. Holcman and I. Kupka, *Singular perturbation for the first eigenfunction and blow-up analysis*, Forum Math., 18 (2006), pp. 445–518.
21. P.-E. Jabin and G. Raoul, *On selection dynamics for competitive interactions*, J. Math. Biol., 63 (2011), pp. 493–517.
22. K.-Y. Lam, *Stability of Dirac concentrations in an integro-PDE model for evolution of dispersal*, Calc. Var. Partial Differential Equations, 56 (2017), p. 32. Paper No. 79.
23. ———, *Dirac-concentrations in an integro-pde model from evolutionary game theory*, Discrete Contin. Dyn. Syst. Ser. B, 24 (2019), pp. 737–754.
24. K.-Y. Lam and Y. Lou, *An integro-PDE model for evolution of random dispersal*, J. Funct. Anal., 272 (2017), pp. 1755–1790.
25. ———, *The principal Floquet bundle and the dynamics of fast diffusing communities*. Manuscript submitted for publication, 2022.
26. K.-Y. Lam, Y. Lou, and B. Perthame, *A Hamilton–Jacobi approach to evolution of dispersal*, 2022. arXiv:2205.05534 [math.AP].
27. T. Lorenzi and C. Pouchol, *Asymptotic analysis of selection-mutation models in the presence of multiple fitness peaks*, Nonlinearity, 33 (2020), pp. 5791–5816.
28. A. Lorz, S. Mirrahimi, and B. Perthame, *Dirac mass dynamics in multidimensional nonlocal parabolic equations*, Comm. Partial Differential Equations, 36 (2011), pp. 1071–1098.
29. P. Magal and G. F. Webb, *Mutation, selection, and recombination in a model of phenotype evolution*, Discrete Contin. Dynam. Systems, 6 (2000), pp. 221–236.
30. S. Mirrahimi, *A Hamilton–Jacobi approach to characterize the evolutionary equilibria in heterogeneous environments*, Math. Models Methods Appl. Sci., 27 (2017), pp. 2425–2460.
31. S. Mirrahimi and J.-M. Roquejoffre, *A class of Hamilton–Jacobi equations with constraint: uniqueness and constructive approach*, J. Differential Equations, 260 (2016), pp. 4717–4738.
32. S. Nordmann and B. Perthame, *Dynamics of concentration in a population structured by age and a phenotypic trait with mutations. Convergence of the corrector*, J. Differential Equations, 290 (2021), pp. 223–261.
33. S. Nordmann, B. Perthame, and C. Taing, *Dynamics of concentration in a population model structured by age and a phenotypical trait*, Acta Appl. Math., 155 (2018), pp. 197–225.
34. B. Perthame, *Transport equations in biology*, Frontiers in Mathematics, Birkhäuser Verlag, Basel, 2007.

35. B. Perthame and G. Barles, *Dirac concentrations in Lotka–Volterra parabolic PDEs*, Indiana Univ. Math. J., 57 (2008), pp. 3275–3301.
36. B. Perthame and P. E. Souganidis, *Rare mutations limit of a steady state dispersal evolution model*, Math. Model. Nat. Phenom., 11 (2016), pp. 154–166.
37. A. Sasaki and S. Ellner, *The evolutionarily stable phenotype distribution in a random environment*, Evolution, 49 (1995), pp. 337–350.

Appendices

Appendix A
The Fixed Point Index

Abstract In this appendix, we recall the Leray–Schauder degree, and derive the closely related notion of the fixed point index. Our treatment follows Amann [1].

A.1 Properties of the Leray–Schauder Degree

Let X be a Banach space, and let

$$\begin{aligned}\mathcal{A}_{LS} = \{(f, U) :\ & U \text{ is an open subset of } X,\ f : \overline{U} \to X \text{ is compact and continuous,} \\ & f(x) \neq x \quad \text{for any } x \in \partial U\}.\end{aligned}$$

Theorem A.1.1 *There exists a unique function that assigns to each* $(f, U) \in \mathcal{A}_{LS}$ *an integer* $\deg(I - f, U, 0)$ *with the following properties:*

(Normalization) $\deg(I, U, 0) = 1$ *if* $0 \in U$.

(Additivity) If U_1, U_2 *are open, disjoint subsets of* U *such that* $f(x) \neq x$ *in* $\overline{U} \setminus (U_1 \cup U_2)$, *then*

$$\deg(I - f, U, 0) = \deg(I - f, U_1, 0) + \deg(I - f, U_2, 0).$$

(Homotopy invariance) Suppose $h : [0, 1] \times \overline{U} \to X$ *is (i) continuous, (ii) for each* $\tau \in [0, 1]$, $x \mapsto h(\tau, x)$ *is a compact mapping from* $\overline{U}$ *to* X, *(iii)* $x \neq h(\tau, x)$ *in* $[0, 1] \times \partial U$. *Then*

$$\deg(I - h(0, \cdot), U, 0) = \deg(I - h(1, \cdot), U, 0).$$

(Excision) If A *is a closed set contained in* $\overline{U}$ *and* $f(x) \neq x$ *in* $A \cup (\partial U)$, *then*

$$\deg(I - f, U, 0) = \deg(I - f, U \setminus A, 0).$$

Such a function $\deg(I - f, U, 0)$ *is called the Leray–Schauder degree of* $I - f$.

K. -Y. Lam, Y. Lou, *Introduction to Reaction-Diffusion Equations*, Lecture Notes on Mathematical Modelling in the Life Sciences, https://doi.org/10.1007/978-3-031-20422-7

See, e.g. [4] for the existence and uniqueness of the Leray–Schauder degree.

A.2 The Fixed Point Index

In many applications, one is interested in a nonlinear map f that is defined on a certain convex subset K of a Banach space X, e.g., the family of nonnegative functions in $C^0(\overline{\Omega})$ or $L^p(\Omega)$ for some bounded domain Ω in $\mathbb{R}^N$. Frequently, the fixed points of the nonlinear map lie on the boundary of the convex subset. In fact, sometimes these convex subsets have no interior points at all. In these instances, the Leray–Schauder degree is not immediately applicable. If the convex subset K is a retract of X, then it is possible to define the fixed point index. The fixed point index is equivalent to the Leray–Schauder degree, but is less well known. In this subsection, we will define the fixed point index in terms of the Leray–Schauder degree.

Definition A.2.1 Let X be a Banach space. A nonempty subset K of X is called a *retract* of X if there exists a continuous map $r : X \to K$ such that $r(x) = x$ for $x \in K$. In this case, r is called a *retraction*.

Remark A.2.2 1. (Exercise) Every retract of X is a closed subset of X.
2. It is proved by Dugundji that every nonempty closed convex subset of a Banach space X is a retract of X [2, 3].
3. The set $K = \{u_0 \in C(\overline{\Omega}) : u_0 \geq 0 \text{ in } \Omega\}$ is a retract of $C(\overline{\Omega})$ with the corresponding retraction being $r(u_0) = \max\{u_0, 0\}$. The same holds with $C(\overline{\Omega})$ replaced by $L^p(\Omega)$.

Definition A.2.3 Let K be a retract of some Banach space X. We say that U is an open (resp. closed) subset of K if $U = K \cap \tilde{U}$ for some open (resp. closed) subset $\tilde{U}$ of X. We define the boundary $\partial_K(U)$ of U relative to K by $\partial_K(U) = \partial\tilde{U} \cap K$.

A triple (f, U, K) is called admissible if K is a retract of X, U is an open subset relative to K, and $f : \overline{U} \to K$ is a compact, continuous mapping such that $f(x) \neq x$ in $\partial_K(U)$.

Theorem A.2.4 *There exists a unique function i that assigns to each admissible (f, U, K) an integer $i(f, U, K)$ with the following properties:*

(Normalization) If $f : \overline{U} \to K$ is a constant map from $\overline{U}$ into U, i.e. $f(x) \equiv x_0$ for some $x_0 \in U$, then $i(f, U, K) = 1$.

(Additivity) Let U_1, U_2 be disjoint subsets of U that are open relative to K. If $f(x) \neq x$ in $\overline{U} \setminus (U_1 \cup U_2)$, then

$$i(f, U, K) = i(f, U_1, K) + i(f, U_2, K),$$

with the convention $i(f, U_k, K) := i(f|_{\overline{U_k}}, U_k, K)$, for $k = 1, 2$.

(Homotopy invariance) Suppose $h : [0, 1] \times \overline{U} \to X$ is (i) continuous, (ii) for each $\tau \in [0, 1]$, $x \mapsto h(\tau, x)$ is a compact mapping from $\overline{U}$ to X, (iii) $x \neq h(\tau, x)$ in $[0, 1] \times \partial_K(U)$. Then

$$i(h(0,\cdot),U,K) = i(h(1,\cdot),U,K).$$

(Permanence) If K_1 *is a retract of* K *and* $f(\overline{U}) \subset K_1$*, then*

$$i(f,U,K) = i(f,U \cap K_1, K_1),$$

with the convention $i(f,U \cap K_1,K_1) := i(f|_{\overline{U \cap K_1}}, U \cap K_1, K_1)$.

(Excision) If A *is a closed subset of* $\overline{U}$*, and* $f(x) \neq x$ *in* $A \cup (\partial_K(U))$*, then*

$$i(f,U,K) = i(f,U \setminus A,K).$$

Such a function $i(f,U,K)$ *is called the fixed point index of* f *over* U *with respect to the retract* K *of* X*.*

Proof The proof is taken from [1, Theorem 11.1]. We first observe that if $K = X$, then the normalization, additivity homotopy invariance, and excision properties are precisely the properties which characterize the Leray–Schauder degree. Hence, $i(f,U,X)$ is uniquely determined by

$$i(f,U,X) = \deg(I - f,U,0).$$

Suppose now that K is an arbitrary retract of X and denote by $r : X \to K$ the retraction. Then by the permanence property,

$$i(f,U,K) = i(f \circ r, r^{-1}(U), X) = \deg(I - f \circ r, r^{-1}(U), 0).$$

Hence $i(f,U,K)$ is again uniquely determined. This proves uniqueness of the fixed point index.

To show existence, we will define the fixed point index in terms of the Leray–Schauder degree. By the uniqueness proof, we are led to define

$$i(f,U,K) = \deg(I - f \circ r_0, r_0^{-1}(U), 0), \tag{A.1}$$

where $r_0 : X \to K$ is an arbitrary retraction.

We need to show that the above definition is independent of the choice of retraction $r_0 : X \to K$. Let $r_1 : X \to K$ be another retraction, and let $V := r_0^{-1}(U) \cap r_1^{-1}(U)$. Since $r_j^{-1}(U) \setminus V$ belongs to $X \setminus K$, and the range of f is K, we have $(f \circ r_j)(x) \neq x$ in $r_j^{-1}(U) \setminus V$. By the excision property of the Leray–Schauder degree, we have

$$\deg(I - f \circ r_j, r_j^{-1}(U), 0) = \deg(I - f \circ r_j, V, 0) \quad \text{for } j = 0,1.$$

We now define $h : [0,1] \times \overline{V} \to K$ by

$$h(\tau,x) = r_0((1-\tau)f(r_0(x)) + \tau f(r_1(x))).$$

We claim that $h(\tau,x) \neq x$ in $[0,1] \times \partial V$. Indeed, suppose to the contrary that $h(\tau,x) = x$ for some $0 \leq \tau \leq 1$ and $x \in \partial V$. Since h takes values in K, we have

$x \in (\partial V) \cap K = \partial_K(U)$ and hence $r_j(x) = x$ for $j = 0, 1$. The equality $h(\tau, x) = x$ then becomes $f(x) = x$ for some $x \in \partial_K(U)$, which is impossible since (f, U, K) is admissible.

Hence, we can apply the homotopy invariance property of the Leray–Schauder degree to get

$$\deg(I - f \circ r_0, V, 0) = \deg(I - f \circ r_1, V, 0).$$

Consequently, the definition $i(f, U, K)$ according to (A.1) is independent of the particular choice of retraction r_0.

Finally, we verify the remaining permanence property. Let K be a retract of X with retraction $r_0 : X \to K$, and Y be a retract of K with retraction $r_2 : K \to Y$, then Y is a retract of X with retraction $r_3 = r_2 \circ r_0$. Let $f : \overline{U} \to X$ such that $f(\overline{U}) \subset Y$ be given. Indeed, we can proceed similarly as before. First, we define $V = r_3^{-1}(U \cap Y) \cap r_0^{-1}(U)$, and observe that by excision,

$$\begin{cases} i(f, U, K) = \deg(I - f \circ r_0, r_0^{-1}(U), 0) = \deg(I - f \circ r_0, V, 0), \\ i(f, U \cap Y, Y) = \deg(I - f \circ r_3, r_3^{-1}(U \cap Y), 0) = \deg(I - f \circ r_3, V, 0). \end{cases}$$

Next, we take $h(\tau, x) = r_3((1-\tau) f(r_0(x)) + \tau f(r_3(x)))$ and observe that $h(\tau, x) \neq x$ in ∂V, so that homotopy invariance of the degree implies that

$$i(f, U, K) = \deg(I - f \circ r_0, V, 0) = \deg(I - f \circ r_3, V, 0) = i(f, U \cap Y, Y).$$

This completes the proof of the permanence property. □

Corollary A.2.5 *The following statements hold.*

1. $i(f, \emptyset, K) = 0$.
2. *If* $i(f, U, K) \neq 0$, *then* f *has at least one fixed point in* U.

Proof The first assertion follows from additivity:

$$i(f, \emptyset, K) = i(f, \emptyset, K) + i(f, \emptyset, K).$$

To prove the second assertion, let (f, U, K) be admissible and let $f(x) \neq x$ in $\overline{U}$. By taking $K = \overline{U}$, the excision property and Corollary A.2.5 yields $i(f, U, K) = i(f, \emptyset, K) = 0$. □

References

1. H. Amann, *Fixed point equations and nonlinear eigenvalue problems in ordered Banach spaces*, SIAM Rev., 18 (1976), pp. 620–709.
2. J. Dugundji, *An extension of Tietze's theorem*, Pacific J. Math., 1 (1951), pp. 353-367.
3. J. Dugundji, *Topology*, Allyn and Bacon, Inc., Boston, Mass., 1966.
4. L. Nirenberg, *Topics in nonlinear functional analysis*, vol. 6 of Courant Lecture Notes in Mathematics, New York University, Courant Institute of Mathematical Sciences, New York; American Mathematical Society, Providence, RI, 2001. Revised reprint of the 1974 original.

Appendix B
The Krein–Rutman Theorem

Abstract We present a self-contained treatment of the Krein–Rutman theorem. Our treatment follows mainly the lecture note of Nussbaum [13], with some modifications. We first prove the Krein–Rutman theorem for positive, compact and homogeneous maps on a cone (see Corollaries B.4.4 and B.4.5), then we derive the classical Krein–Rutman theorem for positive compact linear operators.

B.1 Introduction

It has long been recognized that the full Krein–Rutman theorem can be proved by using fixed point theory, e.g. using the Schauder fixed point theorem. Here we present a fixed point index argument due to Nussbaum [13]. Since the fixed point index (for cones) is directly defined from the Leray-Schauder degree, the argument we present here can be considered a method for obtaining the Krein–Rutman theorem from the Leray-Schauder degree. For simplicity, we restrict ourselves to compact operators, but most of these results can be generalized to α-contractions, provided that the spectral radius is strictly greater than the radius of the essential spectrum. We refer to [10] for recent results and open problems.

B.2 Cones and Orderings

Definition B.2.1 Let K be a subset of a Banach space X.

1. The set K is a *cone* if (i) K is closed and convex, (ii) $\mu K \subset K$ for all $\mu \geq 0$, and (iii) $K \cap (-K) = \{0\}$.
2. K is a *total cone* if it is a cone and $K - K = \{x - y : x, y \in K\}$ is dense in X.
3. K is a *solid cone* if it is a cone with nonempty interior.

Note that a cone K is total if it is solid.

K. -Y. Lam, Y. Lou, *Introduction to Reaction-Diffusion Equations*, Lecture Notes on Mathematical Modelling in the Life Sciences, https://doi.org/10.1007/978-3-031-20422-7

Definition B.2.2 The cone K in X induces a partial ordering of X. For $x, y \in X$, we write

$$x \leq y \qquad (\text{resp.} \qquad x < y, \qquad x \ll y)$$

if

$$y - x \in K \quad (\text{resp. } y - x \in K \setminus \{0\}, \quad y - x \in \operatorname{Int} K).$$

We say that X is an ordered Banach space with order induced by the cone K.

Remark B.2.3 Since any cone $K \subset X$ is convex, Dugundji's theorem [2, 3] asserts that any closed cone K in X is a retract of X, so the fixed point index $i(f, U, K)$ is well-defined for any relatively open subset U in K, and any compact and continuous map $f : \overline{U} \to K$ such that $f(x) \neq x$ in $\overline{U} \setminus U$.

We give several examples of cones.

1. For $X = \mathbb{R}^N$, $K = [0, \infty)^N$.
2. Let X be a metric space. The Banach space $X = C(X; \mathbb{R})$ of all continuous real-valued functions can be regarded as an ordered Banach space with the order induced by the cone of non-negative functions $K = \{u \in C(X) : u(x) \geq 0 \text{ for } x \in X\}$. Furthermore, K is a solid, and total cone.
3. Let Ω be a bounded open set in $\mathbb{R}^N$, and let $X = L^p(\Omega)$. Then $K = \{u \in L^p(\Omega) : u \geq 0 \text{ a.e. in } \Omega\}$ is a total cone. But it is not a solid cone.
4. Let Ω be a smooth bounded domain in $\mathbb{R}^N$. Fix $u_e \in C(\overline{\Omega})$ such that for some $0 < \alpha < \beta$, we have

$$\alpha \operatorname{dist}(x, \partial\Omega) \leq u_e(x) \leq \beta \operatorname{dist}(x, \partial\Omega) \quad \text{in } \Omega.$$

Define

$$C_e(\overline{\Omega}) = \{u \in C(\overline{\Omega}; \mathbb{R}) : \alpha u_e \leq u \leq \beta u_e \text{ in } \Omega, \text{ for some } \alpha, \beta \in \mathbb{R}\}$$

endowed with the norm

$$\|u\|_e = \inf\{\alpha > 0 : -\alpha u_e \leq u \leq \alpha u_e \text{ in } \Omega\}.$$

Then $X = C_e(\overline{\Omega})$ is an ordered Banach space with the order induced by the cone

$$K = \{u \in C(\overline{\Omega}; \mathbb{R}) : \alpha u_e \leq u \leq \beta u_e \text{ in } \Omega, \text{ for some } \alpha, \beta \geq 0\}.$$

Furthermore, K is a solid, and total cone.

Next, we consider maps and linear operators.

Definition B.2.4 Let X be a real Banach space and let $T : X \to X$ be a bounded linear operator.

1. We denote the spectral radius of T by Gelfand's formula

$$r(T) = \lim_{n\to\infty} \|T^n\|^{1/n}, \quad \text{where } \|T^n\| = \sup_{\|x\|\le 1} \|T^n x\|. \tag{B.1}$$

2. We define the complexification of X by $\tilde{X} = \{x + iy : x, y \in X\}$ with

$$\|x + iy\| = \sup_{0\le\theta\le 2\pi} \|(\cos\theta)x + (\sin\theta)y\|,$$

and $\tilde{T}$ to be the usual linear extension of T to $\tilde{X}$:

$$\|\tilde{T}\| = \|T\| \quad \text{so that} \quad r(\tilde{T}) = r(T). \tag{B.2}$$

Equation (B.2) implies that[1]

$$r(T) = \sup\{|z| : z \in \sigma(\tilde{T})\},$$

where $\sigma(\tilde{T})$ denotes the spectrum, which is the complement of the resolvent set of $\tilde{T}$. Since we will also discuss the Krein–Rutman theorem associated with maps $f : K \to K$, we define the monotone property for maps.

Definition B.2.5 Let X be an ordered Banach space with order induced by the cone K.

1. Let C be a subset of X. A map $f : C \to X$ is *monotone* (with respect to the partial order generated by K) if

$$f(x) \le f(y) \quad \text{whenever } x \le y.$$

It is called *strongly monotone* if K is a solid cone and

$$f(x) \ll f(y) \quad \text{whenever } x < y.$$

2. A map $f : K \to K$ is *positively homogeneous with degree one* if

$$f(tx) = tf(x) \quad \text{for all } x \in K, \text{ and } t \ge 0,$$

and we also define $\|f\|_K = \sup\{\|f(x)\| : x \in K, \|x\| \le 1\}$.

3. The *Bonsall cone spectral radius* is given by

$$\tilde{r}_K(f) = \limsup_{n\to\infty} \|f^n\|_K^{1/n}. \tag{B.3}$$

[1] If T is compact, then the supremum can be replaced by a maximum.

In particular, a linear operator $T : X \to X$ is said to be *monotone* (resp. *strongly monotone*) if

$$T(K) \subset K \qquad (\text{resp.} \quad T(K \setminus \{0\}) \subset \operatorname{Int} K).$$

B.3 The Classical Krein–Rutman theorem

Here we state the classical Krein–Rutman theorem [8] for compact positive linear operators.

Theorem B.3.1 (Krein and Rutman, weak form) *Suppose that X is a real ordered Banach space with order induced by the cone K, and $T : X \to X$ is a compact linear operator such that $T(K) \subset K$. If $r = r(T) > 0$ and K is a total cone, then there exist $x_0 \in K \setminus \{0\}$ and $f_0 \in K^* \setminus \{0\}$ such that*

$$Tx_0 = rx_0 \quad \text{and} \quad T^* f_0 = r f_0,$$

where T^* is the adjoint of T, and K^* is the adjoint cone

$$K^* = \{f \in X^* : f(x) \geq 0 \quad \text{for all } x \in K\}.$$

If K is a solid cone in X and T is strongly monotone, then a stronger version of the theorem holds.

Theorem B.3.2 (Krein and Rutman, strong form) *Suppose that X is a real ordered Banach space with order induced by the cone K. If K is a solid cone and $T : X \to X$ is a compact linear operator which is strongly monotone, i.e. $T(K \setminus \{0\}) \subset \operatorname{Int} K$, then*

1. *$r(T) > 0$ and it is a geometrically simple eigenvalue of T.*
2. *There exists an $\epsilon > 0$ such that $|\lambda| < r(T) - \epsilon$ for all eigenvalues $\lambda \in \mathbb{C} \setminus \{r(T)\}$.*
3. *If $\lambda > 0$ is an eigenvalue with an eigenvector in $K \setminus \{0\}$, then $\lambda = r(T)$.*

The proofs of Theorems B.3.1 and B.3.2 can be found at the end of this section, as a consequence of the corresponding result for positively homogeneous maps (Corollaries B.4.4 and B.4.5).

Remark B.3.3 Under the assumptions of Theorem B.3.2, the principal eigenvalue $r(T)$ is, in fact, algebraically simple, i.e. $\operatorname{Ker}(r(T)I - T) = \operatorname{Ker}(r(T)I - T)^2$. This is a consequence of Theorem B.3.4 below.

We note that $r(T)$ is called the principal eigenvalue of T. Before we end this section, we derive a useful consequence of Theorem B.3.2 concerning the inhomogeneous equation

$$\lambda x - Tx = h \quad \text{in } X. \tag{B.4}$$

Theorem B.3.4 *Under the hypotheses of Theorem B.3.2, the following statements hold.*

1. *If* $\lambda > r(T)$ *and* $h \in K \setminus \{0\}$, *then* (B.4) *has a unique solution* x. *Moreover,* $x \in \text{Int}\, K$.
2. *If* $\lambda \leq r(T)$ *and* $h \in K \setminus \{0\}$, *then* (B.4) *has no solution in* K.
3. *If* $\lambda \geq r(T)$ *and* $h \in K \setminus \{0\}$, *then* (B.4) *has no solution in* $X \setminus K$.

Proof (of Theorem B.3.4) Let $r = r(T)$. By Theorem B.3.2, $r > 0$ and there exists an $x_0 \in \text{Int}\, K$ such that $Tx_0 = rx_0$.

We first prove assertion 1. Since $\lambda > r(T)$, we see that λ belongs to the resolvent set of T, so that $x = (\lambda I - T)^{-1} h$ exists and is unique. Note that

$$T(tx_0 + x) < rtx_0 + \lambda x < \lambda(tx_0 + x) \qquad \text{for all } t \geq 0. \tag{B.5}$$

Define

$$B := \{t' \geq 0 : t'x_0 + x \in \text{Int}\, K\}.$$

Since $x_0 \in \text{Int}\, K$, we see that B contains all sufficiently large t and is nonempty.

Next, we claim that $tx_0 + x \notin \partial K$ for any $t \geq 0$. Suppose $tx_0 + x \in \partial K$, then it follows from (B.5) that $tx_0 + x \in K \setminus \{0\}$, so that $T(tx_0 + x) \gg 0$. Then (B.5) implies that $tx_0 + x \in \text{Int}\, K$, which is a contradiction.

Hence, $tx_0 + x \notin \partial K$ for any $t \geq 0$. We can then deduce that $B = [0, \infty)$, and in particular that $x \in \text{Int}\, K$. This proves assertion 1.

Next, we prove assertion 2. Suppose to the contrary that $\lambda \leq r(T)$ and $\lambda x - Tx > 0$ for some $x \in K$. Then $x \neq 0$, so that $\lambda x > Tx \gg 0$. This means $\lambda > 0$ and $x \in \text{Int}\, K$. Hence, $tx - x_0 \in \text{Int}\, K$ for all t sufficiently large. Now,

$$T(tx - x_0) < r(tx - x_0) \qquad \text{for all } t > 0, \tag{B.6}$$

and we observe again that $tx - x_0 \notin \partial K$ for all $t > 0$. This implies that $tx - x_0 \in \text{Int}\, K$ for all $t > 0$. Letting $t \to 0$, we obtain $-x_0 \in K$, which is absurd.

Next, we prove assertion 3. Suppose to the contrary that (B.4) holds for some $\lambda \geq r(T) > 0$, $x \in X \setminus K$ and $h \in K \setminus \{0\}$. By assertion 1, we have $\lambda = r(T)$. Since $x_0 \in \text{Int}\, K$ and $x \in X \setminus K$, we may scale x_0 so that $x_0 + x \in \partial K$. Also, $r(T)x - Tx \neq 0$ implies $x \neq -x_0$. Now the strong monotonicity of T implies

$$r(x_0 + x) > T(x_0 + x) \gg 0,$$

which means $x_0 + x \in \text{Int}\, K$. This is a contradiction. □

B.4 The Generalized Krein–Rutman theorem for Homogeneous Maps

Theorem B.4.1 *Suppose that X is an ordered Banach space with order induced by the cone K. Let* $f : K \to K$ *be a compact, continuous, monotone map which is positively homogeneous of degree one. Assume that there exists a* $u \in K$ *such that* $\{\|f^j(u)\|\}_{j=1}^{\infty}$ *is unbounded, where* f^j *is the composition of f with itself j times. Then there exist* $x \in K$ *and* $t \geq 1$ *such that* $\|x\| = 1$ *and*

$$f(x) = tx. \tag{B.7}$$

Assume, in addition, that $f(y) \neq y$ *for all* $y \in K \setminus \{0\}$*, then* $i(f, U, K) = 0$ *for any neighborhood U of* 0 *which is a relatively open subset of K.*

Proof We will first prove two separate claims in Steps 1 and 2.

Step 1. If (i) there exists a u such that $\{\|f^j(u)\|\}_{j=1}^{\infty}$ is unbounded and (ii) $f(x) \neq x$ for all $x \in K$ such that $\|x\| = 1$, then $i(f, U, K) = 0$ for any $U \subset K$ open relative to K which contains 0.

By the positive homogeneity of f, we have $f(x) \neq x$ for all $x \in K \setminus \{0\}$. By the compactness of f, there exists a $\delta > 0$ such that

$$\inf\{\|x - f(x)\| : x \in \partial_K(U)\} > \delta.$$

Because f is positively homogeneous of degree one, we may replace u by $\delta u/\|u\|$, and assume without loss of generality that $\|u\| \leq \delta$.

Define $g(x) = f(x) + u$ and consider the homotopy $h(\tau, x) = f(x) + \tau u$ for $\tau \in [0, 1]$. It is clear that

$$h(\tau, x) = f(x) + \tau u \neq x \quad \text{for } (\tau, x) \in [0, 1] \times \partial_K(U),$$

so we can conclude that

$$i(f, U, K) = i(g, U, K). \tag{B.8}$$

To complete Step 1, it suffices to prove that $g(x) \neq x$ for all $x \in \overline{U}$, so that $i(g, U, K) = 0$ by Corollary A.2.5.

Since $g(x) \neq x$ on $\partial_K(U) = \overline{U} \setminus U$, we assume to the contrary that $g(x) = x$ for some $x \in U$. Then,

$$x = f(x) + u \geq u.$$

We can iterate and deduce

$$x = g(x) = f(x) + u \geq f(x) \geq f(u).$$

By induction, we deduce that

$$x \geq f^m(u) \quad \text{for all } m \in \mathbb{N}. \tag{B.9}$$

Define $a_m = \|f^m(u)\|$. By the hypothesis of $\limsup_{m\to\infty} a_m = +\infty$, we may choose a subsequence a_{m_i} such that

$$a_{m_i} \geq i \quad \text{and} \quad a_{m_i} = \max_{1\leq j\leq m_i} \{a_j\}.$$

Define

$$V := \left\{ \frac{f^{m_i-1}(u)}{\|f^{m_i}(u)\|} : i \geq 1 \right\}.$$

Then V is a bounded set, since

$$\|f^{m_i-1}(u)\| = a_{m_i-1} \leq a_{m_i} = \|f^{m_i}(u)\|.$$

Since f maps bounded sets into precompact sets[2], the set

$$f(V) = \{v_{m_i} : i \geq 1\} \quad \text{where} \quad v_{m_i} = \frac{f^{m_i}(u)}{\|f^{m_i}(u)\|},$$

is precompact. We can take a further subsequence in $w_j = v_{m_{i_j}}$ in $f(V)$ such that $w_j \to w$ in K and $\|w\| = 1$. Then (B.9) gives

$$\left(a_{m_{i_j}}\right)^{-1} x - w_j \in K,$$

and by taking limits as $j \to \infty$, we derive $-w \in K$, which is a contradiction as $w \in K \setminus \{0\}$.

Step 2. If $f(x) \neq tx$ for all $\|x\| = 1$ and all $t \geq 1$, then $i(f, B, K) = 1$, where $B = \{x \in K : \|x\| < 1\}$.

Consider the homotopy

$$h(\tau, x) = \tau f(x) \quad \text{for } (\tau, x) \in [0, 1] \times \overline{B}.$$

By assumption, $h(\tau, x) \neq x$ for $x \in \partial_K(B)$ and $\tau \in [0, 1]$. Now,

$$i(f, B, K) = i(h(1, \cdot), B, K) = i(h(0, \cdot), B, K),$$

where the second and third equalities are consequences of homotopy invariance and the normalization property of the fixed point index, respectively.

Step 3. We are now in position to conclude the proof. Suppose there is a $u \in K$ such that $\{\|f^j(u)\|\}$ is unbounded, then Steps 1 and 2 combined imply that there exist an $x \in K$ with $\|x\| = 1$ and $t \geq 1$ such that $f(x) = tx$. (Otherwise the fixed point index $i(f, B, K)$ is both zero and one.) The last assertion is already proved in Step 1. □

[2] We say that a subset S of a Banach space is precompact if the closure of S is compact.

Motivated by the condition in Theorem B.4.1, we also define

$$r_K(f) = \sup_{x\in K,\ \|x\|\le 1} \left(\limsup_{n\to\infty} \|f^n(x)\|^{1/n} \right). \tag{B.10}$$

The quantity $r_K(f)$ is referred to as the cone spectral radius of $f : K \to K$ in [10]. The following result is an immediate consequence of Theorem B.4.1.

Lemma B.4.2 *Let X be an ordered Banach space with order induced by the cone $K \subset X$, and let $f : K \to K$ be under the same assumptions as in Theorem B.4.1. If $r_K(f) > 0$, then there exists an $x \in K \setminus \{0\}$ such that $f(x) = r_K(f)x$.*

Proof Let $r = r_K(f)$. For each $\lambda \in (0, r)$, there exists a $u \in K$ such that $\{\|(f/\lambda)^j(u)\|\}_{j=1}^n$ is unbounded. It follows from Theorem B.4.1 that

$$f(x_\lambda) = t_\lambda x_\lambda \quad \text{for some } x_\lambda \in K,\ \|x_\lambda\| = 1 \text{ and } t_\lambda \in [\lambda, r].$$

Since f is compact, we may take a sequence $\lambda \to r$ such that x_λ converges to some $x \in K$ with $\|x\| = 1$. This proves that $f(x) = rx$ for some nonzero $x \in K$. □

Next, we discuss the relationship of $r_K(f)$ with the Bonsall cone spectral radius $\tilde{r}_K(f)$. The following result, due to Mallet-Paret and Nussbaum [9, Theorem 2.3], says that $r_K(f) = \tilde{r}_K(f)$ for a quite general class of maps; see also [14, Theorem 4.3]. We note that if f is a linear map, then the following result is a consequence of Lemma B.4.9.

Lemma B.4.3 ([9, Theorem 2.3]) *Suppose that X is a normed space with order induced by a cone K, and let $f : K \to K$ be continuous, compact and positively homogeneous of degree one. Then $r_K(f) = \tilde{r}_K(f)$, where $r_K(f)$ and $\tilde{r}_K(f)$ are given in* (B.10) *and* (B.3) *respectively.*

Proof Obviously $r_K(f) \le \tilde{r}_K(f)$, so it suffices to show the reverse inequality. Henceforth, we assume $r_K(f)$ is finite, otherwise there is nothing to prove. Assume to the contrary that $r_K(f) < \lambda < \tilde{r}_K(f)$ for some λ. Letting $g(x) = \frac{1}{\lambda} f(x)$, then $r_K(g) < 1$ and hence

$$\lim_{n\to\infty} g^n(x) = 0 \quad \text{for every } x \in K. \tag{B.11}$$

Let $B = \{x \in K : \ \|x\| < 1\}$ and $Q = \overline{g(B)}$. Then Q is compact since $\frac{1}{\lambda} f$ is a compact map.

Next, fix an arbitrary $x \in Q$. Observe from (B.11) that there exists an integer $n(x)$ such that $g^{n(x)-1}(x) \in B$, and hence $g^{n(x)}(x) \in B$. By continuity, there exists a neighborhood $U_x \subset K$ of x, which is open relative to K, such that

$$g^{n(x)-1}(U_x) \subset B \quad \text{and hence } g^{n(x)}(U_x) \subset Q.$$

By compactness of the set Q, there exists a finite set of points $\{x_i\}_{i=1}^k \subset Q$ such that

$$Q \subset \bigcup_{i=1}^{k} U_{x_i}.$$

Let us set $n_0 = \max_{1 \le i \le k} n_i$, where $n_i = n(x_i)$. Roughly speaking, this means that for each $x \in Q$, $g^n(x) \in Q$ for some $n \le n_0$. Precisely, define

$$Q_0 = \bigcup_{i=0}^{n_0-1} g^i(Q).$$

We claim that $g(Q_0) \subset Q_0$, i.e. Q_0 is forward-invariant. Based on the form of Q_0, it is sufficient to prove that $g^{n_0}(Q) \subset Q_0$. To this end, let y be a point of $g^{n_0}(Q)$, i.e. $y = g^{n_0}(x)$ for some $x \in Q$. Then $x \in U_{x_i}$ for some i, and so $g^{n_i}(x) \in Q$. Then

$$y = g^{n_0-n_i}(g^{n_i}(x)) \in g^{n_0-n_i}(Q).$$

Since $0 \le n_0 - n_i < n_0$, it follows that $y \in Q_0$. This establishes the claim.

Finally, notice that for $x \in B$, we have $g(x) \in Q \subset Q_0$, and hence $g^n(x) \in Q_0$ for all $n \ge 1$. Since Q_0 is compact and hence bounded, it follows that $\|g^n\|_K$ is bounded independent of $n \ge 1$. This implies $\tilde{r}_K(g) \le 1$, or that $\tilde{r}_K(f) \le \lambda$. This is a contradiction. □

We are now in position to prove the Krein–Rutman theorems for maps.

Corollary B.4.4 *Suppose that X is an ordered Banach space with order induced by the cone K, and let $f : K \to K$ be a compact, continuous map which is positively homogeneous of degree one and monotone. If $\tilde{r}_K(f) > 0$, then there exists an $\tilde{x} \in K$ with $\|\tilde{x}\| = 1$ such that*

$$f(\tilde{x}) = \tilde{r}\tilde{x} \quad \textit{where} \quad \tilde{r} = \tilde{r}_K(f).$$

Next, we see that if K has nonempty interior and f is strongly monotone, then the eigenvalue t given in Theorem B.4.1 is geometrically simple.

Corollary B.4.5 *Suppose that X is an ordered Banach space with order induced by the cone K. If K is a solid cone and $f : K \to K$ a compact, continuous map which is positively homogeneous of degree one and strongly monotone, then $\tilde{r}_K(f) > 0$ and there exists an $\tilde{x} \in \operatorname{Int} K$ such that*

$$f(\tilde{x}) = \tilde{r}\tilde{x} \quad \textit{where} \quad \tilde{r} = \tilde{r}_K(f).$$

Furthermore, whenever $f(x') = t'x'$ for some $t' > 0$ and $x' \in K \setminus \{0\}$, it must hold that $x' \in \operatorname{span}\{\tilde{x}\}$ and $t' = \tilde{r}_K(f)$.

Remark B.4.6 Under the assumptions of Corollary B.4.5, the principal eigenvalue $\tilde{r}_K(f)$ is, in fact, algebraically simple, i.e. $\operatorname{Ker}(\tilde{r}_K(f)I - f) = \operatorname{Ker}((\tilde{r}_K(f)I - f)^2)$. This is a consequence of Lemma B.4.8 below.

Remark B.4.7 See Problem B.5.7 for a cooperative system where the Krein–Rutman theorems for maps (Corollaries B.4.4 and B.4.5) is needed. See, e.g. [4, 6] for applications of these results to questions arising in ecology.

We also refer to [10] for recent results and open problems in the more general setting involving two cones $K \subset D$ in a Banach space X, where $f : K \to K$ is a positively homogeneous map which preserves the order induced by the cone D.

Proof (of Corollaries B.4.4 and B.4.5) First, we show Corollary B.4.4. By Lemma B.4.3, $r_K(f) = \tilde{r}_K(f) > 0$. Denote by $\tilde{r}$ their common value, then Lemma B.4.2 implies that $f(\tilde{x}) = \tilde{r}\tilde{x}$ for some nonzero $\tilde{x}$.

Next, we assume K has nonempty interior and f is strongly monotone and show Corollary B.4.5. Clearly, $\tilde{x} \in \operatorname{Int} K$ since $f(K \setminus \{0\}) \subset \operatorname{Int} K$.

Suppose now $f(x') = r'x'$ for some $x' \in K \setminus \{0\}$, we claim that $r' = \tilde{r}$ and $x' \in \operatorname{span}\{\tilde{x}\}$. To this end, suppose to the contrary that $x' \notin \operatorname{span}\{\tilde{x}\}$. Since $\tilde{r} = \tilde{r}_K(f)$, we must have $r' \leq \tilde{r}$. By the strong monotonicity property, we again have $x' \in \operatorname{Int} K$, so that $x' - \tau\tilde{x} \in \operatorname{Int} K$ for small positive τ. We observe that

$$\tilde{\tau} := \sup\{\tau \in [0, \infty) : x' - \tau\tilde{x} \in \operatorname{Int} K\}$$

is finite, since otherwise we deduce $-\tilde{x} \in K$, which is impossible. Since $x' \notin \operatorname{span}\{\tilde{x}\}$, we have $x' \neq \tilde{\tau}x$. Applying f to the inequality $\tilde{\tau}\tilde{x} < x'$, the strong monotonicity property implies

$$\tilde{\tau}\tilde{x} = \frac{1}{\tilde{r}} f(\tilde{\tau}\tilde{x}) \ll \frac{1}{\tilde{r}} f(x') = \frac{r'}{\tilde{r}} x' \leq x'.$$

This is in contradiction with the maximality of $\tilde{\tau}$. Hence, $r' = \tilde{r}$ and $x' \in \operatorname{span}\{\tilde{x}\}$. This completes the proof. □

Lemma B.4.8 *Under the assumption of Corollary B.4.5, for* $0 \leq \lambda \leq \tilde{r}_K(f)$ *and* $h \in K \setminus \{0\}$*, the inhomogeneous equation*

$$\lambda x - f(x) = h \tag{B.12}$$

has no solution in K.

Proof Let $\tilde{r} = \tilde{r}_K(f) > 0$ and $\tilde{x} \in \operatorname{Int} K$ be given by Corollary B.4.5. Suppose to the contrary that $\lambda x - f(x) > 0$ for some x. First observe that $x \neq 0$ and $\lambda x > f(x) \gg 0$. This means λ is a positive real number and $x \gg 0$. In particular, $tx - \tilde{x} \gg 0$ for all positive and large real number t. Let

$$t^* = \inf\{t > 0 : tx - \tilde{x} \in K\},$$

which is well-defined and $t^* \geq 0$. We note that $t^* > 0$, since $\tilde{x} \gg 0$. Hence, $t^*x - \tilde{x} \in \partial K$ for some $t^* > 0$. Now, $t^*x - \tilde{x} \neq 0$, otherwise

$$0 = \tilde{r}\tilde{x} - f(\tilde{x}) = \tilde{r}t^*x - f(t^*x) \geq t^*(\lambda x - f(x)) > 0,$$

which is absurd. Hence, $t^*x - \tilde{x} \in \partial K \setminus \{0\}$, but then

$$0 \ll f(t^* x) - f(\tilde{x}) \le \tilde{r}(t^* x - \tilde{x}),$$

which implies that $t^* x - \tilde{x} \notin \partial K$. This is a contradiction. □

Next, we turn our attention to compact linear operators $T : X \to X$. The following result should be compared with Lemma B.4.3.

Lemma B.4.9 *Suppose that* X *is a real ordered Banach space with order induced by the cone* K. *If* K *is a total cone and* $T : X \to X$ *is a compact linear map such that* $r(T) > 0$ *and* $T(K) \subset K$, *then*

$$\tilde{r}_K(T) = r_K(T) = r(T).$$

Proof Clearly, $r_K(T) \le \tilde{r}_K(T) \le r(T)$. Thus we only need to prove $r_K(T) \ge r(T)$.

Suppose, for the moment, that

$$\limsup_{n\to\infty} \frac{\|T^n x\|}{\|T^n\|} = \delta > 0 \quad \text{for some } x \in X. \tag{B.13}$$

Because K is total, there exists a y satisfying $\|x - y\| < \frac{\delta}{2}$ and such that $y = v - w$ for some $v, w \in K$. Then

$$\limsup_{n\to\infty} \frac{\|T^n y\|}{\|T^n\|} \ge \frac{\delta}{2} > 0.$$

Since $y = v - w$, by taking $u = v$ or w, we have

$$\limsup_{n\to\infty} \frac{\|T^n u\|}{\|T^n\|} > 0 \quad \text{for some } u \in K.$$

Hence,

$$r_K(T) \ge \limsup_{n\to\infty} \|T^n u\|^{1/n} \ge \lim_{n\to\infty} \|T^n\|^{1/n} = r(T).$$

Since we always have $r_K(T) \le r(T)$, we have $r_K(T) = r(T)$ and the lemma is proved.

It remains to prove (B.13). Assume to the contrary that

$$\limsup_{n\to\infty} \frac{\|T^n x\|}{\|T^n\|} = 0 \quad \text{for all } x \in X.$$

By the compactness of $T(B)$ where B is the unit ball in X, there exists a finite collection of balls $\{B_j\}_{j=1}^m$ with center x_j and radius $r(T)/4$ such that

$$T(B) \subset \bigcup_{j=1}^m B_j.$$

Choose $N_1 \in \mathbb{N}$ such that

$$\frac{\|T^n x_j\|}{\|T^n\|} < \frac{r(T)}{4} \quad \text{for } n \geq N_1,\ 1 \leq j \leq m.$$

Then for any $x \in B$, we have $Tx \in B_j$ for some j so that

$$\begin{aligned}\|T^{n+1}x\| &\leq \|T^n(Tx - x_j)\| + \|T^n x_j\| \\ &\leq \|T^n\| \left(\frac{r(T)}{4} + \frac{r(T)}{4}\right) \quad \text{for } n \geq N_1.\end{aligned}$$

Taking supremum in $x \in B$, we deduce that $\|T^{n+1}\| \leq \frac{r(T)}{2}\|T^n\|$ for $n \geq N_1$, and hence

$$\|T^n\| \leq \left(\frac{r(T)}{2}\right)^{n-N_1} \|T^{N_1}\| \quad \text{for } n \geq N_1.$$

Taking the n-th root on both sides, and letting $n \to \infty$, we obtain $r(T) \leq \frac{r(T)}{2}$, which is a contradiction to $r(T) > 0$. □

Theorem B.3.1 follows directly from Corollary B.4.4 and Lemma B.4.9.

Proof (of Theorem B.3.1) Since K is a total cone, and $r(T) > 0$, Lemma B.4.9 asserts that $\tilde{r}_K(T) = r(T) > 0$. It then follows from Corollary B.4.4 that there exists an $\tilde{x} \in K$ with $\|\tilde{x}\| = 1$ such that $T\tilde{x} = r(T)\tilde{x}$.

It remains to show the existence of an $f_0 \in K^* \setminus \{0\}$ such that $T^* f_0 = r(T) f_0$. First, it is classical that $r = r(T^*) = r(T) > 0$. Next, we claim that

$$\tilde{r}_{K^*}(T^*) = r > 0. \tag{B.14}$$

Indeed, let $\tilde{x}$ be the eigenvector of T as above, then K and $\{-\tilde{x}\}$ are two disjoint convex subsets for which one is closed and one is compact. By the Hahn–Banach theorem [1, Theorem 1.7], the sets K and $\{-\tilde{x}\}$ are strictly separated, i.e. there exists a $g \in K^*$ such that $\langle -\tilde{x}, g\rangle < 0$. Furthermore, one has

$$\langle \tilde{x}, (T^*)^n g\rangle = \langle T^n \tilde{x}, g\rangle = r^n \langle \tilde{x}, g\rangle,$$

which implies $\tilde{r}_{K^*}(T^*) \geq r$. Since the reverse inequality $r = r(T^*) \geq \tilde{r}_K(T^*)$ follows by definition, we have proved (B.14).

Having proved (B.14), the existence of $f_0 \in K^* \setminus \{0\}$ such that $T^* f_0 = r(T) f_0$ follows from Corollary B.4.4. □

Proof (of Theorem B.3.2) To show assertion 1, observe that the restriction of T to K is continuous, compact, positively homogeneous of degree one, and strongly monotone. By Corollary B.4.5, we have $\tilde{r}_K(T) > 0$ and there exists an $\tilde{x} \in \operatorname{Int} K$ such that $T\tilde{x} = \tilde{r}_K(T)\tilde{x}$.

Next, observe that $r(T) \geq \tilde{r}_K(T) > 0$, so that Lemma B.4.9 implies $\tilde{r}_K(T) = r_K(T) = r(T) > 0$. In particular,

$$T\tilde{x} = r(T)\tilde{x} \quad \text{for some } \tilde{x} \in \operatorname{Int} K.$$

By repeating the proof of Corollary B.4.5, we prove that if x' is an eigenvector corresponding to the eigenvalue $r(T)$, then $x' \in \operatorname{span}\{\tilde{x}\}$. (Essentially, we show that $\tilde{x} - \tau x' \in \partial K$ if and only if $x = \tau x'$.) This proves assertion 1.

Next, Corollary B.4.5 implies that $r(T)$ is the only eigenvalue that has a positive eigenvector. This proves assertion 3.

To show assertion 2, we first assume without loss of generality that $r(T) = 1$. We need to show that there exists an $\epsilon > 0$ such that $|\lambda| < 1 - \epsilon$ for any eigenvalue $\lambda \in \mathbb{C}$ that is distinct from 1.

It is enough to show that $|\lambda| < 1$ for each eigenvalue $\lambda \in \mathbb{C}$ of $\tilde{T}$, where $\tilde{T}$ is the complexification of T. This is because $\tilde{T}$, being a compact linear operator, has finitely many eigenvalues in $\{z \in \mathbb{C} : 1/2 < |z| < 1\}$.

Suppose $\tilde{T}w = \lambda w$ for some $w \in \tilde{X} \setminus \{0\}$ and $\lambda \in \mathbb{C}$ such that $|\lambda| = 1$. We will show that the only possibility is $\lambda = 1$; i.e. we will rule out $\lambda = -1$ and $\lambda = \sigma + i\tau$ for some $\tau \neq 0$.

Suppose $\lambda = -1$. Observe that T^2 is compact, linear and $T^2(K \setminus \{0\}) \subset \operatorname{Int} K$, and that $T^2\tilde{x} = \tilde{x}$ for some $\tilde{x} \in \operatorname{Int} K$. By assertions 1 and 3, we deduce that $r(T^2) = 1$ is a simple eigenvalue with eigenfunction $\tilde{x}$. However, $T^2 w = T(-w) = w$, and we deduce from assertion 1 that $w \in \operatorname{span}\{\tilde{x}\}$, but then this is in contradiction with $Tw = -w$. Hence, $\lambda \neq -1$.

Consider now the case $\lambda = \sigma + i\tau$ with $\tau \neq 0$ and $\sigma^2 + \eta^2 = 1$. Then $w = u + iv$ and

$$Tu = \sigma u - \tau v, \quad \text{and} \quad Tv = \tau u + \sigma v. \tag{B.15}$$

We observe that u and v must be linearly independent since $\tau \neq 0$. Let $X_1 = \operatorname{span}\{u, v\}$. Then (B.15) implies that $T(X_1) \subset X_1$. We claim that $K_1 := K \cap X_1 = \{0\}$.

Suppose to the contrary that $x_0 \in X_1$ for some $x_0 \in K \setminus \{0\}$. Then $x_1 = Tx_0 \in X_1 \cap \operatorname{Int} K$. Hence, $X_1 \cap \operatorname{Int} K$ is nonempty. This implies that K_1 is a solid cone in X_1. By assertion 1 applied to $T_1 := T\big|_{X_1}$, we deduce that T has an eigenvector in $K_1 \setminus \{0\}$. By assertion 3, that eigenvector can only be $\tilde{x}$, i.e. $\tilde{x} \in X_1$. Hence, $\tilde{x} = \alpha u + \beta v$. By applying T and using (B.15), we can show that $\alpha = \beta = 0$, and hence $\tilde{x} = 0$, this is a contradiction. Therefore, $K_1 = \{0\}$.

From $\operatorname{span}\{u, v\} \cap K = \{0\}$ we deduce that the set

$$\Sigma := \{(\xi, \eta) \in \mathbb{R}^2 : \tilde{x} + \xi u + \eta v \in K\}$$

is bounded and closed. Roughly speaking, $T\tilde{x} = \tilde{x}$, T is a rotation in Σ. Since T is strongly monotone, it maps Σ to its interior, which implies that $|\lambda| < 1$.

Precisely, since $\tilde{x} \in \operatorname{Int} K$, the supremum $M := \sup\{\xi^2 + \eta^2 : (\xi, \eta) \in \Sigma\}$ is positive, and is achieved at some $(\xi_0, \eta_0) \in \Sigma$. Let $z_0 = \tilde{x} + \xi_0 u + \eta_0 v$. Then $z_0 \in K \setminus \{0\}$ and $Tz_0 \in \operatorname{Int} K$. Therefore we can find $\alpha \in (0, 1)$ such that $Tz_0 \geq \alpha\tilde{x}$, i.e.

$$(1-\alpha)\tilde{x} + (\xi_1 u + \eta_1 v) \geq 0, \tag{B.16}$$

where

$$\begin{pmatrix} \xi_1 \\ \eta_1 \end{pmatrix} = \begin{pmatrix} \sigma & \tau \\ -\tau & \sigma \end{pmatrix} \begin{pmatrix} \xi_0 \\ \eta_0 \end{pmatrix} = \begin{pmatrix} \sigma\xi_0 + \tau\eta_0 \\ \sigma\eta_0 - \tau\xi_0 \end{pmatrix}.$$

Clearly

$$\xi_1^2 + \eta_1^2 = (\sigma^2 + \tau^2)(\xi_0^2 + \eta_0^2) = M|\lambda|^2.$$

By (B.16), we find that $(\xi_1, \eta_1)/(1-\alpha) \in \Sigma$ and hence

$$\xi_1^2 + \eta_1^2 \leq M(1-\alpha)^2,$$

that is,

$$|\lambda|^2 \leq (1-\alpha)^2,$$

and hence $|\lambda| < 1$. We have now proved that if λ is an eigenvalue of T with modulus 1, then necessarily $\lambda = 1$. Since T is compact, there are only finitely many eigenvalues with modulus greater than $1/2 > 0$. Hence, there exists an $\epsilon > 0$ such that $|\lambda| < r(T) - \epsilon$ for any eigenvalue $\lambda \in \mathbb{C} \setminus \{r(T)\}$. The proof of assertion 2 is now complete. □

B.5 Further Reading

In Lemma B.4.2, we followed the strategy of [13] to deduce the existence of a positive eigenfunction corresponding to $r_K(f)$ from Theorem B.4.1. The latter is a special case of [12, Theorem 2.1]. Lemma B.4.3 is due to Mallet-Paret and Nussbaum [9, Theorem 2.3]; see also [14, Theorem 4.3] for a generalization.

The original article by Krein and Rutman [8] contained theorems concerning eigenvalues of nonlinear, cone-preserving operators. The weak Krein–Rutman theorem for maps in this article (Corollary B.4.4) is a special case of the results in [9, 10], where it is generalized to noncompact maps. Also, in the case of compact maps, the completeness of X or K is not necessary [14, Theorem 7.8]. These generalizations can be useful for other biological situations where the solutions enjoy fewer regularity properties than reaction-diffusion models. For the main text of this monograph, the classical Krein–Rutman theorem for linear operators is sufficient. However, homogeneous maps appears naturally in biological applications in the context of sexually reproducing populations [5, 6, 7], or populations with internal nutrient storage [4, 11]. See also Problem B.5.7, which is motivated by sexually reproducing populations.

Problems

B.5.1 Let X be an ordered Banach space with the order $\leq_K$ induced by a cone $K \subset X$. Let $\{x_n\}$ and $\{y_n\}$ be sequences such that

$$0 \leq_K x_n \leq_K y_n \quad \text{for all } n \in \mathbb{N}.$$

Show that $x_n \to 0$ if $y_n \to 0$. [Hint: If $X = \mathbb{R}$ and $K = [0, \infty)$, then this is the usual squeeze theorem. Use the formal definition of a cone to prove the general case.]

B.5.2 Let X be an ordered Banach space with the order $\leq_K$ induced by a cone $K \subset X$. Let x_n be an increasing sequence, i.e. $x_1 \leq_K x_2 \leq_K x_3 \leq_K \cdots$. Show that $\{x_n\}$ is convergent if and only if $\{x_n\}$ is precompact.

In Problems B.5.3 to B.5.6, we will generalize Theorem B.4.1, Lemma B.4.3, and Corollaries B.4.4 and B.4.5 to the following setting: Let $C \subset K$ be two cones in the Banach space X. Suppose $f : C \to C$ is compact, positively homogeneous of degree one, and is monotone with respect to the order induced by the cone K.

B.5.3 (Generalize Theorem B.4.1) Show that if there exists a $u \in C$ such that $\{\|f^j(u)\|\}_{j=1}^{\infty}$ is unbounded, then there exist $x \in C$ and $t \geq 1$ such that $\|x\| = 1$ and $f(x) = tx$.

B.5.4 (Generalize Lemma B.4.3) Show that if $\tilde{r}_C(f) > 0$, then $r_C(f) = \tilde{r}_C(f)$.

B.5.5 (Generalize Corollary B.4.4) Show that if $\tilde{r}_C(f) > 0$, then there exists an $\tilde{x} \in C$ with $\|\tilde{x}\| = 1$ such that $f(\tilde{x}) = \tilde{r}_C(f)\tilde{x}$.

B.5.6 (Generalize Corollary B.4.5) Show that if C and K both have nonempty interior in X, and $f : C \to C$ is strongly monotone with respect to the order induced by the cone K, i.e.

$$f(x) \ll_K f(\overline{x}) \quad \text{if} \quad x, \overline{x} \in C \text{ and } x <_K \bar{x},$$

then $\tilde{r}_C(f) > 0$ and there exists an $\tilde{x} \in \operatorname{Int} K$ such that

$$f(\tilde{x}) = \tilde{r}\tilde{x} \quad \text{where} \quad \tilde{r} = \tilde{r}_C(f),$$

and that whenever $f(x') = t'x'$ for some $t' > 0$ and $x' \in C \setminus \{0\}$, it must hold that $x' \in \operatorname{span}\{\tilde{x}\}$ and $t' = \tilde{r}_C(f)$.

B.5.7 Let Ω be a bounded domain in $\mathbb{R}^N$ with smooth boundary $\partial\Omega$. Consider the following system of parabolic equations:

$$\begin{cases} \partial_t u = d_1 \Delta U + a_1 \frac{uv}{u+v} + b_1 u + c_1 v & \text{in } \Omega \times (0, \infty), \\ \partial_t v = d_2 \Delta v + a_2 \frac{uv}{u+v} + c_2 u + b_2 v & \text{in } \Omega \times (0, \infty), \\ \mathbf{n} \cdot \nabla u = \mathbf{n} \cdot \nabla v = 0 & \text{on } \partial\Omega \times (0, \infty), \\ u(x, 0) = u_0(x), \quad v(x, 0) = v_0(x) & \text{in } \Omega, \end{cases} \tag{B.17}$$

where $d_i, a_i, b_i, c_i \in C^{\alpha,\alpha/2}(\overline{\Omega} \times [0,\infty))$ are T-periodic in T. Moreover, $d_i, c_i > 0$ and $a_i \geq 0$ in $\overline{\Omega} \times [0,\infty)$. Define

$$X = C(\overline{\Omega}; \mathbb{R}^2), \quad K = C(\overline{\Omega}; \mathbb{R}^2_+),$$

and consider the map $f : K \to K$ given by

$$(u_0, u_0) \mapsto (u(\cdot, T), u(\cdot, T)).$$

Show that f satisfies the hypotheses of Corollary B.4.5. Hence, the Bonsall cone spectral radius $\tilde{r} = \tilde{r}_K(f)$ is positive, and there exists $(\tilde{u}, \tilde{v}) \in \operatorname{Int} K$ such that $f(\tilde{u}, \tilde{v}) = \tilde{r}(\tilde{u}, \tilde{v})$. In other words, let $\tilde{\lambda} = \frac{1}{T} \log \tilde{r}$, then

$$\begin{cases} \partial_t u = d_1 \Delta U + a_1 \frac{uv}{u+v} + b_1 u + c_1 v + \tilde{\lambda} u & \text{in } \Omega \times (0,\infty), \\ \partial_t v = d_2 \Delta v + a_2 \frac{uv}{u+v} + c_2 u + b_2 v + \tilde{\lambda} v & \text{in } \Omega \times (0,\infty), \\ \mathbf{n} \cdot \nabla u = \mathbf{n} \cdot \nabla v = 0 & \text{on } \partial\Omega \times (0,\infty) \end{cases}$$

has a positive T-periodic solution.

References

1. H. Brezis, *Functional analysis, Sobolev spaces and partial differential equations*, Universitext, Springer, New York, 2011.
2. J. Dugundji, *An extension of Tietze's theorem*, Pacific J. Math., 1 (1951), pp. 353–367.
3. J. Dugundji, *Topology*, Allyn and Bacon, Inc., Boston, Mass., 1966.
4. S.-B. Hsu, K.-Y. Lam, and F.-B. Wang, *Single species growth consuming inorganic carbon with internal storage in a poorly mixed habitat*, J. Math. Biol., 75 (2017), pp. 1775–1825.
5. W. Jin, H. L. Smith, and H. R. Thieme, *Persistence and critical domain size for diffusing populations with two sexes and short reproductive season*, J. Dynam. Differential Equations, 28 (2016), pp. 689–705.
6. W. Jin and H. Thieme, *An extinction/persistence threshold for sexually reproducing populations: the cone spectral radius*, Discrete Contin. Dyn. Syst. Ser. B, 21 (2016), pp. 447–470.
7. W. Jin and H. R. Thieme, *Persistence and extinction of diffusing populations with two sexes and short reproductive season*, Discrete Contin. Dyn. Syst. Ser. B, 19 (2014), pp. 3209–3218.
8. M. G. Kreĭn and M. A. Rutman, *Linear operators leaving invariant a cone in a Banach space*, Amer. Math. Soc. Translation, 1950 (1950), p. 128.
9. J. Mallet-Paret and R. D. Nussbaum, *Eigenvalues for a class of homogeneous cone maps arising from max-plus operators*, Discrete Contin. Dyn. Syst., 8 (2002), pp. 519–562.
10. ———, *Generalizing the Krein–Rutman theorem, measures of noncompactness and the fixed point index*, J. Fixed Point Theory Appl., 7 (2010), pp. 103–143.
11. H. Nie, S.-B. Hsu, and F.-B. Wang, *Steady-state solutions of a reaction-diffusion system arising from intraguild predation and internal storage*, J. Differential Equations, 266 (2019), pp. 8459–8491.
12. R. D. Nussbaum, *Eigenvectors of nonlinear positive operators and the linear Kreĭn–Rutman theorem*, in Fixed point theory (Sherbrooke, Que., 1980), vol. 886 of Lecture Notes in Math., Springer, Berlin-New York, 1981, pp. 309–330.

13. ———, *The fixed point index and some applications*, vol. 94 of Séminaire de Mathématiques Supérieures [Seminar on Higher Mathematics], Presses de l'Université de Montréal, Montreal, QC, 1985.
14. H. R. Thieme, *Discrete-time population dynamics on the state space of measures*, Math. Biosci. Eng., 17 (2020), pp. 1168–1217.

Appendix C
Subhomogeneous Dynamics

Abstract We discuss a threshold type result concerning strictly subhomogeneous maps, and its continuous-time counterpart. These results are often applicable to models of a single species, or a group of cooperating species. Examples include the diffusive logistic equation discussed in Chapter 5, and the single phytoplankton model discussed in Chapter 8.

Let X be an ordered Banach space with the order induced by a solid cone K. If $x, \overline{x} \in X$, then we write

$$x \leq_K \overline{x} \quad (\text{resp.} \quad x <_K \overline{x}, \quad x \ll_K \overline{x})$$

provided that

$$\overline{x} - x \in K \quad (\text{resp.} \quad \overline{x} - x \in K \setminus \{0\}, \quad \overline{x} - x \in \operatorname{Int} K).$$

C.1 Subhomogeneous Maps

In this section, we discuss a threshold type result concerning strictly subhomogeneous maps. As before, let X be an ordered Banach space with order generated by a solid cone K. Next, let $C \subset K$ be a closed cone (possibly equal to K) and $S : C \to C$ be a continuous map. We first state our assumptions on the map S.

(S1) $S(0) = 0$ and S is strongly monotone with respect to K, i.e.

$$S(x_1) \ll_K S(x_2) \quad \text{if} \quad x_1, x_2 \in C, \ \text{and} \ x_1 <_K x_2;$$

(S2) $\{S^n(x)\}_{n=1}^{\infty}$ is precompact for each $x \in K$;
(S3) S is strictly subhomogeneous, i.e.

$$\lambda S(x) <_K S(\lambda x) \quad \text{for all } x \in C \cap \operatorname{Int} K, \ 0 < \lambda < 1;$$

K. -Y. Lam, Y. Lou, *Introduction to Reaction-Diffusion Equations*, Lecture Notes on Mathematical Modelling in the Life Sciences, https://doi.org/10.1007/978-3-031-20422-7

(S4) There exists a compact map $f : C \to C$ such that

$$\lim_{\lambda \to 0+} \frac{1}{\lambda} S(\lambda x) = f(x) \quad \text{pointwise for each } x \in C;$$

We may also replace (S4) by the following.

(S5) $S : C \to C$ is compact and is Fréchet differentiable at zero.

Remark C.1.1 (S5) implies (S4). Indeed, if S is compact and differentiable at zero, it follows that $DS(0) : X \to X$ is compact.

In the problems section, we give some examples of population models that satisfy (S1)–(S3) and one of (S4) and (S5).

Lemma C.1.2 *Suppose $S : C \to C$ is a continuous mapping satisfying* (S1)–(S4), *and let $f : C \to C$ be given by* (S4). *Then*

1. *f is positively homogeneous of degree one, i.e. $f(\lambda x) = \lambda f(x)$ for $x \in C$, $\lambda \geq 0$;*
2. *f is monotone with respect to K;*
3. *$f(x) >_K S(x)$ for all $x \in C \setminus \{0\}$;*
4. *$f(x) \gg_K 0$ for all $x \in C \setminus \{0\}$;*
5. *$\tilde{r}_C(f) > 0$, and there exists an $\tilde{x} \in C \cap \operatorname{Int} K$ such that $f(\tilde{x}) = \tilde{r}_C(f)\tilde{x}$.*

Here $\tilde{r}_C(f)$ is the Bonsall cone spectral radius given in (B.3).

Proof Assertion 1 follows directly from the definition of f in (S4). For assertion 2, let $x_1 <_K x_2$. It follows from the monotonicity of S that

$$f(x_1) = \lim_{\lambda \to 0} \frac{1}{\lambda} S(\lambda x_1) \leq_K \lim_{\lambda \to 0} \frac{1}{\lambda} S(\lambda x_1) = f(x_2).$$

Next, we observe that by (S3),

$$f(x) \geq_K \frac{1}{\lambda} S(\lambda x) >_K S(x) \quad \text{for all } 0 < \lambda < 1.$$

In particular,

$$f(x) >_K S(x) \gg_K 0 \quad \text{for all } x \in C \setminus \{0\}. \tag{C.1}$$

Hence, there exist $x_1 \in C \setminus \{0\}$ and $c_1 > 0$ such that $f(x_1) \geq_K c_1 x_1$. We claim that $\tilde{r}_C(f) > 0$. Otherwise,

$$\tilde{r}_C(f/c_1) = \frac{1}{c_1} \tilde{r}_C(f) = 0,$$

and

$$x_1 \leq_K (f/c_1)^n (x_1) \to 0.$$

This implies $x_1 \in C \cap (-K)$, which means $x_1 = 0$, which is a contradiction. Hence, $\tilde{r}_C(f) > 0$. This proves assertions 3 and 4.

Finally, we apply Corollary B.4.4 (see Problem B.5.4) to obtain $\tilde{x} \in C \setminus \{0\}$ such that $f(\tilde{x}) = \tilde{r}_C(f)\tilde{x}$. By (C.1), $\tilde{x} \gg_K 0$. This proves assertion 5. □

Theorem C.1.3 *Suppose $S : C \to C$ is a continuous mapping satisfying* (S1)–(S4). *Then the following dichotomy holds:*

1. *If $\tilde{r}_C(f) \le 1$, then $S^n(x) \to 0$ for each $x \in C$.*
2. *If $\tilde{r}_C(f) > 1$, then S has a unique fixed point $x^* \in C \cap \operatorname{Int} K$, and $S^n(x) \to x^*$ for each $x \in C \setminus \{0\}$.*

Here $\tilde{r}_C(f) > 0$ is the Bonsall spectral radius of the function $f : C \to C$ given by (S4).

If we replace hypothesis (S4) by (S5), we obtain a result contained in [14, Chap. 2].

Corollary C.1.4 *Suppose $S : C \to C$ is a continuous mapping satisfying* (S1)–(S3) *and* (S5). *Then the following dichotomy holds:*

1. *If $r(DS(0)) \le 1$, then $S^n(x) \to 0$ for each $x \in C$.*
2. *If $r(DS(0)) > 1$, then S has a unique fixed point $x^* \in C \cap \operatorname{Int} K$, and $S^n(x) \to x^*$ for each $x \in C \setminus \{0\}$.*

Here $DS(0) : X \to X$ is the Fréchet derivative of S at zero, and $r(DS(0))$ is its spectral radius.

Before giving the proof of Theorem C.1.3, we give a corollary that is easier to use.

Corollary C.1.5 *Suppose $S : C \to C$ is a continuous mapping satisfying* (S1)–(S3). *If, in addition, one of* (S4), (S5) *holds, then exactly one of the following alternatives hold:*

1. *S has no fixed points in $C \cap \operatorname{Int} K$. In this case $S^n(x) \to 0$ for each $x \in C \setminus \{0\}$.*
2. *There exists an $x_0 \in C \setminus \{0\}$ such that $S(x_0) \ge_K x_0$. In this case, S has a unique fixed point $x^* \in C \cap \operatorname{Int} K$, and $S^n(x) \to x^*$ for each $x \in C \setminus \{0\}$.*

Proof (of Corollary C.1.5) First, suppose S has no fixed points in $C \cap \operatorname{Int} K$, then alternative 1 of Theorem C.1.3 must hold, i.e., $\tilde{r}_C(f) \le 1$ and 0 attracts all solutions.

Next, suppose $S(x_0) \ge_K x_0$ for some $x_0 \in C \setminus \{0\}$, then $S^n(x_0) \ge_K S^1(x_0) \gg_K 0$, so $S^n(x_0) \not\to 0$. This implies alternative 2 of Theorem C.1.3 holds. □

Now, we give the proof of Theorem C.1.3.

Proof (of Theorem C.1.3) Suppose $\tilde{r}_C(f) \le 1$, then by items 1, 3 and 5 of Lemma C.1.2, there exists an $\tilde{x} \in C \cap \operatorname{Int} K$ such that

$$S(t\tilde{x}) <_K f(t\tilde{x}) \le_K t\tilde{x} \quad \text{for all } t > 0. \tag{C.2}$$

We claim that S has no fixed points in $C \setminus \{0\}$. Suppose to the contrary that $S(x_0) = x_0$ for some $x_0 \in C \setminus \{0\}$. Then (C.2) implies $x_0 \notin \operatorname{span}\{\tilde{x}\}$. Since $\tilde{x} \in \operatorname{Int} K$ and $-x_0 \notin K$, we may choose a minimal $t_0 > 0$ such that $t_0\tilde{x} - x_0 \in \partial K \setminus \{0\}$, so that

$$x_0 = S(x_0) \ll_K S(t_0\tilde{x}) < t_0\tilde{x}.$$

This is a contradiction. Hence S has no non-zero fixed points in C.

Next, observe from (C.2) that $S^n(t\tilde{x})$ is strictly decreasing in n. In the absence of nonzero fixed points, we must have $S^n(t\tilde{x}) \to 0$, for each $t > 0$. Hence, given $x \in C$, choose t such that $0 \leq_K x \leq_K t\tilde{x}$. By the monotonicity of S, we have $0 \leq_K S^n(x) \leq_K S^n(t\tilde{x})$. We can then let $n \to \infty$ to deduce that $S^n(x) \to 0$. This proves assertion 1.

Suppose now that $\tilde{r}_C(f) > 1$. By Lemma C.1.2, there exists an $\tilde{x} \in C \cap \operatorname{Int} K$ such that

$$f(\tilde{x}) = \tilde{r}_C(f)\tilde{x} \gg_K \tilde{x}.$$

By definition of f, there exists $0 < \tilde{\lambda} < 1$ such that

$$S(\tilde{\lambda}\tilde{x}) \geq_K \tilde{\lambda}\tilde{x}. \tag{C.3}$$

In particular, $S^n(\tilde{\lambda}\tilde{x})$ is increasing in n. By (S3), $S^n(\tilde{\lambda}\tilde{x}) \to x^*$ for some $x^* \in C \cap \operatorname{Int} K$. By arguing as above, x^* is the unique fixed point in $C \cap \operatorname{Int} K$.

Next, observe that (S3) implies that

$$S(\lambda x^*) >_K \lambda x^* \quad \text{for } 0 < \lambda < 1, \quad S(\lambda x^*) <_K \lambda x^* \quad \text{for } \lambda > 1.$$

In particular, $S^n(\lambda x^*)$ is increasing (resp. decreasing) in n if $0 < \lambda < 1$ (resp. $\lambda > 1$). Hence, by (S2) they must converge to a fixed point of S as $n \to \infty$ (see Problem C.3.1). Since x^* is the unique positive fixed point, we have

$$S^n(\lambda x^*) \to x^* \quad \text{for any } \lambda > 0.$$

Given $x \in C \cap \operatorname{Int} K$, we claim that $S^n(x) \to x^*$. Indeed, choose $\underline{c} < 1 < \overline{c}$ such that $\underline{c}x^* \leq_K x \leq_K \overline{c}x^*$, then $S^n(\underline{c}x^*) \leq_K S^n(x) \leq_K S^n(\overline{c}x^*)$ for all $n \geq 1$. Letting $n \to \infty$, we have $S^n(x) \to x^*$.

In general, given $x \in C \setminus \{0\}$, we have $S(x) \in C \cap \operatorname{Int} K$, so it again follows that $S^n(x) \to x^*$. This fixed point x^* must then be unique in $C \setminus \{0\}$. This proves assertion 2. □

C.2 Subhomogeneous Semiflows

We now formulate a continuous-time version of the results in the previous section. Let C be a closed cone satisfying $C \subset K$, and let $\Phi : [0, \infty) \times C \to C$ be a continuous semiflow. We write $\Phi_t(x) = \Phi(t, x)$. The semiflow properties are (i) $\Phi_0(x) = x$ for all $x \in C$, and (ii) $\Phi_t \circ \Phi_s = \Phi_{t+s}$ for all $t, s \geq 0$.

In this section, we discuss a threshold type result concerning strictly subhomogeneous semiflows. As before, let X be an ordered Banach space with order generated by a solid cone K. We first state the assumptions on the semiflow $\Phi_t : C \to C$.

(S1') For each $t > 0$, $\Phi_t(0) = 0$ and Φ_t is strongly monotone with respect to K, i.e.

$$\Phi_t(x_1) \ll_K \Phi_t(x_2) \quad \text{if} \quad x_1, x_2 \in C \text{ and } x_1 <_K x_2;$$

(S2') For each $x \in C$, $\{\Phi_t(x) : t > 0\}$ is precompact;
(S3') Φ_t is strictly subhomogeneous, i.e.

$$\lambda\Phi_t(x) <_K \Phi_t(\lambda x) \quad \text{for all } x \in C \cap \operatorname{Int} K,\ 0 < \lambda < 1,\ t > 0;$$

(S4') For each $t > 0$, there exists a compact map $f_t : C \to C$ such that

$$\lim_{\lambda \to 0+} \frac{1}{\lambda}\Phi_t(\lambda x) = f_t(x) \quad \text{for each } x \in C;$$

(S5') For each $t > 0$, $\Phi_t : C \to C$ is compact and Fréchet differentiable at zero for each $t > 0$.

Note that (S5') implies (S4').

Theorem C.2.1 *Suppose $\Phi_t : C \to C$ is a continuous semiflow satisfying* (S1')–(S4'). *Then the following dichotomy holds:*

1. *If $\tilde{r}_C(f_1) \le 1$, then $\Phi_t(x) \to 0$ for each $x \in C$.*
2. *If $\tilde{r}_C(f_1) > 1$, then Φ_t has a unique equilibrium $x^* \in C \cap \operatorname{Int} K$, and $\Phi_t(x) \to x^*$ for each $x \in C \setminus \{0\}$.*

Here $\tilde{r}_C(f_1) > 0$ is the Bonsall spectral radius of the function $f_1 : C \to C$ given by (S4').

Corollary C.2.2 *Suppose $\Phi_t : C \to C$ is a continuous semiflow satisfying* (S1')–(S3') *and* (S5'). *Then the following dichotomy holds:*

1. *If $r(D\Phi_1(0)) \le 1$, then $\Phi_t(x) \to 0$ for each $x \in C$;*
2. *If $r(D\Phi_1(0)) > 1$, then Φ_t has a unique equilibrium $x^* \in C \cap \operatorname{Int} K$, and $\Phi_t(x) \to x^*$ for each $x \in C \setminus \{0\}$.*

Here $D\Phi_1(0) : X \to X$ is the Fréchet derivative of Φ_1 at zero, and $r(D\Phi_1(0))$ is its spectral radius.

Proof (of Theorem C.2.1) Let $S = \Phi_1$. If $\tilde{r}_C(f_1) \le 1$, then it follows from Theorem C.1.3 that $S^n(x) \to 0$ for all x. By continuous dependence of parameter, we have $\Phi_t(x) \to 0$ for all x. In fact, we observe that $\tilde{r}_C(f_{2^{-m}}) \le 1$ for some m implies $\Phi_t(x) \to 0$ as $t \to \infty$.

If $\tilde{r}_C(f_1) > 1$, then it follows that $\tilde{r}_C(f_{2^{-m}}) > 1$ for all $m \ge 1$. By Theorem C.1.3, for each $m \ge 1$, the map $S_m = \Phi_{2^{-m}}$ has a unique positive fixed point $x_m^* \in C \cap \operatorname{Int} K$. Since $S_{m+1} \circ S_{m+1} = S_m$, we deduce that $x^* = x_m^*$ is independent of m, i.e. it is an equilibrium of Φ_t. Since

$$\lim_{n \to \infty} \Phi_{n/2^m}(x) = \lim_{n \to \infty} (S_m)^n(x) = x^* \quad \text{for each } x \in C \setminus \{0\},$$

it is straightforward to see that $\Phi_t(x) \to x^*$ for each $x \in C \setminus \{0\}$. □

Similarly, we have the following useful corollary.

Corollary C.2.3 *Suppose* $\Phi_t : C \to C$ *is a continuous semiflow satisfying* (S1')–(S3'). *If, in addition, either (S4') or (S5') holds, then exactly one of the following alternatives hold:*

1. Φ_t *has no equilibrium points in* $C \cap \operatorname{Int} K$. *In this case* $\Phi_t(x) \to 0$ *for each* $x \in C \setminus \{0\}$.
2. *There exists an* $x_0 \in C \setminus \{0\}$ *such that* $\Phi_t(x_0) \geq_K x_0$ *for all* $t > 0$. *In this case, S has a unique equilibrium* $x^* \in C \cap \operatorname{Int} K$, *and* $\Phi_t(x) \to x^*$ *for each* $x \in C \setminus \{0\}$.

C.3 Further Reading

Early results on concave maps can be traced to the work of Krasnoselskii [4, 5, 6]. See also Thieme [11]. Hirsch [1] and Smith [9] proved convergence for monotone and concave maps, which was subsequently extended to global convergence for subhomogeneous and strongly monotone maps by Takáč [10]. Further progress was obtained by Hirsch [2] and Zhao [13]. See also Jiang [3] and Wang [12]. We refer the interested readers to the monographs [14, Chap. 2] and [7] in which a more complete account of the discrete-time results is presented.

Corollary C.1.4 can be found in [13, Theorem 2.3], which is adequate for most applications. Theorem C.1.3 is a simple extension which can be applied to subhomogeneous maps S for which the limit $f(x) = \lim_{\lambda \to 0} \frac{1}{\lambda} S(\lambda x)$ exists, but fails to be continuously differentiable at zero. See Problem C.3.3 for an example where Theorem C.1.3 is applied to a parabolic problem with T-periodic coefficients, for which the associated map S is not differentiable at zero.

Problems

C.3.1 Let X be an ordered Banach space with the order $\leq_K$ induced by a cone $K \subset X$, and let $S : K \to K$ be a map. Suppose there is an x_0 such that $S(x_0) \geq_K x_0$ and that $\{S^n(x_0)\}_{n=1}^{\infty}$ is precompact, show that $S^n(x_0)$ converges to a fixed point of S as $n \to \infty$.

C.3.2 Consider the logistic equation

$$\begin{cases} \partial_t u = \Delta u + u g(x,t,u) & \text{in } \Omega \times (0,\infty), \\ \mathbf{n} \cdot \nabla u = 0 & \text{on } \partial\Omega \times (0,\infty), \\ u(x,0) = u_0(x) & \text{in } \Omega, \end{cases} \tag{C.4}$$

where Ω is a bounded smooth domain, and $g \in C^\infty(\overline{\Omega} \times [0,\infty) \times [0,\infty))$ satisfies

$$\begin{cases} u \mapsto g(x,t,u) \text{ is strictly decreasing,} \\ \text{there exists } M > 0 \text{ such that } g(x,t,s) < 0 \text{ for } s \geq M. \end{cases}$$

Let $X = C(\overline{\Omega})$, $C = K = \{u_0 \in X : u_0 \geq 0\}$.

(a) If $g(x,t,s)$ is T-periodic in time, then the map $S : C \to C$ given by

$$S(u_0) = u(\cdot, T)$$

satisfies (S1)–(S3) and (S5).

(b) If $g(x,t,s)$ is independent of time, then the semiflow $\Phi_t : C \to C$ given by

$$\Phi_t(u_0) = u(\cdot, t)$$

satisfies (S1')–(S3') and (S5').

C.3.3 Let Ω be a bounded domain in $\mathbb{R}^N$ with smooth boundary $\partial\Omega$. Consider the following system of parabolic equations:

$$\begin{cases} \partial_t U = d_1 \Delta U + a_1 \frac{UV}{U+V} + b_1 U + c_1 V - U^2 & \text{in } \Omega \times (0,\infty), \\ \partial_t V = d_2 \Delta V + a_2 \frac{UV}{U+V} + c_2 U + b_2 V - V^2 & \text{in } \Omega \times (0,\infty), \\ \mathbf{n} \cdot \nabla U = \mathbf{n} \cdot \nabla V = 0 & \text{on } \partial\Omega \times (0,\infty), \\ U(x,0) = U_0(x), \quad V(x,0) = V_0(x) & \text{in } \Omega, \end{cases} \tag{C.5}$$

where $d_i, a_i, b_i, c_i \in C^{\alpha,\alpha/2}(\overline{\Omega} \times [0,\infty))$ are T-periodic in T. Moreover, $d_i, c_i > 0$ and $a_i \geq 0$ in $\overline{\Omega} \times [0,\infty)$. Define

$$X = C(\overline{\Omega}; \mathbb{R}^2), \quad C = K = C(\overline{\Omega}; \mathbb{R}^2_+),$$

and consider the map $S : C \to C$ given by

$$(U_0, V_0) \mapsto (U(\cdot, T), V(\cdot, T)).$$

Verify the hypotheses (S1)–(S4).
[Hint: For (S2), use the fact that $S(M, M) \leq_K (M, M)$ for all $M \gg 1$. For (S4'), observe that $f(\cdot) = \lim_{\lambda \to 0+} \frac{1}{\lambda} S(\lambda \cdot)$ exists. In fact, $f : C \to C$ is given by

$$f(u_0, v_0) = (u(\cdot, T), v(\cdot, T)),$$

where (u, v) is the solution of (B.17) with initial data (u_0, v_0). By Problem B.5.7, f is compact and monotone with respect to C. Note that (S5) does not hold.]

C.3.4 The following model of a single phytoplankton population was proposed by Shigesada and Okubo (see Chapter 8 for the biological background) [8].

$$\begin{cases} \partial_t u = \partial_{xx} u + g(e^{-\int_0^x u(y,t)\,dy})u - du & \text{for } x \in [0, L],\ t > 0, \\ \partial_x u = 0 & \text{for } x = 0, L,\ t > 0, \\ u(x, 0) = u_0(x), \end{cases} \tag{C.6}$$

where L, d are positive constants and $g(s) = \frac{as}{b+s}$ for some $a, b > 0$. Define

$$X = C([0, L]; \mathbb{R}), \quad C = C([0, L]; \mathbb{R}_+),$$

and

$$K = \{u_0 \in C([0, L]) : \int_0^x u_0(y)\,dy \geq 0 \text{ for all } x \in [0, L]\}.$$

It is proved in Chapter 8 (Propositions 8.1.1 and 8.2.2) that (C.6) generates a semiflow $\Phi_t : C \to C$, which is strongly monotone with respect to the order induced by the cone K. In this case, C is a proper subset of K. Show that Φ_t satisfies (S1')–(S3') and (S5').

References

1. M. W. Hirsch, *The dynamical systems approach to differential equations*, Bull. Amer. Math. Soc. (N.S.), 11 (1984), pp. 1–64.
2. ———, *Positive equilibria and convergence in subhomogeneous monotone dynamics*, in Comparison methods and stability theory (Waterloo, ON, 1993), vol. 162 of Lecture Notes in Pure and Appl. Math., Dekker, New York, 1994, pp. 169–188.
3. J. F. Jiang, *Sublinear discrete-time order-preserving dynamical systems*, Math. Proc. Cambridge Philos. Soc., 119 (1996), pp. 561–574.
4. M. A. Krasnosel'skii, *Positive solutions of operator equations*, Noordhoff, Groningen, (1964).
5. ———, *The operator of translation along the trajectories of differential equations*, Translations of Mathematical Monographs, Vol. 19, American Mathematical Society, Providence, R.I., 1968. Translated from the Russian by Scripta Technica.
6. M. A. Krasnosel'skii and V. Y. Stetsenko, *Toward a theory of equation with concave operations*, Siberian Mathematical Journal, 10 (1969), pp. 405–410.
7. U. Krause, *Positive dynamical systems in discrete time*, vol. 62 of De Gruyter Studies in Mathematics, De Gruyter, Berlin, 2015. Theory, models, and applications.
8. N. Shigesada and A. Okubo, *Analysis of the self-shading effect on algal vertical distribution in natural waters*, J. Math. Biol., 12 (1981), pp. 311–326.
9. H. L. Smith, *Cooperative systems of differential equations with concave nonlinearities*, Nonlinear Anal., 10 (1986), pp. 1037–1052.
10. P. Takáč, *Asymptotic behavior of discrete-time semigroups of sublinear, strongly increasing mappings with applications to biology*, Nonlinear Anal., 14 (1990), pp. 35–42.
11. H. R. Thieme, *On a class of Hammerstein integral equations*, Manuscripta Math., 29 (1979), pp. 49–84.
12. Y. Wang, *Convergence to periodic solutions in periodic quasimonotone reaction-diffusion systems*, J. Math. Anal. Appl., 268 (2002), pp. 25–40.
13. X.-Q. Zhao, *Global attractivity and stability in some monotone discrete dynamical systems*, Bull. Austral. Math. Soc., 53 (1996), pp. 305–324.
14. ———, *Dynamical systems in population biology*, CMS Books in Mathematics/Ouvrages de Mathématiques de la SMC, Springer, Cham, second ed., 2017.

Appendix D
Existence of Connecting Orbits

Abstract We prove the Dancer–Hess lemma concerning the existence of a heteroclinic orbit between two ordered equilibria.

While monotone methods and comparison arguments have long been applied in the study of differential equations [8], their synthesis with dynamical systems originate from the works of Hirsch [3, 4, 5, 6] and Matano [9, 10, 11]. An important result was the Dancer–Hess lemma, which asserts the existence of a connecting orbit between two ordered equilibria. Used in conjunction with the comparison principle, the existence of a heteroclinic orbit implies that the long time dynamics are determined by the presence/absence of equilibria, to a certain extent.

Let X be an ordered Banach space with the order induced by a positive cone K. If $x, \overline{x} \in X$, then we write

$$x \leq_K \overline{x} \quad \text{(resp.} \quad x <_K \overline{x}, \quad x \ll_K \overline{x})$$

provided

$$\overline{x} - x \in K \quad \text{(resp.} \quad \overline{x} - x \in K \setminus \{0\}, \quad \overline{x} - x \in \operatorname{Int} K).$$

D.1 Discrete-Time Monotone Dynamical Systems

Definition D.1.1 Let C be a convex subset of X, and consider a map $S : C \to C$.

1. S is said to be monotone if $x <_K \bar{x}$ implies $S(x) \leq_K S(\bar{x})$.
2. S is said to be strictly monotone if $x <_K \bar{x}$ implies $S(x) <_K S(\bar{x})$.
3. S is said to be strongly monotone if $x <_K \bar{x}$ implies $S(x) \ll_K S(\bar{x})$.

An element $x \in C$ is called a *subequilibrium* (resp. a *strict subequilibrium*) for the fixed point equation $S(u) = u$ provided $x \leq_K S(x)$ (resp. $x <_K S(x)$). An element $y \in C$ is called superequilibrium or strict superequilibrium if the above inequalities are reversed. Assume that u_1, u_2 are respectively subequilibrium and superequilibrium such that $u_1 < u_2$, and that

K. -Y. Lam, Y. Lou, *Introduction to Reaction-Diffusion Equations*, Lecture Notes on Mathematical Modelling in the Life Sciences, https://doi.org/10.1007/978-3-031-20422-7

$$\tilde{I} := [u_1, u_2] = \{x \in C : u_1 \leq_K x \leq_K u_2\}.$$

Then S maps $\tilde{I}$ into itself.

Finally, we say that S is order compact if the set $S([x_1, x_2])$ has compact closure in X for every $x_1 <_K x_2$.

Lemma D.1.2 *Suppose $S : C \to C$ is order compact and strictly monotone. Assume that $u_1, u_2 \in C$ are respectively strict subequilibrium and strict superequilibrium such that $u_1 < u_2$. Define the sequences $\{x_n\}$ and $\{y_n\}$ by*

$$x_n = S^n(u_1) \quad \text{and} \quad y_n = S^n(u_2) \quad \text{for } n \in \mathbb{N}.$$

Then $\{x_n\}$ is an increasing sequence converging to the minimal fixed point $\underline{E}$ in $\tilde{I}$, while $\{y_n\}$ is a decreasing sequence converging to the maximal fixed point $\overline{E}$ in $\tilde{I}$. For each n, the elements x_n and y_n are respectively strict sub- and superequilibria.

Proof Clearly, $x_{n+1} \geq_K x_n$ for all $n \geq 0$. We prove the convergence of the sequence $\{x_n\}$. Suppose it is not convergent. By the compactness of the mapping S, we can find two convergent subsequences n_k and n'_k such that $x_{n_k} \to x$ and $x'_{nk} \to x'$ with $x \neq x'$. Moreover, we can assume without loss of generality that

$$n_k \leq n'_k \leq n_{k+1} \quad \text{for all } k.$$

By the monotonicity of the whole sequence $\{x_n\}$, we have

$$x_{n_k} \leq_K x_{n'_k} \leq_K x_{n_{k+1}} \quad \text{for all } k.$$

Letting $k \to \infty$, we deduce that

$$x \leq_K x' \leq_K x, \quad \text{i.e.} \quad x - x' \in K \cap (-K),$$

where we used the closedness of K. Since $K \cap (-K) = \{0\}$, this implies $x = x'$. This is a contradiction. Thus $\underline{E} = \lim_{n\to\infty} x_n$ is well-defined.

By letting $n \to \infty$ in $x_{n+1} = S(x_n)$, we deduce that $\underline{E} = S(\underline{E})$, i.e. $\underline{E}$ is a fixed point. Similarly, $\overline{E} = \lim_{n\to\infty} y_n$ is well-defined and is a fixed point.

To see that $\underline{E}$ (resp. $\overline{E}$) is the minimal (resp. maximal) fixed point in $\tilde{I}$, let $E^* \in \tilde{I}$ be an arbitrary fixed point, then $u_1 <_K E^* <_K u_2$ as u_i are not fixed points. By the order-preserving property of S, we have

$$x_n = S^n(u_1) \leq_K E^* \leq_K S^n(u_2) = y_n.$$

Letting $n \to \infty$, we have $\underline{E} \leq_K E^* \leq_K \overline{E}$. This completes the proof. □

The next result, due to [7, Proposition 2.1], is a version of a result of Dancer and Hess [1, Proposition]; see also [2, Proposition 2.1].

Lemma D.1.3 (Existence of a connecting orbit) *Let $u_1, u_2 \in C$ be isolated fixed points of the strictly monotone continuous mapping $S : C \to C$, such that $u_1 <_K u_2$. Define*

$$\tilde{I} := [u_1, u_2]_K = \{x \in C : u_1 \leq_K x \leq_K u_2\}.$$

Suppose that $S(\tilde{I})$ has compact closure in $\tilde{I}$, and that there is a finite set of fixed points $\Theta \subset \tilde{I} \setminus \{u_1, u_2\}$ (possibly empty) such that every $\theta \in \Theta$ satisfies

(i) *$S^n(x) \not\to \theta$ as $n \to \infty$ for each $x \in \tilde{I}$ such that $x <_K \theta$ or $x >_K \theta$;*
(ii) *θ is an isolated fixed point of S, and $i(S, B_\delta(E_0), \tilde{I}) = 0$ for all small $\delta > 0$. Here $i(\cdot,\cdot,\cdot)$ refers to the fixed point index defined in Appendix A.*

Then at least one of the following holds:

(a) *S has at least one fixed point in $\tilde{I} \setminus (\{u_1, u_2\} \cup \Theta)$.*
(b) *There is an entire orbit $\{z_n\}_{n\in\mathbb{Z}}$ of S joining u_1 to u_2 and satisfying*

$$u_1 <_K z_n <_K z_{n+1} <_K u_2 \quad \text{for } n \in \mathbb{Z}, \quad \lim_{n\to\infty} z_n = u_2, \quad \lim_{n\to-\infty} z_n = u_1.$$

(c) *There is an entire orbit $\{y_n\}_{n\in\mathbb{Z}}$ of S joining u_2 to u_1 and satisfying*

$$u_1 <_K y_{n+1} <_K y_n <_K u_2 \quad \text{for } n \in \mathbb{Z}, \quad \lim_{n\to\infty} y_n = u_1, \quad \lim_{n\to-\infty} y_n = u_2.$$

Remark D.1.4 If S has no fixed points in $\tilde{I} \setminus \{u_1, u_2\}$ (i.e. Θ is empty), the proposition is proved in [1, Proposition 1] (see also [2, Proposition 2.1]). The existence of a connecting orbit allowing for the presence of boundary equilibria θ was given in [7]. See [7, Proposition 2.1] and the paragraph that follows, where conditions (i) and (ii) are mentioned.

Remark D.1.5 In fact, Hsu et al. [7] showed that, in place of (i) and (ii), one may alternatively assume that θ is an extreme point of $\tilde{I}$ and is ejective. We say that θ is an extreme point of $\tilde{I}$ if $\theta = \frac{1}{2}(x + y)$, for some $x, y \in \tilde{I}$, implies $x = y = \theta$. We say that θ is ejective if there exists a neighborhood U of θ in C such that for each $x \in C \setminus \{\theta\}$, there is an integer $n = n(x)$ such that $S^n(x) \in X \setminus U$. While (i) and the first part of (ii) clearly hold for ejective points, the verification of the latter part of (ii) requires the application of more advanced theory in [12, 13]. We formulated the proposition in terms of (i) and (ii), which is frequently applicable (see Lemmas E.1.4 and E.1.5), as an alternative of using the more advanced theory.

Proof For simplicity, we assume Θ contains a single fixed point θ. The general case follows by a similar argument.

Suppose that S has no fixed point other than u_1, u_2 and those in Θ, we will show that either (b) or (c) holds.

Define, for $\lambda \in [0, 1]$ and $x \in I$, the mappings

$$S_\lambda(x) := \lambda S(x) + (1 - \lambda)u_2, \quad \tilde{S}_\lambda(x) := \lambda S(x) + (1 - \lambda)u_1.$$

Step 1. If $S_\lambda(w) = w$ for some $\lambda \in [0,1]$ and $w \in \tilde{I} \setminus \{u_1, u_2, \theta\}$, then w is a strict superequilibrium. Similarly, if $\tilde{S}_\lambda(v) = v$ for some $\lambda \in [0,1]$ and $v \in \tilde{I} \setminus \{u_1, u_2\}$, then v is a strict subequilibrium.

Indeed, note that $\lambda \neq 0$ as $w \neq u_2$, and that $\lambda \neq 1$ since w is not a fixed point. So we can use $u_2 >_K w$ to get

$$w = S_\lambda(w) = \lambda S(w) + (1-\lambda)u_2 >_K \lambda S(w) + (1-\lambda)w.$$

This gives $w >_K S(w)$, i.e. w is a strict superequilibrium. The proof of the second part of Step 1 is analogous and is omitted.

Next, define

$$B_\epsilon(a) := \{x \in \tilde{I} : \|x-a\| < \epsilon\} \text{ and } \partial B_\epsilon(a) := \partial_{\tilde{I}}(B_\epsilon(a)) = \{x \in \tilde{I} : \|x-a\| = \epsilon\}.$$

Step 2. For each $\epsilon > 0$ sufficiently small, at least one of the following holds.

(i') There exist $\lambda \in [0,1]$ and $w \in \partial B_\epsilon(u_2)$ such that $S_\lambda(w) = w$.
(ii') There exist $\lambda \in [0,1]$ and $v \in \partial B_\epsilon(u_1)$ such that $\tilde{S}_\lambda(v) = v$.

Fix any ϵ small enough so that the closure of $B_\epsilon(u_1)$, $B_\epsilon(u_2)$ and $B_\epsilon(\theta)$ are disjoint. Suppose to the contrary that

$$\begin{cases} S_\lambda(w) \neq w & \text{for all } \lambda \in [0,1],\ w \in \partial B_\epsilon(u_2), \\ \tilde{S}_\lambda(v) \neq v & \text{for all } \lambda \in [0,1],\ v \in \partial B_\epsilon(u_1). \end{cases}$$

The first statement implies that

$$i(S, B_\epsilon(u_2), \tilde{I}) = i(S_1, B_\epsilon(u_2), \tilde{I}) = i(S_0, B_\epsilon(u_2), \tilde{I}) = 1,$$

where the first inequality is by definition of S_λ, the second equality is by homotopy invariance, and the last one is from the normalization property. Note that the fixed point index is well-defined since $\tilde{I}$ is convex and S maps $\tilde{I}$ into itself. Similarly, we can show that $i(S, B_\epsilon(u_1), \tilde{I}) = 1$.

Since $B_\epsilon(u_j)$ $(j = 1, 2)$ and $B_\epsilon(\theta)$ are open subsets of $\tilde{I} \setminus U$ with disjoint closure, and

$$S(x) \neq x \quad \text{for} \quad x \in \tilde{I} \setminus [B_\epsilon(u_1) \cup B_\epsilon(u_2) \cup B_\epsilon(\theta)],$$

the additivity of the fixed point index implies

$$1 = i(S, \tilde{I}, \tilde{I}) = i(S, B_\epsilon(u_1), \tilde{I}) + i(S, B_\epsilon(u_2), \tilde{I}) + i(S, B_\epsilon(\theta), \tilde{I}) = 2,$$

where $i(S, B_\epsilon(\theta), \tilde{I}) = 0$ follows from the hypothesis (ii). This is a contradiction.

Step 3. Take a sequence $\epsilon_n \to 0$. Then at least one of the two statements of Step 2 holds for infinitely many ϵ_n. We will pass to a subsequence and proceed assuming that statement (i') of Step 2 holds for all $\epsilon_n \to 0$, and derive alternative (c) of the statement of the lemma. (If statement (ii') of Step 2 holds for infinitely many ϵ_n, one can analogously derive alternative (b).)

In this case, there exist $\epsilon_n \to 0$ and a sequence of strict superequilibria $w_n \in \partial B_{\epsilon_n}(u_2)$. Indeed, there exist a $\lambda_{\epsilon_n} \in [0, 1]$ and $w_n \in \partial B_{\epsilon_n}(u_2)$ such that $S_{\lambda_{\epsilon_n}}(w_n) = w_n$, and so w_n is a strict superequilibrium.

Step 4. Let $\delta_0 > 0$ be such that $\overline{B}_{\delta_0}(u_1)$, $\overline{B}_{\delta_0}(u_2)$, and $\overline{B}_{\delta_0}(\theta)$ are pairwise disjoint. By continuity of S, there exists $\delta_1 \in (0, \delta_0)$ such that $\|S(x) - u_2\| \le \delta_0$ if $\|x - u_2\| < \delta_1$. By the previous Step 3 there is a strict superequilibrium w_1 such that $\|w_1 - u_2\| < \delta_1$. By Lemma D.1.2, $u_2 > w_1 > S(w_1) > S^2(w_1) > \cdots \to u_1$ in $\tilde{I}$, as the monotone semiorbit must converge to a fixed point and cannot converge to θ or u_2. Let $n(1) \in \mathbb{N}$ be chosen such that $\delta_1 \le \|S^{n(1)}(w_1) - u_2\| \le \delta_0$, and note that $n(1) \ge 1$.

Next, there exists a $\delta_2 \in (0, \delta_1)$ such that $\|S(z) - u_2\| \le \delta_1$ if $\|z - u_2\| < \delta_2$. By Step 3 again, there is a strict superequilibrium w_2 such that $\|w_2 - u_2\| < \delta_2$. Hence $u_2 > w_2 > S(w_2) > S^2(w_2) > \cdots \to u_1$ in $\tilde{I}$. We can choose again $n(2) \ge 2$ such that $\delta_1 \le \|S^{n(2)}(w_2) - u_2\| \le \delta_0$.

Continuing in this fashion, we obtain a sequence of strict superequilibria $\{S^{n(k)}(w_k)\}_{k=1}^{\infty} \subset \overline{B}_{\delta_0}(u_2) \setminus B_{\delta_1}(u_2)$ such that $n(k) \ge k$.

Since S is order compact, the set $S(\tilde{I})$ is relatively compact. Therefore, there exists a subsequence $\{S^{n(k')}(w_{k'})\}_{k'}$ converging to some $y_0 \in \tilde{I}$ such that $\delta_1 \le \|y_0 - u_2\| \le \delta_0$. By passing to a further subsequence, we may assume as well that $\{S^{n(k')-1}(w_{k'})\}_{k'}$ converges in $\tilde{I}$ to some limit which we denote by y_{-1}. Note that $S(y_{-1}) = y_0$, and $u_1 < y_0 < y_{-1} < u_2$ (since neither y_0, y_{-1} can be equal to the fixed points u_1, u_2).

Recursively, we can obtain a sequence $\{y_k : k = 0, -1, -2, \ldots\}$ of strict superequilibria such that $y_k > y_{k+1}$ and $S(y_k) = y_{k+1}$ for all $k < 0$. By compactness, the sequence y_k converges to some fixed point $y_{-\infty} \in \overline{B}_{\delta_0}(u_2)$ as $k \to -\infty$. Since u_2 is the only fixed point in $\overline{B}_{\delta_0}(u_2)$, we have $y_k \to u_2$ as $k \to -\infty$.

Finally, for $k \ge 1$, define $y_k = S^k(y_0)$, then we obtain an entire orbit $\{y_k\}_{k=-\infty}^{\infty}$ which is strictly decreasing. By Lemma D.1.2, y_k converges to some fixed point as $k \to \infty$. Since $y_k \not\to \theta$ by the hypothesis (iii), the only possibility is that $y_k \to u_1$ as $k \to \infty$. □

Lemma D.1.6 (Order Interval Trichotomy) *In addition to the hypotheses of Lemma D.1.3, assume that C has nonempty interior in X and $u_1 \ll_K u_2$. If $u \in \tilde{I} \setminus \{u_1, u_2, \theta\}$ and $Su = u$ implies $u_1 \ll_K u \ll_K u_2$, then precisely one of the alternatives (a)–(c) of Lemma D.1.3 holds.*

Proof First, we show that (b) and (c) are incompatible. Assume to the contrary that both (b) and (c) hold. Then $y_n - z_n \to u_2 - u_1 \in \operatorname{Int} K$ as $n \to -\infty$, so that $z_{n_0} \ll_K y_{n_0}$ for some n_0. But then

$$z_n = S^{n-n_0}(z_{n_0}) <_K S^{n-n_0}(y_{n_0}) = y_n \quad \text{for all } n \ge n_0.$$

Letting $n \to \infty$, we deduce that $u_2 \le_K u_1$, which is a contradiction.

It remains to show that (a) and (b) are incompatible. Since (a) holds, the hypothesis implies the existence of a fixed point u such that $u_1 \ll_K u \ll_K u_2$. Now, take n_1 such that $z_{n_1} \ll_K u$, then by applying S^n on both sides, and letting $n \to \infty$, we deduce that $u_2 \leq_K u$, which is a contradiction. The incompatibility of (a) and (c) is analogous and we omit the proof. □

Lemma D.1.7 *Let P be a compact subset of X, then P contains a maximal (minimal) element.*

Proof By Zorn's lemma, it suffices to show that every totally ordered subset of P has an upper bound in P.

Let Q be a totally ordered subset of P. Then Q is precompact, so for each $\epsilon > 0$, there exists a finite subset A_ϵ of Q such that $\overline{Q} \subset \cup_{p \in A_\epsilon} B_\epsilon(p)$. Let $p'_\epsilon = \max A_\epsilon$, which exists since A_ϵ is finite and totally ordered.

Next, we use compactness of P to choose a sequence $\epsilon = \epsilon_n \to 0$ such that $p'_{\epsilon_n} \to p^*$ for some $p^* \in P$.

We claim that p^* is an upper bound of Q. Indeed, for any $q \in Q$ and $n \in \mathbb{N}$, there exists $\hat{p}_n \in A_{\epsilon_n}$ such that

$$\|q - \hat{p}_n\| < \epsilon_n, \quad \text{and} \quad \hat{p}_n \leq p'_{\epsilon_n}.$$

Letting $n \to \infty$, we have $q \leq p^*$. □

D.2 Continuous-Time Monotone Dynamical Systems

We now formulate a continuous-time version of the results in the previous section. Assume that $\Phi : [0, \infty) \times C \to C$ is a continuous semiflow. We write $\Phi_t(x) = \Phi(t, x)$. The semiflow properties are (i) $\Phi_0(x) = x$ for all $x \in C$, and (ii) $\Phi_t \circ \Phi_s = \Phi_{t+s}$ for all $t, s \geq 0$.

Lemma D.2.1 (Existence of a connecting orbit) *Suppose $\Phi_t : C \to C$ is order compact and strictly monotone for each $t > 0$. Let $u_1 <_K u_2$ be equilibrium points of the semiflow Φ, and let $\tilde{I} := [u_1, u_2]_K = \{x \in C : u_1 \leq_K x \leq_K u_2\}$. Suppose further that there is a finite set of equilibria $\Theta \subset \tilde{I} \setminus \{u_1, u_2\}$ (possibly empty) such that every $\theta \in \Theta$ satisfies*

(i) *$\Phi_t(x) \not\to \theta$ as $t \to \infty$ for each $x \in \tilde{I}$ such that $x <_K \theta$ or $x >_K \theta$.*
(ii) *There is a $\delta > 0$ such that for each $t \in (0, \delta]$, Φ_t has no fixed points in $\overline{B}_\delta(\theta) \setminus \{\theta\}$ and $i(\Phi_t, B_\delta(\theta), \tilde{I}) = 0$, where $B_\delta(\theta) = \{x \in \tilde{I} : \|x - \theta\| < \delta\}$.*

Then at least one of the following holds:

(a) *Φ has an equilibrium point in $\tilde{I}$ distinct from u_1, u_2, and θ.*
(b) *There is a strictly increasing entire orbit $\{\gamma(s)\}_{s \in \mathbb{R}}$ connecting u_1 to u_2, i.e. $\Phi_t(\gamma(s)) = \gamma(t + s)$ for any $t \geq 0$ and $s \in \mathbb{R}$, and*

$$\gamma(t_1) <_K \gamma(t_2) \quad \textit{for } t_1 < t_2, \quad \textit{and} \quad \gamma(-\infty) = u_1, \quad \gamma(\infty) = u_2.$$

(c) *There is a strictly decreasing entire orbit* $\{\gamma(s)\}_{s\in\mathbb{R}}$ *connecting* u_2 *to* u_1, *i.e.* $\Phi_t(\gamma(s)) = \gamma(t+s)$ *for any* $t \geq 0$ *and* $s \in \mathbb{R}$, *and*

$$\gamma(t_1) >_K \gamma(t_2) \quad \textit{for } t_1 < t_2, \qquad \textit{and} \qquad \gamma(-\infty) = u_2, \quad \gamma(\infty) = u_1.$$

Remark D.2.2 Suppose $\Phi_t : C \to C$ is order compact and strictly monotone for each $t > 0$. Let $u_1 <_K u_2$ be equilibrium points of the semiflow Φ, and let $\tilde{I} := [u_1, u_2]_K = \{x \in C : u_1 \leq_K x \leq_K u_2\}$. If $\tilde{I} \setminus \{u_1, u_2\}$ contains no equilibrium points of Φ, then either (b) or (c) of Lemma D.2.1 holds.

Remark D.2.3 If θ is an extreme point of $\tilde{I}$ and θ is ejective, i.e. there exists an open neighborhood U of θ in $\tilde{I}$ such that for each $x \in U$, there exists $t = t(x)$ such that $\Phi_t(x) \notin U$, then one can verify conditions (i) and (ii). See [7, Proof of Proposition 2.4] and [12, 13].

Proof For simplicity, we prove the case when Θ contains a single equilibrium θ. The general case follows from the same argument.

For each $k \geq 1$ we consider the discrete dynamical system $\{(\Phi_{2^{-k}})^n\}_{n\geq 1} = \{\Phi_{n2^{-k}}\}_{n\geq 1}$. Suppose $\tilde{I} \setminus \{u_1, u_2, \theta\}$ contains no equilibrium points of Φ. There are two cases:

(I) There is a sequence $k_1 < k_2 < \cdots \to \infty$ such that for $k = k_i$, the family of discrete dynamical systems $\{(\Phi_{2^{-k}})^n\}_{n\geq 1}$ has no fixed points in $\tilde{I}$ other than u_1, u_2, θ.

(II) For all large k, the family of discrete dynamical systems $\{(\Phi_{2^{-k}})^n\}_{n\geq 1}$ has additional fixed points in $\tilde{I}$ other than u_1, u_2, θ.

Proof of case (I). For each $j \geq 1$ (i.e. $k = k_j$), we may apply Lemma D.1.3 to obtain a discrete entire orbit $\{y_m^{(j)}\}_{m\in\mathbb{Z}}$ of the map $\Phi_{2^{-k_j}}$ connecting u_1 and u_2. Note that $\Phi_{2^{-k_j}}(y_m^{(j)}) = y_{m+1}^{(j)}$. By passing to a sequence in j, we may assume that $\{y_m^{(j)}\}_{m\in\mathbb{Z}}$ is either strictly decreasing for all j, or strictly increasing for all j. Suppose they are decreasing for all j, and connect u_2 to u_1 as m varies from $-\infty$ to ∞. By performing a translation if necessary, we may also assume that

$$\delta_1 \leq \|y_0^{(j)} - u_2\| \leq \delta_0 \quad \text{and} \quad \sup_{m\leq 0} \|y_m^{(j)} - u_2\| \leq \delta_0,$$

where $\delta_1 < \delta_0$ are positive numbers chosen as in the proof of Lemma D.1.3, so that

$$\begin{cases} B_{\delta_0}(u_2) \text{ is disjoint with the neighborhood } U \text{ of } \theta, \\ u_2 \text{ is the only equilibrium point in } B_{\delta_0}(u_2), \\ \|\Phi_t(z) - u_2\| \leq \delta_0 \text{ for } 0 \leq t \leq 1 \text{ and } \|z - u_2\| \leq \delta_1. \end{cases}$$

Denote, for each j,

$$\gamma^{(j)}\left(\frac{m}{2^{k_j}}\right) = y_m^{(j)} \qquad \text{for } m \in \mathbb{Z}.$$

By compactness and the diagonal process, we may take $j \to \infty$ (via a subsequence), so that for each dyadic number s (i.e. $s = n/2^k$ for some $n \in \mathbb{Z}$ and $k \in \mathbb{N}$),

$$\tilde{\gamma}(s) = \lim_{j\to\infty} \gamma^{(j)}(s)$$

is well-defined such that

$$\begin{cases} \tilde{\gamma}(0) \in \tilde{I} \setminus \{u_1, u_2, \theta\}, & \text{and} \quad \tilde{\gamma}(t) \in \overline{B}_{\delta_0}(u_2) \quad \text{for } t \le 0, \\ \Phi_t(\tilde{\gamma}(s)) = \tilde{\gamma}(t+s) & \text{for any dyadic number } t, s \text{ with } t \ge 0. \end{cases}$$

Observe that $\tilde{\gamma}(s)$ is decreasing, since given any two dyadic numbers $t_1 < t_2$, there exist k, n_1, n_2 such that $t_i = n_i/2^k$ for $i = 1, 2$. Since the discrete orbits $\{y_m^{(j)}\}$ are decreasing, we deduce that

$$\tilde{\gamma}(t_1) = \lim_{j\to\infty} \gamma^{(j)}(t_1) \ge_K \lim_{j\to\infty} \gamma^{(j)}(t_2) = \tilde{\gamma}(t_2).$$

Next, we take $\gamma : \mathbb{R} \to \tilde{I}$ to be the unique continuous extension of $\tilde{\gamma}$ to all of $\mathbb{R}$. Since $\gamma(s)$ is also monotone decreasing, it must converge to some equilibrium points as $s \to \pm\infty$. Now, $\gamma(s) \in \overline{B}_{\delta_0}(u_2)$ for all $s \le 0$, and u_2 is the only equilibrium in $\overline{B}_{\delta_0}(u_2)$, so $\gamma(-\infty) = u_2$. Since $\gamma(-\infty) = u_2 > \gamma(0)$, we deduce that γ is strictly decreasing, and $\gamma(\infty) = u_1$. Here we used the fact that there are no monotone trajectories converging to θ. This completes the proof of case (I).

Proof of case (II). Suppose for all $k \gg 1$, $\Phi_{2^{-k}}$ has fixed points in $\tilde{I} \setminus \{u_1, u_2, \theta\}$.

We claim that for each $\epsilon > 0$, there exists a $k_0 \ge 1$ such that for all $k \ge k_0$, the map $\Phi_{2^{-k}}$ does not have fixed points in $\tilde{I} \setminus (B_\epsilon(u_1) \cup B_\epsilon(u_2) \cup \{\theta\})$. Suppose not, then there is a fixed $\epsilon > 0$ and a sequence $k_j \to \infty$ and a sequence $\theta_j \in \tilde{I} \setminus (B_\epsilon(u_1) \cup B_\epsilon(u_2) \cup \{\theta\})$ such that

$$\Phi_{2^{-k_j}}(\theta_j) = \theta_j \quad \text{and} \quad \lim_{j\to\infty} \theta_j = \tilde{\theta} \quad \text{for some} \quad \tilde{\theta} \in \tilde{I} \setminus \{u_1, u_2\}.$$

However, it is easy to see that $\Phi_t(\tilde{\theta}) = \tilde{\theta}$ for any dyadic number $t > 0$ and, by continuity, all real numbers $t > 0$. Thus $\tilde{\theta}$ is an equilibrium point of Φ in $\tilde{I} \setminus \{u_1, u_2\}$, and it must hold that $\tilde{\theta} = \theta$. Hence, we have $\theta_j \to \theta$ as $j \to \infty$. But this is impossible in view of hypothesis (ii). This shows that for large k, there is no fixed point of $\Phi_{2^{-k}}$ in $\tilde{I} \setminus (B_\epsilon(u_1) \cup B_\epsilon(u_2) \cup \{\theta\})$.

Let $\delta_0 > \delta_1 > 0$ be as in the proof of case (i). Fix $\epsilon \in (0, \delta_1)$ and k_0 large enough so that for $k \ge k_0$, the discrete map $\Phi_{2^{-k}}$ has no fixed point in $\tilde{I} \setminus (B_\epsilon(u_1) \cup B_\epsilon(u_2) \cup \{\theta\})$. By Zorn's lemma (see Lemma D.1.7), we can choose a minimal fixed point $u_2^{(k)}$ of $\Phi_{2^{-k}}$ in $B_\epsilon(u_2)$ and a maximal fixed point $u_1^{(k)}$ of $\Phi_{2^{-k}}$ in $B_\epsilon(u_1) \cap [u_1, u_2^{(k)}]_K$. Note that

$$\begin{cases} u_1^{(k)} <_K u_2^{(k)}, \quad u_i^{(k)} \to u_i \quad \text{as} \quad k \to \infty \quad \text{for } i = 1, 2, \\ [u_1^{(k)}, u_2^{(k)}]_K \setminus \{u_1^{(k)}, u_2^{(k)}, \theta\} \quad \text{contains no fixed points of } \Phi_{2^{-k}}. \end{cases}$$

Moreover, for each $k \ge k_0$, Lemma D.1.3 gives an entire orbit $\{y_m^{(k)}\}_{m\in\mathbb{Z}}$ connecting $u_1^{(k)}$ and $u_2^{(k)}$. Again, assume we are in the case that they are all decreasing. We again perform a translation such that $y_0^{(k)} \in \overline{B}_{\delta_0}(u_2) \setminus B_{\delta_1}(u_2)$. Then we again use compactness and a diagonal process to obtain a decreasing orbit $\gamma : \mathbb{R} \to \tilde{I} \setminus U$ such that $\gamma(0) \in \tilde{I} \setminus \{u_1, u_2, \theta\}$, and argue as before that it must be strictly decreasing, and satisfies $\gamma(-\infty) = u_2$ and $\gamma(\infty) = u_1$. □

Lemma D.2.4 (Order Interval Trichotomy) *In addition to the hypotheses of Lemma D.2.1, assume that C has nonempty interior in X and* $u_1 \ll_K u_2$. *If* $u \in \tilde{I} \setminus \{u_1, u_2, \theta\}$ *and* $Su = u$ *implies* $u_1 \ll_K u \ll_K u_2$, *then precisely one of the alternatives (a)–(c) of Lemma D.2.1 holds.*

Proof This is analogous to the proof of Lemma D.1.6 and is omitted. □

References

1. E. N. Dancer and P. Hess, *Stability of fixed points for order-preserving discrete-time dynamical systems*, J. Reine Angew. Math., 419 (1991), pp. 125–139.
2. P. Hess, *Periodic-parabolic boundary value problems and positivity*, vol. 247 of Pitman Research Notes in Mathematics Series, Longman Scientific & Technical, Harlow; copublished in the United States with John Wiley & Sons, Inc., New York, 1991.
3. M. W. Hirsch, *Systems of differential equations which are competitive or cooperative. I. Limit sets*, SIAM J. Math. Anal., 13 (1982), pp. 167–179.
4. ———, *The dynamical systems approach to differential equations*, Bull. Amer. Math. Soc. (N.S.), 11 (1984), pp. 1–64.
5. ———, *Stability and convergence in strongly monotone dynamical systems*, J. Reine Angew. Math., 383 (1988), pp. 1–53.
6. ———, *Positive equilibria and convergence in subhomogeneous monotone dynamics*, in Comparison methods and stability theory (Waterloo, ON, 1993), vol. 162 of Lecture Notes in Pure and Appl. Math., Dekker, New York, 1994, pp. 169–188.
7. S. B. Hsu, H. L. Smith, and P. Waltman, *Competitive exclusion and coexistence for competitive systems on ordered Banach spaces*, Trans. Amer. Math. Soc., 348 (1996), pp. 4083–4094.
8. E. Kamke, *Zur Theorie der Systeme gewöhnlicher Differentialgleichungen. II*, Acta Math., 58 (1932), pp. 57–85.
9. H. Matano, *Asymptotic behavior and stability of solutions of semilinear diffusion equations*, Publ. Res. Inst. Math. Sci., 15 (1979), pp. 401–454.
10. ———, *Existence of nontrivial unstable sets for equilibriums of strongly order-preserving systems*, J. Fac. Sci. Univ. Tokyo Sect. IA Math., 30 (1984), pp. 645–673.
11. H. Matano, *Strongly order-preserving local semidynamical systems—theory and applications*, in Semigroups, theory and applications, Vol. I (Trieste, 1984), vol. 141 of Pitman Res. Notes Math. Ser., Longman Sci. Tech., Harlow, 1986, pp. 178–185.
12. R. D. Nussbaum, *Periodic solutions of some nonlinear autonomous functional differential equations*, Ann. Mat. Pura Appl. (4), 101 (1974), pp. 263–306.
13. ———, *The fixed point index and some applications*, vol. 94 of Séminaire de Mathématiques Supérieures [Seminar on Higher Mathematics], Presses de l'Université de Montréal, Montreal, QC, 1985.

Appendix E
Abstract Competition Systems in Ordered Banach Spaces

Abstract In this appendix, we present selected theorems concerning abstract competitive systems of two species. The proof will be self-contained, since the necessary tools have been developed in earlier appendices. We also prove an improved trichotomy result (Theorems E.1.15 and E.2.13) for competitive systems satisfying an additional hypothesis. This novel hypothesis is satisfied by a large class of competitive systems, including the diffusive Lotka–Volterra systems that are discussed in Chapter 7, as well as the two-species phytoplankton system in Chapter 8. We first develop the result for discrete-time competition systems in Section E.1. The continuous-time analogue is contained in Section E.2.

For $i = 1, 2$, let X_i be a Banach space with positive cone X_i^+ with nonempty interior $\mathrm{Int}\, X_i^+$. The same symbol for the partial orders generated by the cones X_i^+ are used. If $x_i, \overline{x}_i \in X_i$, then we write

$$x_i \leq \overline{x}_i \quad (\text{resp.} \quad x_i < \overline{x}_i, \quad x_i \ll \overline{x}_i)$$

provided that

$$\overline{x}_i - x_i \in X_i^+ \quad (\text{resp.} \quad \overline{x}_i - x_i \in X_i^+ \setminus \{0\}, \quad \overline{x}_i - x_i \in \mathrm{Int}\, X_i^+).$$

Let $X = X_1 \times X_2$, $X^+ = X_1^+ \times X_2^+$. Then X^+ is a cone in X with nonempty interior $\mathrm{Int}\, X^+ = \mathrm{Int}\, X_1^+ \times \mathrm{Int}\, X_2^+$. The solid cone X^+ generates the order relations $\leq$, $<$, $\ll$ in X in the usual way. In particular, if $x = (x_1, x_2)$, and $\bar{x} = (\bar{x}_1, \bar{x}_2)$, then $x \leq \bar{x}$ if and only if $x_i \leq \bar{x}_i$ for $i = 1, 2$. For our purposes, the more important cone is $K = X_1^+ \times (-X_2^+)$, with nonempty interior given by $\mathrm{Int}\, K = \mathrm{Int}\, X_1^+ \times (-\mathrm{Int}\, X_2^+)$. It generates the partial order relations $\leq_K$, $<_K$, $\ll_K$. In this case,

$$x \leq_K \overline{x} \quad \text{iff} \quad x_1 \leq \bar{x}_1 \quad \text{and} \quad \bar{x}_2 \leq x_2.$$

A similar statement holds with $<_K$ (resp. $\ll_K$) replacing $\leq_K$ and $<$ (resp. $\ll$) replacing $\leq$.

K. -Y. Lam, Y. Lou, *Introduction to Reaction-Diffusion Equations*, Lecture Notes on Mathematical Modelling in the Life Sciences, https://doi.org/10.1007/978-3-031-20422-7

We consider dynamical systems with state space being a solid cone C of X^+, where $C = C_1 \times C_2$ and $C_i \subset X_i^+$ are cones with non-empty interior $\operatorname{Int} C_i$, for $i = 1, 2$. If $x <_K y$, then $[x, y]_K = \{z \in C : x \leq_K z \leq_K y\}$.

This chapter is adapted from [1, 2, 4], where the case $C_i = X_i^+$ for $i = 1, 2$ was considered. In this section, we modify their arguments to the slightly more general situation for maps $S : C \to C$ which preserve the competitive order generated by the cone K. The situation when C_i is a proper subset of X_i^+ arises, for example, in the phytoplankton population model [3]. For the continuous-time situation, sharper results are contained in [11], where bi-stability is considered, and X_i^+ can have empty interior; see also [7]. Under some additional assumptions, we will obtain an improvement on the trichotomy result; see Theorems E.1.15 and E.2.13. We first develop the result for discrete-time competition systems in Section E.1. The continuous-time analogue is treated in Section E.2.

E.1 Discrete-Time Competition Systems

Let $S : C \to C$ be a continuous map and denote by S^n the n-fold composition of S. The following hypotheses on S are meant to capture the essence of competition between two adequate competitors, i.e. ones which can persist in the absence of competition.

(H1) S is order compact and strictly monotone with respect to $<_K$. That is,

$$x <_K \bar{x} \quad \text{implies} \quad S(x) <_K S(\bar{x}).$$

(H2) For $i = 1, 2$, there exist maps $\eta : C \to X$ and $f_i : C_i \to C_i$ such that

$$S(x_1, x_2) = (f_1(x_1), f_2(x_2)) + \eta(x_1, x_2),$$

where $\|\eta(x)\|/\|x\| \to 0$ as $\|x\| \to 0$ and f_i is compact, positively homogeneous of degree one, strongly monotone with respect to the order generated by C_i, and the Bonsall cone spectral radius[1] satisfies $\tilde{r}_{C_i}(f_i) > 1$.

(H3) $S(C_1 \times \{0\}) \subset C_1 \times \{0\}$. There exists an $\hat{x}_1 \in \operatorname{Int} C_1$ such that $E_1 = (\hat{x}_1, 0)$ is a fixed point of S, and $S^n((x_1, 0)) \to (\hat{x}_1, 0)$ for every $x_1 \in C_1 \setminus \{0\}$. The symmetric conditions hold for S on $\{0\} \times C_2$, where the fixed point is denoted by $E_2 = (0, \hat{x}_2)$.

(H4) If $x, y \in C$ satisfy $x <_K y$ and at least one of x, y belongs to $\operatorname{Int} C$, then $S(x) \ll_K S(y)$. If $x = (x_1, x_2) \in C$ satisfies $x_i \neq 0$ for $i = 1, 2$, then $S(x) \in \operatorname{Int} C$.

We have introduced the boundary fixed points of S:

$$E_0 = (0, 0), \quad E_1 = (\hat{x}_1, 0), \quad E_2 = (0, \hat{x}_2).$$

[1] $\tilde{r}_{C_i}(f_i)$ is the Bonsall cone spectral radius of f_i with respect to C_i; see Definition B.2.5.

We say that a fixed point E_* is positive if it belongs to the interior of C. The ordered interval $I = [E_2, E_1]_K = [(0, \hat{x}_2), (\hat{x}_1, 0)]_K$ will play an important role.

Recall that S is order compact if $S([(0, x_2), (x_1, 0)]_K)$ has compact closure in X for every $(x_1, x_2) \in C$. Biologically interpreted, $x = (x_1, x_2)$ and $\bar{x} = (\bar{x}_1, \bar{x}_2)$ represent two alternative states of the competition system, in which the state of the first population is given by the first components x_1 or $\bar{x}_1$, and the state of the second population is given by the second component. The relation $x \leq_K \bar{x}$ says that the second species has an advantage in the state x relative to the state $\bar{x}$, since $\bar{x}_2 \leq x_2$ and $x_1 \leq \bar{x}_1$. The order preserving property (H1) says that this relative advantage is being propagated, or preserved, into the future by the dynamical system.

Remark E.1.1 (Condition (H2) concerning the instability of E_0.) Suppose S is continuously differentiable in I, then (H3) implies that

$$DS(0,0)[y_1, y_2] = (T_1 y_1, T_2 y_2)$$

for some compact linear operator T_i which is monotone with respect to C_i (i.e. $T_i(C_i) \subset C_i$). In this case, the condition (H2) holds if and only if $T_i : C_i \to C_i$ is strongly monotone and $r(T_i) > 1$ for $i = 1, 2$. See Lemma B.4.9.

Remark E.1.2 Suppose S is continuously differentiable in I, and write $S(x_1, x_2) = (S_1(x_1, x_2), S_2(x_1, x_2))$. A sufficient condition for (H2)–(H3) is given by the following.

1. $S(C_1 \times \{0\}) \subset C_1 \times \{0\}$ and $S(\{0\} \times C_2) \subset \{0\} \times C_2$.
2. For $i = 1, 2$, $D_i S(0,0) : X_i \to X_i$ is compact and strongly monotone with respect to the cone C_i, and $r(D_i S(0,0)) > 1$.
3. $x_1 \mapsto S_1(x_1, 0)$ is strictly monotone and strongly subhomogeneous, i.e.

$$\begin{cases} x_1, \bar{x}_1 \in C_1, \ \text{ and } \ x_1 < \bar{x}_1 \ \Rightarrow \ S_1(x_1, 0) < S_1(\bar{x}_1, 0) \\ \lambda S_1(x_1, 0) \ll S_1(\lambda x_1, 0) \quad \text{for } 0 < \lambda < 1, \ \text{ and } \ x_1 \in C_1 \cap \text{Int} X_1^+, \end{cases}$$

And a similar condition holds for $x_2 \mapsto S_2(0, x_2)$.

Indeed, conditions 1 and 2 imply (H2), as discussed in Remark E.1.1. Moreover, conditions 1, 2 and 3 together imply (H3), which follow from a result in subhomogeneous dynamical systems. See Theorem C.1.3.

Remark E.1.3 We observe that $E_2 \ll_K S^2(x) \ll_K E_1$ for all $x = (x_1, x_2) \in I \setminus \{E_1, E_2\}$ such that $x_i \neq 0$ for all $i = 1, 2$. Indeed, $S(x) \gg_C 0$ by the second part of (H4). Since $E_2 <_K S(x) <_K E_1$, we have $E_2 \ll_K S^2(x) \ll_K E_1$ by the first part of (H4).

Lemma E.1.4 *Suppose* (H3) *holds, then* $S^n(x) \not\to E_0$ *for each* $x \in C$ *such that* $x <_K E_0$ *or* $x >_K E_0$.

Proof Suppose $x = (x_1, x_2) \in C$ is such that $x <_K E_0$ or $x >_K E_0$, then $x \neq E_0$ and either $x_1 = 0$ or $x_2 = 0$. In this case, (H3) implies that $S^n(x) \to E_1$ or E_2. □

Lemma E.1.5 *Suppose* (H2) *holds, then there exists a* $\delta > 0$ *such that* $E_0 = (0,0)$ *is the unique fixed point of* S *in* $\overline{B}_\delta(E_0) = \{x \in I : \|x\| \le \delta\}$, *and* $i(S, B_\delta(E_0), I) = 0$.

Proof Fix $y = (y_1, y_2) \in I$ such that $y_i \neq 0$ for $i = 1, 2$, and define

$$h(\tau, x) = \tau S(x) + (1-\tau)y \quad \text{for } \tau \in [0,1], x = (x_1, x_2) \in I.$$

It is enough to show that there exists a $\delta > 0$ sufficiently small, such that $h(\tau, x) \neq x$ for every $\tau \in [0,1]$ and $0 < \|x\| \le \delta$. Indeed, if this is true, then E_0 is the unique fixed point in $\overline{B}_\delta(E_0) = \{x \in C : \|x\| \le \delta\}$, and $y \notin \overline{B}_\delta(E_0)$. By homotopy invariance,

$$i(S, B_\delta(E_0), I) = i(h(0,\cdot), B_\delta(E_0), I) = 0.$$

Suppose to the contrary that there is no such δ. Then to each $n \in \mathbb{N}$ we find $\tau_n \in [0,1]$ and $x_n = (x_{1,n}, x_{2,n}) \in C \setminus \{E_0\}$ with $\delta_n := \|x_n\| > 0$ and $\delta_n \to 0$, such that $x_n = h(\tau_n, x_n)$. Setting $(p_n, q_n) = \frac{1}{\delta_n}(x_{1,n}, x_{2,n})$, we infer from (H2) that

$$\begin{cases} p_n - \tau_n f_1(p_n) = \frac{1-\tau_n}{\delta_n} y_1 + o(1), \\ q_n - \tau_n f_2(q_n) = \frac{1-\tau_n}{\delta_n} y_2 + o(1). \end{cases}$$

Using the boundedness and compactness of the maps f_i, and the fact that the left-hand side is bounded in norm, we conclude that $\tau_n \to 1$, and (passing to a subsequence if necessary)

$$(p_n, q_n) \to (p_\infty, q_\infty) \quad \text{and} \quad \frac{1-\tau_n}{\delta_n} \to \alpha$$

for some $\|(p_\infty, q_\infty)\| = 1$ and some real number $\alpha \ge 0$. In particular,

$$p_\infty - f_1(p_\infty) = \alpha y_1 \quad \text{for some } p_\infty \in C_1 \setminus \{0\}.$$

Since $\tilde{r}_{C_1}(f_1) > 1$, we invoke Lemma B.4.8 to deduce that $p_\infty = 0$ and $\alpha = 0$. In the same way, $\tilde{r}_{C_2}(f_2) > 1$ implies that $q_\infty = 0$. This is a contradiction since $\|(p_\infty, q_\infty)\| = 1$. □

Remark E.1.6 The condition (H2) can be replaced by the weaker assumption that E_0 is an extreme and ejective point with respect to I. In this case the conclusions of every result in this chapter continue to hold. See Remark D.1.5.

Theorem E.1.7 *Let* (H1)–(H4) *hold and let* $I = [E_2, E_1]_K$. *Then the omega limit set* $\omega(x) \subset I$ *for every* $x \in C$. *Suppose, in addition, that* S *has no fixed points in* $I \setminus \{E_0, E_1, E_2\}$, *then for each* $x \in C \setminus \{E_0\}$,

$$S^n(x) \to E_1 \quad \text{or} \quad S^n(x) \to E_2 \quad \text{as } n \to \infty.$$

Moreover, exactly one of the following holds:

1. $S^n(x) \to E_1$ *as* $n \to \infty$ *for every* $x = (x_1, x_2) \in I$ *with* $x_i \neq 0$, $i = 1, 2$.
2. $S^n(x) \to E_2$ *as* $n \to \infty$ *for every* $x = (x_1, x_2) \in I$ *with* $x_i \neq 0$, $i = 1, 2$.

Proof First, we show that $\omega(x) \subset I$ for every $x \in C$. Since $S(I) \subset I$, it suffices to consider $x = (x_1, x_2) \notin I$. If one of the components of x is trivial, then (H3) implies $S^n(x) \to E_1$ or E_2. It remains to consider $x = (x_1, x_2)$ such that $x_i \neq 0$ for $i = 1, 2$. Then $(0, x_2) \leq_K (x_1, x_2) \leq_K (x_1, 0)$ and hence

$$S^n(0, x_2) \leq_K S^n(x_1, x_2) \leq_K S^n(x_1, 0) \quad \text{for all } n \geq 0. \tag{E.1}$$

Let $y \in \omega(x)$, i.e. there exists a subsequence $\{n_j\}$ such that $S^{n_j}(x) \to y$. Then we observe from (H3) and (E.1) that $E_2 \leq_K y \leq_K E_1$, which means $y \in I$. This proves that $\omega(x) \subset I$ for every $x \in C$.

Henceforth, we assume S has no fixed points in $I \setminus \{E_0, E_1, E_2\}$. By Lemmas E.1.4 and E.1.5, we can now apply Lemma D.1.3, which implies that at least one of (b) or (c) in the statement of Lemma D.1.3 holds.

Step 1. We claim that (b) and (c) in the statement of Lemma D.1.3 (with $u_1 = E_2$ and $u_2 = E_1$) cannot simultaneously hold.

For in such a case two connecting orbits $\{y_n\}$ and $\{z_n\}$ exist such that

$$y_n - z_n \to E_1 - E_2 \quad \text{as } n \to -\infty, \qquad y_n - z_n \to E_2 - E_1 \quad \text{as } n \to +\infty.$$

Since $E_1 - E_2 \in \operatorname{Int} K$, there exists an n_0 such that $y_{n_0} - z_{n_0} \in \operatorname{Int} K$. As S is strictly monotone, we get

$$z_n = S^{n-n_0}(z_{n_0}) <_K S^{n-n_0}(y_{n_0}) = y_n \quad \text{for all } n \geq n_0.$$

Letting $n \to \infty$, we deduce that $E_1 \leq_K E_2$. This is a contradiction.

Step 2. Suppose that assertion (b) (resp. (c)) in Lemma D.1.3 holds, then

$$S^n(x) \to E_1 \quad (\text{resp. } S^n(x) \to E_2)$$

for all $x = (x_1, x_2) \in I$, $x_i \neq 0$ for $i = 1, 2$.

Indeed, suppose (b) holds, then there exists an entire orbit $\{z_n\}_{n \in \mathbb{Z}}$ such that $z_n \to E_2$ as $n \to -\infty$ and $z_n \to E_1$ as $n \to \infty$. Fix an $x = (x_1, x_2) \in I \setminus \{E_1, E_2\}$ such that $x_i \neq 0$ for $i = 1, 2$. By the compactness of the trajectory, it suffices to show that $\omega(x) \subset \{E_1\}$. By Remark E.1.3, we have $E_2 \ll_K S^2(x) \ll_K E_1$. It is therefore possible to choose n_0 such that

$$z_{n_0} \leq_K S^2(x) \leq_K E_1. \tag{E.2}$$

For $n \geq n_0$, applying S^{n-2} to (E.2), we have $z_{n_0+n-2} \leq_K S^n(x) \leq_K E_1$. Since $z_{n+n_0-2} \to E_1$ as $n \to \infty$, we have $\omega(x) \subset \{E_1\}$, i.e. $S^n(x) \to E_1$ as desired. The case when assertion (c) in Lemma D.1.3 holds is analogous.

Step 3. It remains to show that if $x \in C \setminus I$, then either $S^n(x) \to E_1$ or $S^n(x) \to E_2$.

We fix an $x = (x_1, x_2) \in C \setminus I$. If $x_1 = 0$ (resp. $x_2 = 0$), then (H3) says that $S^n(x) \to E_2$ (resp. $S^n(x) \to E_1$) and we are done. Hence we assume $x_i \neq 0$ for $i = 1, 2$, and divide into two cases:

(i) $S^{n_0}(x) \in I$ for some n_0.
(ii) $S^n(x) \notin I$ for all n.

In the case (i), $E_2 <_K S^{n_0}(x) <_K E_1$ since $S^n(x) \subset \operatorname{Int} C$ for all n by (H4). By (H4) again, $E_2 \ll_K S^{n_0+1}(x) \ll_K E_1$, and we can repeat Step 2 to show that $S^n(x) \to E_2$ in case (b) of Lemma D.1.3 holds, or $S^n(x) \to E_1$ in case (c) of Lemma D.1.3 holds.

In case (ii), we first observe that $E_0 \notin \omega(x)$. Indeed, for δ small enough, we have $B_\delta(E_0) \subset I$ (this follows from the fact that $\hat{x}_i \in \operatorname{Int} C_i$ for $i = 1, 2$). Hence in case (ii), we have $S^n(x) \notin B_\delta(E_0)$ for all $n \geq 0$.

We claim that

$$\omega(x) \subset [(C_1 \times \{0\}) \cup (\{0\} \times C_2)] \setminus B_\delta(E_0). \tag{E.3}$$

To prove (E.3), suppose to the contrary that there exists $z = (z_1, z_2) \in \omega(x)$ such that $z_i \neq 0$ for $i = 1, 2$. By Remark E.1.3, we have $E_2 \ll_K S^2(z) \ll_K E_1$. Since $\omega(x)$ is forward invariant[2], $S^2(z) \in \omega(x)$, so we can find $n_0 \in \mathbb{N}$ such that $E_2 \ll_K S^{n_0}(x) \ll_K E_1$. But this does not happen in case (ii). We have proved (E.3).

Since $\omega(x)$ is connected, we have

$$\omega(x) \in (C_1 \times \{0\}) \setminus B_\delta(E_0) \quad \text{or} \quad \omega(x) \in (\{0\} \times C_2) \setminus B_\delta(E_0).$$

By hypothesis (H3), there exists an $i \in \{1, 2\}$ such that

$$\omega(x) \subset \cup_{n \geq 1} S^{-n}(B_\epsilon(E_i)) \quad \text{for each } \epsilon > 0.$$

Since $\omega(x)$ is invariant (i.e. $S(\omega(x)) = \omega(x)$), we have $\omega(x) \subset B_\epsilon(E_i)$ for each $\epsilon > 0$, i.e. $\omega(x) = \{E_i\}$ for $i = 1$ or $i = 2$, i.e. $S^n(x) \to E_1$ or E_2 as $n \to \infty$. This completes the proof. □

Corollary E.1.8 *Let* (H1)–(H4) *hold. Then S has a positive fixed point if one of the following holds.*

- *Both E_1 and E_2 are locally asymptotically stable relative to I.*
- *Both E_1 and E_2 are unstable relative to I.*
- *There is a point $x \in X$ such that $\omega(x) \cap \operatorname{Int} C$ is nonempty.*

Proof If S has no positive fixed points, then by Theorem E.1.7, there is a connecting orbit connecting E_1 and E_2 and that $\omega(x) \subset \{E_1, E_2\}$ for any $x \in C \setminus \{0\}$. This is incompatible with any of the three conditions in the statement of the corollary. □

In general, the nonexistence of a positive fixed point does not imply that one of E_1, E_2 attracts all trajectories; see Problem 7.5.1. To prove that E_1 (resp. E_2) attracts all trajectories, one needs to impose further conditions to guarantee that the boundary fixed point is repelling, i.e. any trajectory in $\operatorname{Int} C$ does not converge to E_2 (resp. E_1). Suppose S is continuously differentiable in a neighborhood of E_1 and E_2, we write

[2] $S(z') \in \omega(x)$ if $z' \in \omega(x)$.

$$S(x_1,x_2) = (S_1(x_1,x_2), S_2(x_1,x_2)) \in X_1 \times X_2.$$

One way is to impose linear instability in one of the boundary equilibria; see [2].

Proposition E.1.9 *Let* (H1)–(H4) *hold, and assume that S has no fixed points in $I \setminus \{E_0, E_1, E_2\}$. Suppose one of the following conditions hold:*

1. *There exist $\delta > 0$, $u_1 \in C_1 \setminus \{0\}$ such that*

$$S_1(tu_1, x_2) \geq tu_1 \quad \text{for all } 0 \leq t \leq 1 \text{ and } (0,x_2) \in B_\delta(E_2). \tag{E.4}$$

2. *$S : C \to C$ is continuously differentiable in a neighborhood of E_2, such that $D_1S_1(E_2) : X_1 \to X_1$ is compact and strongly positive w.r.t. C_1 and that $r(D_1S_1(E_2)) > 1$.*

Then $S^n(x) \to E_1$ for all $x = (x_1,x_2) \in C$, $x_i \neq 0$.

Proof Step 1. First, we deduce that condition 2 implies condition 1. By (H3), we deduce that $D_2S_1(0,x_2) = 0$ for all $x_2 \in C_2$. Since D_1S_1 is strongly monotone with respect to C_1 (i.e. $D_1S_1(E_2)(C_1 \setminus \{0\}) \subset \operatorname{Int} C_1$) and $r(D_1S_1(E_2)) > 1$, the Krein–Rutman Theorem asserts that there exists a $u_1 \in \operatorname{Int} C_1$ such that $D_1S_1(E_2)[u_1] = r(D_1S_1(E_2))u_1 \gg u_1$. By continuity, we can find a $\delta > 0$ such that

$$D_1S_1(x_1,x_2)[u_1] \geq u_1 \quad \text{for } (x_1,x_2) \in B_{2\delta}(E_2).$$

Next, we normalize u_1 so that $(tu_1,x_2) \in B_{2\delta}(E_2)$ for all $t \in [0,1]$ and $(0,x_2) \in B_\delta(E_2)$. Then, using the fact that $S_1(0,x_2) \equiv 0$, (E.4) can be verified as follows:

$$S_1(tu_1,x_2) = \int_0^1 D_1S_1(\tau t u_1, x_2)[tu_1]\, d\tau \geq tu_1.$$

Hence condition 1 holds in both cases.

Step 2. In view of Theorem E.1.7, it suffices to show that $S^n(x_1,x_2) \not\to E_2$ if $x_i \neq 0$ for $i = 1,2$. Suppose to the contrary that $S^n(x_1,x_2) \to E_2$. Without loss of generality, we may assume that $x \in \operatorname{Int} C$ and $\|S^n(0,x_2) - E_2\| < \delta$ for all $n \geq 0$. Let $u_1 \in C_1 \setminus \{0\}$ be given by condition 1, and choose $0 < t \leq 1$ such that $tu_1 \leq x_1$. By repeated applications of (E.4), we have

$$S_1^n(x_1,x_2) \geq S_1^n(tu_1,x_2) \geq tu_1 \quad \text{for all } n \geq 0.$$

Since $S_1^n(x_1,x_2) \to 0$, we may let $n \to \infty$ to deduce that $0 \geq tu_1$ for some $t > 0$, i.e. $-u_1 \in X_1^+$. This is in contradiction with $u_1 \in C_1 \setminus \{0\} \subset X_1^+ \setminus \{0\}$. □

When the two species are mutually invasible, i.e. E_1 and E_2 are both unstable, then Theorem E.1.7 yields the existence of at least one positive fixed point.

Corollary E.1.10 *Let* (H1)–(H4) *hold. Suppose both E_1 and E_2 are linearly unstable (i.e. at least one of the two conditions of Proposition E.1.9 holds for E_2, and a symmetric statement holds for E_1), then there exists at least one positive fixed point E_*.*

Proof By Proposition E.1.9, for each $i = 1, 2$ and $x = (x_1, x_2) \in C$ with $x_i \neq 0$, we have $S^n(x) \not\to E_i$. It follows from Theorem E.1.7 that S has at least one fixed point in $I \setminus \{E_0, E_1, E_2\}$. □

Next, we seek to further relax the linear instability criterion. The following conditions hold, for example, for two-species Lotka–Volterra systems, as well as certain phytoplankton models.

(H5) S is continuously differentiable in a neighborhood of E_1 and E_2, such that

- The map $DS(E_2) : X \to X$ is compact and satisfies $r(D_2S_2(E_2)) < 1$. If $r(DS(E_2)) \geq 1$, then

$$DS(E_2)[v] \geq_K v \quad \text{for some } v = (v_1, v_2) \in \text{Int}\, C_1 \times (-\text{Int}\, X_2^+).$$

- The map $DS(E_1) : X \to X$ is compact and satisfies $r(D_1S_1(E_1)) < 1$. If $r(DS(E_1)) \geq 1$, then

$$DS(E_1)[v] \geq_K v \quad \text{for some } v = (v_1, v_2) \in \text{Int}\, X_1^+ \times (-\text{Int}\, C_2).$$

We give a sufficient condition for the first part of (H5) concerning E_2. A symmetric result holds for the second part of (H5) concerning E_1.

Lemma E.1.11 *Suppose S is continuously differentiable in a neighborhood of E_2, $DS(E_2) : X \to X$ is compact, and the following conditions hold.*

1. *$D_2S_2(E_2) : X_2 \to X_2$ is strongly monotone w.r.t. X_2^+, and $r(D_2S_2(E_2)) < 1$.*
2. *$D_1S_1(E_2) : X_1 \to X_1$ is strongly monotone w.r.t. C_1.*
3. *$D_1S_2(E_2)(v_1') \neq 0$ for all $v_1' \in \text{Int}\, C_1$.*

Then the first part of (H5) concerning $DS(E_2)$ holds.

Remark E.1.12 The condition $r(D_2S_2(E_2)) < 1$ means that the fixed point E_2 is linearly stable when S is restricted to $\{0\} \times C_2$. The condition 3 is crucial, if one compares with the counterexample; see Problem 7.5.1.

Proof (of Lemma E.1.11) First, observe that

$$DS(E_2)[v_1, v_2] = (D_1S_1(E_2)[v_1], D_1S_2(E_2)[v_1] + D_2S_2(E_2)[v_2]), \tag{E.5}$$

since $D_2S_1(E_2) \equiv 0$ due to $S(\{0\} \times C_2) \subset \{0\} \times C_2$.

Step 1. We claim that $DS(E_2) : X \to X$ is monotone in $C_1 \times (-X_2^+)$.

Indeed, it is clear that $DS(E_2)$ inherits the monotonicity with respect to $K = X_1^+ \times (-X_2^+)$ from S. To improve it to $C_1 \times (-X_2^+)$, let $v = (v_1, v_2) \in C_1 \times (-X_2^+)$, and observe that

$$\begin{aligned} DS(E_2)[v] &= \lim_{\lambda \to 0} \frac{1}{\lambda}(S(E_2 + \lambda v) - S(E_2)) \\ &= \lim_{\lambda \to 0} \frac{1}{\lambda} (S_1(E_2 + \lambda v), S_2(E_2 + \lambda v) - \hat{x}_2) \in C_1 \times (-X_2^+). \end{aligned}$$

Step 2. Suppose $r := r(DS(E_2)) \geq 1$, then there exists a nonzero vector $v = (v_1, v_2) \in \operatorname{Int} C_1 \times (-\operatorname{Int} X_2^+)$ such that $DS(E_2)[v] = rv$.

It follows from the Krein–Rutman Theorem that there exists such a nonzero $v = (v_1, v_2) \in C_1 \times (-X_2^+)$. We claim that $v_1 \neq 0$. Suppose not, then $v_2 \neq 0$ and we deduce from (E.5) that

$$D_2 S_2(E_2)[v_2] = r v_2.$$

But this is impossible since $r(D_2 S_2(E_2)) < 1 \leq r$. Thus $v_1 \in C_1 \setminus \{0\}$ and is an eigenvector of $D_1 S_1(E_2)$. Since $D_1 S_1(E_2)$ is strongly monotone in C_1, we deduce that $v_1 \in \operatorname{Int} C_1$.

It remains to prove that $v_2 \in (-\operatorname{Int} X_2^+)$. Indeed, from (E.5) we deduce that $v_2 \in -X_2^+$ satisfies

$$r v_2 - D_2 S_2(E_2) v_2 = h,$$

where $h = D_1 S_2(E_2)[v_1] \neq 0$ since $v_1 \in \operatorname{Int} C_1$. Moreover, by Step 1, $DS(E_2)$ is monotone with respect to $C_1 \times (-X_2^+)$, so $h \in (-X_2^+) \setminus \{0\}$. In particular $v_2 \in (-X_2^+) \setminus \{0\}$. It follows from the strong monotonicity of $D_2 S_2(E_2)$ that $v_2 \in (-\operatorname{Int} X_2^+)$. □

We can now prove a sufficient condition for E_2 (resp. E_1) to be repelling.

Remark E.1.13 We give two examples that satisfy (H1)–(H5). Let $f_i : \overline{\Omega} \times \mathbb{R}_+ \times \mathbb{R}_+ \times \mathbb{R}_+ \to \mathbb{R}$ be given such that

(F) f_i is smooth, T-periodic in t. Moreover, f_i strictly decreasing in u and in v.

The first example is a Lotka–Volterra type competition system.

$$\begin{cases} \partial_t u - \mu_1 \Delta u = f_1(x, t, u, v) u & \text{in } \Omega \times (0, \infty), \\ \partial_t v - \mu_2 \Delta v = f_2(x, t, u, v) v & \text{in } \Omega \times (0, \infty), \\ \mathbf{n} \cdot \nabla u = \mathbf{n} \cdot \nabla v = 0 & \text{on } \partial\Omega \times (0, \infty), \\ u(x, 0) = u_0(x), \quad v(x, 0) = v_0(x) & \text{in } \Omega, \end{cases} \tag{E.6}$$

where Ω is a bounded smooth domain in $\mathbb{R}^N$ with outer unit normal vector $\mathbf{n}$ on $\partial\Omega$, Δ is the Laplacian operator, and ∇ is the gradient operator. Let $X = C(\overline{\Omega}; \mathbb{R}^2)$, $X_i^+ = C_i = C(\overline{\Omega}; \mathbb{R}_+)$ for $i = 1, 2$, and $K = X_1^+ \times (-X_2^+)$. If μ_i and f_i satisfy (F) and additionally

$$\mu_i > 0, \quad f_1(x, t, 0, 0) > 0 > \partial_v f_1(x, t, u, 0), \quad f_2(x, t, 0, 0) > 0 > \partial_u f_2(x, t, u, 0)$$

for all $x \in \overline{\Omega}$, $t, u, v \geq 0$, then the mapping $S(u_0, v_0) = (u(\cdot, T), v(\cdot, T))$ satisfies (H1)–(H5). See Lemma 7.1.3 for the proof of the autonomous case.

The second example arises in a nonlocal model where two phytoplankton species compete for light in a water column. Let $X = C([0, L]; \mathbb{R}^2)$,

$$C_i = C([0, L]; \mathbb{R}_+) \quad X_i^+ = \left\{ u_0 \in C([0, L]) : \int_0^x u_0(y)\, dy \geq 0 \text{ for } 0 \leq x \leq L \right\}$$

for $i = 1, 2$, and $K = X_1^+ \times (-X_2^+)$.

$$\begin{cases} \partial_t u - \mu_1 u_{xx} + \alpha_1 u_x = f_1(x, t, \int_0^x u(y, t)\, dy, \int_0^x v(y, t)\, dy) u & \text{in } [0, L] \times (0, \infty), \\ \partial_t v - \mu_2 v_{xx} + \alpha_2 v_x = f_2(x, t, \int_0^x u(y, t)\, dy, \int_0^x v(y, t)\, dy) v & \text{in } [0, L] \times (0, \infty), \\ \mu_1 \partial_x u - \alpha_1 u = \mu_2 \partial_x v - \alpha_2 v = 0 & \text{on } \{0, L\} \times (0, \infty), \\ u(x, 0) = u_0(x), \quad v(x, 0) = v_0(x) & \text{in } [0, L], \end{cases} \tag{E.7}$$

where $\mu_i > 0$ and α_i are real constants. Assume f_i satisfies (F),

$$\partial_x f_i \leq 0, \quad \partial_u f_i < 0, \quad \partial_v f_i < 0$$

and that $f(x, t, u, v) < 0$ for $\min\{u, v\}$ sufficiently large. Then the mapping $S(u_0, v_0) = (u(\cdot, T), v(\cdot, T))$ satisfies (H1)–(H5); see [6].

Lemma E.1.14 *Let* (H1)–(H5) *hold. Assume that*

1. *E_2 is an isolated fixed point.*
2. *There exists a fixed point E^* such that $E_2 \ll_K E^* \leq_K E_1$ and*

$$S^n(x) \to E^* \quad \textit{for any } x = (x_1, x_2) \in [E_2, E^*] \textit{ such that } x_1 \neq 0.$$

Then $S^n(x_1, x_2) \not\to E_2$ whenever $x_i \neq 0$ for $i = 1, 2$.

Proof Step 1. There exist $m \in \mathbb{N}$, $0 < \lambda_1 < 1$ and $\delta_1 > 0$ such that

$$\|S^m(0, x_2) - E_2\| \leq (\lambda_1)^m \|(0, x_2) - E_2\| \quad \text{for } (0, x_2) \in B_{\delta_1}(E_2). \tag{E.8}$$

Indeed, since $r(D_2 S_2(E_2)) < 1$, we can choose m and $0 < \lambda_1 < 1$ such that

$$\|(D_2 S_2(E_2))^m\|^{1/m} < \lambda_1.$$

Recalling that $E_2 = (0, \hat{x}_2)$, we have

$$S^m(0, x_2) - E_2 = \Big(0, (D_2 S_2(E_2))^m [x_2 - \hat{x}_2] + o(\|x_2 - \hat{x}_2\|)\Big),$$

where we used $D(S^m)(E_2) = (DS(E_2))^m$. Hence,

$$\|S^m(0, x_2) - E_2\| \leq (\lambda_1)^m \|(0, x_2) - E_2\|.$$

This proves that (E.8) holds.

Step 2. We claim that $r(DS(E_2)) \geq 1$. If $r(DS(E_2)) < 1$, then E_2 is locally asymptotically stable. This is impossible since E^* attracts points $(x_1, x_2) \in [E_2, E^*]$ with $x_1 \neq 0$.

Step 3. There exist real numbers $\lambda_2 \in (\lambda_1, 1)$, $\delta \in (0, \delta_1]$, and a vector $v = (v_1, v_2) \in \operatorname{Int} C_1 \times (-\operatorname{Int} X_2^+)$ such that

$$S(tv_1, tv_2 + x_2) \geq_K \lambda_2 t(v_1, v_2) + S(0, x_2) \quad \text{if } 0 < t \leq 1, \text{ and } (0, x_2) \in B_\delta(E_2). \tag{E.9}$$

Indeed, let v be given by (H5). Fixing an arbitrary $\lambda_2 \in (\lambda_1, 1)$, we have $DS(E_2)[v] \gg_K \lambda_2 v$. By continuity, there exists a $\delta \in (0, \delta_1]$ such that $DS(x)[v] \geq_K \lambda_2 v$ for $x \in B_{2\delta}(E_2)$. By scaling v, we may assume that $(tv_1, \pm tv_2 + x_2) \in B_{2\delta}(E_2)$ for all $t \in [0, 1]$ and $(0, x_2) \in B_\delta(E_2)$. Therefore,

$$S(tv_1, tv_2 + x_2) - S(0, x_2) = \int_0^1 DS(\tau tv_1, \tau tv_2 + x_2)[tv]\, d\tau \geq_K \lambda_2 tv.$$

This proves (E.9).

Step 4. We claim that $S^n(x_1, x_2) \not\to E_2$ whenever $x_i \neq 0$ for $i = 1, 2$.

Assume to the contrary that $S^n(x_1, x_2) \to E_2$. We may again reduce to the case that $(x_1, x_2) \in \operatorname{Int} C$, and $\|S^n(x_1, x_2) - E_2\| < \delta$ for all $n \geq 0$. By the fact that $E_2 \ll_K E^*$, we may further assume that $S^n(x_1, x_2) \leq_K E^*$ for all n.

By Step 1, and scaling $v = (v_1, v_2)$ if necessary, we may assume in addition that $S^n(0, x_2 - tv_2) \in B_\delta(E_2)$ for all $n \geq 0$ and $0 \leq t \leq 1$.

Fix $0 < t \leq 1$ such that $x_1 \geq tv_1$. By Step 3, we have

$$S(x) \geq_K S(tv_1, x_2) \geq_K \lambda_2 tv + S(0, x_2 - tv_2).$$

Repeating the process, we deduce that, for each n,

$$\begin{aligned} S^n(x) &\geq_K (\lambda_2)^n tv + S^n(0, x_2 - tv_2) \\ &= (\lambda_2)^n t \left[v + \frac{1}{(\lambda_2)^n t} (S^n(0, x_2 - tv_2) - E_2) \right] + E_2 \\ &= (\lambda_2)^n t \,[v + w_n] + E_2. \end{aligned}$$

By taking $n = km$, where m is given by Step 1, we have $\|w_n\| \leq C(\lambda_1/\lambda_2)^n$ if $n = km$ for some integer k. Since $v \in \operatorname{Int} K$, we may send $n = km \to \infty$, to deduce that $S^{km}(x) \gg_K E_2$ for all sufficiently large k. Hence, $E_2 \ll_K S^n(x) \leq_K E^*$ for some n. But that would imply that $S^n(x) \to E^*$ as $n \to \infty$. This is a contradiction. □

We can now prove a stronger version of the trichotomy result in Theorem E.1.7.

Theorem E.1.15 *Let* (H1)–(H5) *hold and let* $I = [E_2, E_1]_K$. *Then the omega limit set* $\omega(x) \subset I$ *for every* $x \in C$, *and the following trichotomy holds:*

1. *S has at least one fixed point in* $I \setminus \{E_0, E_1, E_2\}$.
2. $S^n(x) \longrightarrow E_1$ *for all* $x = (x_1, x_2) \in C$ *such that* $x_i \neq 0$, $i = 1, 2$.
3. $S^n(x) \longrightarrow E_2$ *for all* $x = (x_1, x_2) \in C$ *such that* $x_i \neq 0$, $i = 1, 2$.

Proof Suppose S has no fixed points in $I \setminus \{E_0, E_1, E_2\}$. It follows from Theorem E.1.7 that

$$S^n(x) \longrightarrow E_1 \text{ or } E_2 \quad \text{if } x = (x_1, x_2) \in C,\ x_i \neq 0,\ i = 1, 2. \tag{E.10}$$

Moreover, one of E_1, E_2 attracts all positive orbits in I. Suppose for definiteness that E_1 is the stable fixed point relative to I. It then follows from Lemma E.1.14 (taking $E^* = E_1$) that $S^n(x) \not\rightarrow E_2$ for any $x = (x_1, x_2) \in C$ such that $x_i \neq 0$, $i = 1, 2$. It then follows from (E.10) that $S^n(x) \longrightarrow E_1$ for any $x = (x_1, x_2) \in C$ such that $x_i \neq 0$, $i = 1, 2$. □

If the first alternative of Theorem E.1.15 holds, we can impose a further condition to derive a global attractivity result.

Theorem E.1.16 *Let* (H1)–(H5) *hold, and assume that S has at least one positive fixed point. Assume further that every positive fixed point is locally asymptotically stable, then there is a unique positive fixed point* E^* *such that* $S^n(x) \longrightarrow E^*$ *for every* $x = (x_1, x_2) \in C$ *such that* $x_i \neq 0$ *for* $i = 1, 2$.

Proof Let $E^* = (u^*, v^*)$ be a positive fixed point of S. Observe by (H3) that $E_2 \ll_K E^* \ll_K E_1$. Next, let E' be a maximal equilibrium in $[E_2, E^*] \setminus \{E^*\}$. Such a choice is possible thanks to Lemma D.1.7 and the fact that E^* is isolated, so that the set of equilibria in $[E_2, E^*] \setminus \{E^*\}$ is compact. Then $E' <_K E^*$ and $[E', E^*] \setminus \{E', E^*\}$ contains no equilibrium. It follows from the local asymptotic stability of E^*, and Lemma D.1.3, that there is a strict increasing sequence $\{x_n\}_{n=-\infty}^{\infty} \subset \mathrm{Int}\,[E', E^*]$ such that $S(x_n) = x_{n+1}$ for all n, $\lim_{n\to-\infty} x_n = E'$ and $\lim_{n\to\infty} x_n = E^*$. In particular, E' is a fixed point in $[E_2, E^*]$ which is not locally asymptotically stable. Since every positive fixed point is locally asymptotically stable, it follows that $E' = E_2$. This proves the existence of a connecting orbit $\{x_n\}_{n\in\mathbb{Z}}$ connecting E_2 to E^*. Similarly, there exists a second connecting orbit connecting E_1 to E^*. By comparison, it follows that

$$S(x_1, x_2) \longrightarrow E^* \quad \text{for all } (x_1, x_2) \in I,\ x_i \neq 0,\ i = 1, 2.$$

In particular, both E_1 and E_2 are not linearly stable, i.e. $r(DS(E_i)) \geq 1$ for $i = 1, 2$.

Next, we apply Lemma E.1.14 to conclude that E_2 (and similarly E_1) is repelling.

Finally, given $x = (x_1, x_2) \in C$ such that $x_i \neq 0$, $i = 1, 2$. Theorem E.1.7 implies that the omega limit set of x belongs to $I = [E_2, E_1]$. Since $S^n(x) \not\rightarrow E_i$ for $i = 1, 2$, it follows (as in Step 3 in the proof of Theorem E.1.7) that the omega limit set contains a point $x' = (x_1', x_2') \in I$ such that $x_i' \neq 0$ for both $i = 1, 2$. Then $E_2 \ll_K S^2(x') \ll_K E_1$, and thus $E_2 \ll_K S^n(x) \ll_K E_1$ for some $n \gg 1$. By comparing with the connecting orbits connecting $E_i \rightarrow E^*$, it follows that $S^n(x) \rightarrow E^*$. □

Theorem E.1.17 (Hess and Lazer's compression result) *Let* (H1)–(H4) *hold. Suppose $S : C \to C$ is continuously differentiable in a neighborhood of E_1 and E_2, and the following conditions hold:*

- *$D_2S_2(E_1) : X_1 \to X_1$ is compact and strongly positive w.r.t. C_2 and we have $r(D_2S_2(E_1)) > 1$.*
- *$D_1S_1(E_2) : X_2 \to X_2$ is compact and strongly positive w.r.t. C_1 and we have $r(D_1S_1(E_2)) > 1$.*

Then there exist fixed points E_*, E^*, such that $E_2 \ll_K E_* \leq_K E^* \ll_K E_1$ and

1. *$S^n(x) \to E_*$ for $x = (x_1, x_2) \in [E_2, E_*]$, $x_1 \neq 0$.*
2. *$S^n(x) \to E^*$ for $x = (x_1, x_2) \in [E^*, E_1]$, $x_2 \neq 0$.*
3. *For any $x = (x_1, x_2) \in C$, $x_i \neq 0$, the omega limit set $\omega(x)$ satisfies*

$$\omega(x) \subset [E_*, E^*].$$

If, in addition, $E_* = E^*$, then $S^n(x) \to E^*$ for all $x = (x_1, x_2) \in C$, $x_i \neq 0$.

See Figure E.1 for an illustration.

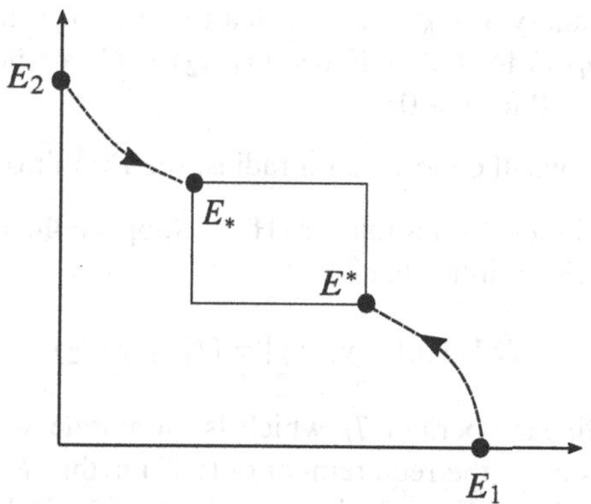

Fig. E.1: A diagram illustrating the compression result (Theorem E.2.15).

Proof We leave this as an exercise. □

Corollary E.1.18 *Assume, in addition to the assumptions of Theorem E.1.17, that every positive fixed point is locally asymptotically stable, then there is a unique positive fixed point E^* such that $S^n(x) \to E^*$ for every $x = (x_1, x_2) \in C$ such that $x_i \neq 0$ for $i = 1, 2$.*

Proof We leave this as an exercise. □

E.2 Continuous-Time Competition Systems

We now formulate a continuous-time version of the results in the previous section. Assume that $\Phi : [0,\infty)\times C \to C$ is a continuous semiflow. We write $\Phi_t(x) = \Phi(t,x)$. The semiflow properties are (i) $\Phi_0(x) = x$ for all $x \in C$, and (ii) $\Phi_t \circ \Phi_s = \Phi_{t+s}$ for all $t, s \geq 0$. The analogous hypotheses to (H1)–(H4) above are given below

(H1') Φ_t is order compact for each $t > 0$ and strictly monotone with respect to $<_K$. That is, $x <_K \bar{x}$ implies $\Phi_t(x) <_K \Phi_t(\bar{x})$.

(H2') For each $t > 0$, there exist maps $\eta : C \to X$ and $f_i : C_i \to C_i$ $(i = 1, 2)$ such that

$$\Phi_t(x_1, x_2) = (f_1(x_1), f_2(x_2)) + \eta(x_1, x_2),$$

where $\|\eta(x)\|/\|x\| \to 0$ as $\|x\| \to 0$ and f_i is compact, positively homogeneous of degree one, strongly monotone with respect to the order generated by C_i, and $\tilde{r}_{C_i}(f_i) > 1$.

(H3') $\Phi_t(C_1 \times \{0\}) \subset C_1 \times \{0\}$ for $t \geq 0$. There exists an $\hat{x}_1 \in \operatorname{Int} C_1$ such that $E_1 = (\hat{x}_1, 0)$ is an equilibrium of Φ_t, and $\Phi_t(x_1, 0) \to (\hat{x}_1, 0)$ as $t \to \infty$ for every $x_1 \in C_1 \setminus \{0\}$. The symmetric conditions hold for Φ_t on $\{0\}\times C_2$, where the equilibrium is denoted by $E_2 = (0, \hat{x}_2)$.

(H4') If $x, y \in C$ satisfy $x <_K y$ and at least one of x, y belongs to $\operatorname{Int} C$, then $\Phi_t(x) \ll_K \Phi_t(y)$ for $t > 0$. If $x = (x_1, x_2) \in C$ satisfies $x_i \neq 0$ for $i = 1, 2$, then $\Phi_t(x) \gg_C 0$ for $t > 0$.

Here $\tilde{r}_{C_1}(f_i)$ is the Bonsall cone spectral radius; see Definition B.2.5.

Remark E.2.1 (A sufficient condition for (H2')) Suppose Φ_t is continuously differentiable in I, then (H3') implies that

$$D\Phi_t(0,0)[y_1, y_2] = (T_1 y_1, T_2 y_2)$$

for some compact linear operator T_i which is monotone with respect to C_i (i.e. $T_i(C_i) \subset C_i$). In this case, the requirement of (H2') is that T_i is strongly monotone and that there is an $\bar{x}_i \in C_i$ such that $T_i(\bar{x}_i) \gg_{C_i} \bar{x}_i$. Indeed, this implies that $\tilde{r}_{C_i}(T_i) = r(T_i) > 1$, where $r(T_i)$ is the usual spectral radius for linear operators, and the equality follows from Lemma B.4.9.

We have introduced the boundary equilibrium points of Φ_t:

$$E_0 = (0,0), \quad E_1 = (\hat{x}_1, 0), \quad E_2 = (0, \hat{x}_2).$$

We say that an equilibrium point E_* is positive if it belongs to the interior of C. The ordered interval $I = [E_2, E_1]_K = [(0, \hat{x}_2), (\hat{x}_1, 0)]_K$ will play an important role.

Lemma E.2.2 *Suppose* (H3') *holds. Then for each $x \in C$ such that $x <_K E_0$ or $x >_K E_0$, $\Phi_t(x) \to E_1$ or $\Phi_t(x) \to E_2$ as $t \to \infty$*

Proof Suppose $x = (x_1, x_2) \in C$ is such that $x <_K E_0$ or $x >_K E_0$, then $x \neq E_0$ and either $x_1 = 0$ or $x_2 = 0$. In this case, (H3') implies that $\Phi_t(x) \to E_1$ or E_2. □

Lemma E.2.3 *Suppose* (H2') *holds. Let $t > 0$ be given and let $S = \Phi_t$. Then there exists a $\delta > 0$ such that $E_0 = (0,0)$ is the unique fixed point of S in $\overline{B}_\delta(E_0) = \{x \in I : \|x\| \le \delta\}$, and $i(S, B_\delta(E_0), I) = 0$.*

Proof See the proof of Lemma E.1.5. □

Remark E.2.4 (H2') can be replaced the weaker assumption that E_0 is an extreme point with respect to I and is an ejective point with respect to the semiflow Φ_t. See Remark D.1.5.

Theorem E.2.5 *Let* (H1')–(H4') *hold. Then the omega limit set $\omega(x)$ of every orbit is contained in $I = [E_2, E_1]_K$. Suppose, in addition, that S has no equilibrium points in $I \setminus \{E_0, E_1, E_2\}$, then for each $x \in C \setminus \{E_0\}$,*

$$\Phi_t(x) \to E_1 \quad or \quad \Phi_t(x) \to E_2 \quad as\ t \to \infty.$$

Moreover, exactly one of the following holds:

1. *$\Phi_t(x) \to E_1$ as $t \to \infty$ for every $x = (x_1, x_2) \in I$ with $x_i \neq 0$, $i = 1, 2$.*
2. *$\Phi_t(x) \to E_2$ as $t \to \infty$ for every $x = (x_1, x_2) \in I$ with $x_i \neq 0$, $i = 1, 2$.*

Proof First, we show that $\omega(x) \subset I$ for every $x \in C$. If one of the components of x is trivial, then (H3') says that $\Phi_t(x) \to E_i$ for some $i = 0, 1, 2$, and we are done. It remains to consider $x = (x_1, x_2)$ such that $x_i \neq 0$ for $i = 1, 2$. Then $(0, x_2) \le_K (x_1, x_2) \le_K (x_1, 0)$ and hence

$$\Phi_t(0, x_2) \le_K \Phi_t(x_1, x_2) \le_K \Phi_t(x_1, 0) \quad \text{for } t \ge 0.$$

Letting $t \to \infty$ we deduce that $\omega(x) \subset I$.

Henceforth, we assume Φ_t has no equilibrium points in $I \setminus \{E_0, E_1, E_2\}$.

By Lemmas E.2.2 and E.2.3, we have verified (i)–(ii) in Lemma D.2.1. Since Φ has no other fixed points in $I \setminus \{E_0, E_1, E_2\}$, either alternative (b) or (c) of Lemma D.2.1 holds, i.e., there exists at least one monotone orbit $\gamma : \mathbb{R} \to I$ connecting E_1 and E_2. By Step 1 in the proof of Theorem E.1.7, either there exists a strictly increasing orbit connecting E_2 to E_1, or a strictly decreasing orbit connecting E_1 to E_2, but not both. We can now argue as in Steps 2 and 3 in the proof of Theorem E.1.7 to conclude. □

Corollary E.2.6 *Let* (H1')–(H4') *hold. Then Φ has a positive equilibrium point if one of the following holds.*

- *(bistability) Both E_1 and E_2 are locally asymptotically stable relative to I.*
- *(mutual invasibility) Both E_1 and E_2 are unstable relative to I.*
- *(persistence) There is a point $x \in X$ such that $\omega(x) \cap \operatorname{Int} C$ is nonempty.*

Proof If Φ has no positive equilibrium points, then by Theorem E.2.5, there is a connecting orbit $\gamma(t)$ connecting E_1 and E_2 and for each $x \in C \setminus \{0\}$, $\Phi_t \to E_1$ or E_2. This is incompatible with any of the three conditions in the statement of the corollary. □

Next, we formulate a sufficient condition to ensure no internal trajectories converge to E_2 (resp. E_1). For this purpose, we write

$$\Phi_t(x_1,x_2) = (\Phi_t^{(1)}(x_1,x_2), \Phi_t^{(2)}(x_1,x_2)),$$

where $\Phi_t^{(i)} : (C_1 \times C_2) \to C_i$ is continuous and order compact.

In general, the nonexistence of positive equilibria does not imply that one of E_1, E_2 attracts all trajectories; see Problem 7.5.1. To prove that E_1 (resp. E_2) attracts all trajectories, one needs to impose further conditions to guarantee that any trajectory in $\operatorname{Int} C$ does not converge to E_2 (resp. E_1). For this purpose, we introduce the following assumption. We suppose Φ_t is continuously differentiable in a neighborhood of E_1 and E_2, and write

$$\Phi_t(x_1,x_2) = (\Phi_t^{(1)}(x_1,x_2), \Phi_t^{(2)}(x_1,x_2)) \in X_1 \times X_2.$$

One way is to impose linear instability in one of the boundary equilibria. This was originally proved in [2].

Proposition E.2.7 *Let* (H1')–(H4') *hold, and assume that Φ_t has no equilibria in $I \setminus \{E_0, E_1, E_2\}$. Suppose one of the following conditions hold:*

1. *There exist $\tau, \delta > 0$, $u_1 \in C_1 \setminus \{0\}$ such that*

$$\Phi_\tau(su_1, x_2) \geq su_1 \quad \text{for all } 0 \leq s \leq 1 \text{ and } (0,x_2) \in B_\delta(E_2). \qquad \text{(E.11)}$$

2. *For some $\tau > 0$, $\Phi_\tau : C \to C$ is continuously differentiable in a neighborhood of E_2, such that $D_1\Phi_\tau^{(1)}(E_2) : X_2 \to X_2$ is compact and strongly positive w.r.t. C_1 and $r(D_1\Phi_\tau^{(1)}(E_2)) > 1$.*

Then $\Phi_t(x) \to E_1$ as $t \to \infty$ for all $x = (x_1,x_2) \in C$, $x_i \neq 0$.

Proof By Theorem E.2.5, it suffices to ensure that $\Phi_t(x) \not\to E_2$ for each $(x_1,x_2) \in C$, $x_i \neq 0$. Suppose to the contrary that $\Phi_t(x) \to E_2$ for some x, then we can repeat the proof of Proposition E.1.9 to derive a contradiction. □

When the two species are mutually invasible, i.e. E_1 and E_2 are both unstable, then Theorem E.2.5 yields the existence of at least one positive fixed point.

Corollary E.2.8 *Let* (H1')–(H4') *hold. Suppose both E_1 and E_2 are linearly unstable (i.e. at least one of the two conditions of Proposition E.2.7 holds for E_2, and a symmetric statement holds for E_1), then the semiflow has at least one positive equilibrium E_*.*

Proof By Proposition E.2.7, for each $i = 1, 2$ and $(x_1,x_2) \in C$ with $x_i \neq 0$, we have $\Phi_t(x_1,x_2) \not\to E_i$. It follows from Theorem E.2.5 that Φ_t has at least one equilibrium in $I \setminus \{E_0, E_1, E_2\}$. □

Next, we seek to further relax the linear instability criterion. The following conditions holds, for example, for two-species Lotka–Volterra systems with diffusion (see Chapter 7), as well as certain phytoplankton models (see Chapter 8).

(H5') For each $t > 0$, Φ_t is continuously differentiable, and the following hold.

- The map $D\Phi_t(E_2) : X \to X$ is compact and $r(D_2\Phi_t^{(2)}(E_2)) < 1$ for all $t > 0$. If $r(D\Phi_\tau(E_2)) \geq 1$ for some τ, then we have

$$D\Phi_\tau(E_2)[v] \geq_K v \quad \text{for some } v = (v_1, v_2) \in \operatorname{Int} C_1 \times (-\operatorname{Int} X_2^+).$$

- The map $D\Phi_t(E_1) : X \to X$ is compact and $r(D_1\Phi_t^{(1)}(E_1)) < 1$ for all $t > 0$. If $r(D\Phi_\tau(E_1)) \geq 1$ for some τ, then we have

$$D\Phi_\tau(E_1)[v] \geq_K v \quad \text{for some } v = (v_1, v_2) \in \operatorname{Int} X_1^+ \times (-\operatorname{Int} C_2).$$

Remark E.2.9 In Chapter 7, we verify that the semiflow Φ_t generated by the two-species diffusive Lotka–Volterra competition system satisfies (H1')–(H5'). As a consequence, for each fixed $T > 0$, the mapping $S = \Phi_T$ verifies (H1)–(H5).

We give a sufficient condition for the first part of (H5') concerning E_2. A symmetric condition holds for the latter part of (H5') concerning E_1.

Lemma E.2.10 *Suppose that Φ_t is continuously differentiable in a neighborhood of E_2, and $D\Phi_t(E_2)$ is compact for all $t > 0$. In addition, suppose the following conditions hold for all $t > 0$.*

1. $D_2\Phi_t^{(2)}(E_2)$ *is strongly monotone w.r.t.* X_2^+, *and* $r(D_2\Phi_t^{(2)}(E_2)) < 1$;
2. $D_1\Phi_t^{(1)}(E_2)$ *is strongly monotone w.r.t.* C_1;
3. $D_1\Phi_t^{(2)}(E_2)(v_1') \neq 0$ *for all* $v_1' \in C_1 \setminus \{0\}$.

Then (H5') holds.

Remark E.2.11 The condition $r(D_2\Phi_t^{(2)}(E_2)) < 1$ means that the equilibrium E_2 is linearly stable when the semiflow is restricted to $\{0\} \times C_2$. The condition 3 is crucial, if one compares with the counterexample; see Problem 7.5.1.

Proof (of Lemma E.2.10) This is analogous to Lemma E.1.11. □

We can now prove a sufficient condition for E_2 (resp. E_1) to be repelling.

Lemma E.2.12 *Let* (H1')–(H5') *hold. Assume that*

(i) E_2 *is an isolated equilibrium.*
(ii) *There exists an equilibrium* $E^* = (x_1^*, x_2^*) \gg_K E_2$ *such that*

$$\Phi_t(x) \to E^* \quad \textit{for any } x = (x_1, x_2) \in [E_2, E^*] \textit{ such that } x_1 \neq 0.$$

Then $\Phi_t(x_1, x_2) \not\to E_2$ *whenever* $x_i \neq 0$ *for* $i = 1, 2$.

Proof By the assumptions, E_2 is not locally asymptotically stable, so that for all $t > 0$, $r(D\Phi_t(E_2)) \geq 1$.

Suppose to the contrary that $\Phi_t(x) \to E_2$ for some $x = (x_1, x_2)$, $x_i \neq 0$, then $S^n(x) \to E_2$, where $S = \Phi_\tau$ for some $\tau > 0$. One can then repeat the proof of Lemma E.1.14 to obtain a contradiction. □

We can now prove a stronger version of the trichotomy result in Theorem E.2.5.

Theorem E.2.13 *Let* (H1')–(H5') *hold and let* $I = [E_2, E_1]_K$. *Then the omega limit set* $\omega(x) \subset I$ *for every* $x \in C$, *and the following trichotomy holds:*

1. Φ_t *has at least one equilibrium in* Int I;
2. $\Phi_t(x) \to E_1$ *for all* $x = (x_1, x_2) \in C$ *such that* $x_i \neq 0$, $i = 1, 2$;
3. $\Phi_t(x) \to E_2$ *for all* $x = (x_1, x_2) \in C$ *such that* $x_i \neq 0$, $i = 1, 2$.

Proof Suppose Φ_t has no equilibrium in Int I, then it follows from (H3') that it has no equilibrium in $I \setminus \{E_0, E_1, E_2\}$. By Theorem E.2.5,

$$\Phi_t(x) \to E_1 \text{ or } E_2, \quad \text{if } x = (x_1, x_2) \in C,\ x_i \neq 0,\ i = 1, 2. \tag{E.12}$$

Moreover, one of E_1, E_2 attracts all positive orbits in I. Suppose for definiteness that E_1 is the stable equilibrium relative to I. It then follows from Lemma E.2.12 (taking $E^* = E_1$) that $\Phi_t(x) \not\to E_2$ for any $x = (x_1, x_2) \in C$ such that $x_i \neq 0$, $i = 1, 2$. It then follows from (E.12) that $\Phi_t(x) \to E_1$ for any $x = (x_1, x_2) \in C$ such that $x_i \neq 0$, $i = 1, 2$. □

If the first alternative of Theorem E.1.15 holds, we can impose a further condition to derive a global attractivity result.

Theorem E.2.14 *Let* (H1')–(H5') *hold, and assume that* S *has at least one positive fixed point. Assume further that every positive fixed point is locally asymptotically stable, then there is a unique positive equilibrium* E_* *such that* $\Phi_t(x) \to E_*$ *for every* $x = (x_1, x_2) \in C$ *such that* $x_i \neq 0$ *for* $i = 1, 2$.

Proof This is analogous to Theorem E.1.16. We omit the details. □

Theorem E.2.15 (Hess and Lazer's compression result, continuous version) *Let* (H1')–(H4') *hold. Suppose* $\Phi_t : C \to C$ *is continuously differentiable in a neighborhood of* E_1 *and* E_2, *and the following conditions hold:*

- *For some* $\tau > 0$, $D_2\Phi_\tau^{(2)}(E_1) : X_1 \to X_1$ *is compact and strongly positive w.r.t.* C_2 *and that* $r(D_2\Phi_\tau^{(2)}(E_1)) > 1$.
- *For some* $\tau > 0$, $D_1\Phi_\tau^{(1)}(E_2) : X_2 \to X_2$ *is compact and strongly positive w.r.t.* C_1 *and that* $r(D_1\Phi_\tau^{(1)}(E_2)) > 1$.

Then there exist fixed points E_*, E^*, *such that* $E_2 \ll_K E_* \leq_K E^* \ll_K E_1$ *and*

1. $\Phi_t(x) \to E_*$ *for* $x = (x_1, x_2) \in [E_2, E_*]$, $x_1 \neq 0$.
2. $\Phi_t(x) \to E^*$ *for* $x = (x_1, x_2) \in [E^*, E_1]$, $x_2 \neq 0$.
3. *For any* $x = (x_1, x_2) \in C$, $x_i \neq 0$, *the omega limit set* $\omega(x)$ *satisfies*

$$\omega(x) \subset [E_*, E^*].$$

If, in addition, $E_* = E^*$, *then* $\Phi_t(x) \to E^*$ *for all* $x = (x_1, x_2) \in C$, $x_i \neq 0$.

See Figure E.1 for an illustration.

Proof We leave this as an exercise. □

Corollary E.2.16 *Assume, in addition to the assumptions of Theorem E.2.15, that every positive equilibrium is locally asymptotically stable, then there is a unique positive equilibrium E^* such that $\Phi_t(x) \to E^*$ as $t \to \infty$ for every $x = (x_1, x_2) \in C$ such that $x_i \neq 0$ for $i = 1, 2$.*

Proof We leave this as an exercise. □

E.3 Further Reading

The abstract treatment of competitive system was initiated by Hess and Lazer [2], who first observed that the dynamics of two-species competition systems have common features regardless of the fine details of the model, or its dimension. Hsu, Waltman and Ellermeyer [5] considered continuous time competitive systems using monotone dynamical systems theory, which was later extended in [4]. Further progress, particularly on the bistable dynamics and saddle point behaviors, was obtained in [7, 9, 11]. Building on the framework of [4], we show that diffusive Lotka–Volterra systems satisfy an additional hypothesis and deduce an improved trichotomy result for long-time dynamics (Theorem 7.1.6). This improves an earlier result in [8]. See [10] for a recent review article.

Problems

E.3.1 Prove Theorem E.1.17 or Theorem E.2.15.

E.3.2 Prove Corollary E.1.18 or Corollary E.2.16.

References

1. P. Hess, *Periodic-parabolic boundary value problems and positivity*, vol. 247 of Pitman Research Notes in Mathematics Series, Longman Scientific & Technical, Harlow; copublished in the United States with John Wiley & Sons, Inc., New York, 1991.
2. P. Hess and A. C. Lazer, *On an abstract competition model and applications*, Nonlinear Anal., 16 (1991), pp. 917–940.
3. S.-B. Hsu, K.-Y. Lam, and F.-B. Wang, *Single species growth consuming inorganic carbon with internal storage in a poorly mixed habitat*, J. Math. Biol., 75 (2017), pp. 1775–1825.
4. S. B. Hsu, H. L. Smith, and P. Waltman, *Competitive exclusion and coexistence for competitive systems on ordered Banach spaces*, Trans. Amer. Math. Soc., 348 (1996), pp. 4083–4094.
5. S.-B. Hsu, P. Waltman, and S. F. Ellermeyer, *A remark on the global asymptotic stability of a dynamical system modeling two species competition*, Hiroshima Math. J., 24 (1994), pp. 435–445.

6. D. Jiang, K.-Y. Lam, Y. Lou, and Z.-C. Wang, *Monotonicity and global dynamics of a nonlocal two-species phytoplankton model*, SIAM J. Appl. Math., 79 (2019), pp. 716–742.
7. J. Jiang, X. Liang, and X.-Q. Zhao, *Saddle-point behavior for monotone semiflows and reaction-diffusion models*, J. Differential Equations, 203 (2004), pp. 313–330.
8. K.-Y. Lam and D. Munther, *A remark on the global dynamics of competitive systems on ordered Banach spaces*, Proc. Amer. Math. Soc., 144 (2016), pp. 1153–1159.
9. X. Liang and J. Jiang, *On the finite-dimensional dynamical systems with limited competition*, Trans. Amer. Math. Soc., 354 (2002), pp. 3535–3554.
10. H. L. Smith, *Monotone dynamical systems: reflections on new advances & applications*, Discrete Contin. Dyn. Syst., 37 (2017), pp. 485–504.
11. H. L. Smith and H. R. Thieme, *Stable coexistence and bi-stability for competitive systems on ordered Banach spaces*, J. Differential Equations, 176 (2001), pp. 195–222.

Index

K. -Y. Lam, Y. Lou, *Introduction to Reaction-Diffusion Equations*, Lecture Notes on Mathematical Modelling in the Life Sciences, https://doi.org/10.1007/978-3-031-20422-7

Printed in the United States
by Baker & Taylor Publisher Services